21世纪高等学校规划教材｜电子信息

自动控制原理
（双语教材）（第2版）

摆玉龙 编著

清华大学出版社
北京

内容简介

本书采用中英双语相结合的方式，全面地阐述了自动控制系统的基本理论与方法。全书共分为8章，主要论述连续控制系统的分析和综合方法，包括各类控制系统数学模型的建立和模型间的等效变换；利用经典控制理论中的时域分析法、根轨迹法和频域分析法分析控制系统性能；应用PID控制和串联校正等对系统进行设计；现代控制理论基础等。同时概述计算机控制系统、过程控制系统和机电一体化系统。书中针对各章内容，适当增加了英文MATLAB应用实验设计，编著了大量英文阅读材料和概念解析，便于双语教学使用。

本书可以作为高等院校"自动控制原理"课程的教材，适用于电气、自动化、电子、信息与通信、计算机、机械等专业，特别适用于有双语教学要求的相关课程，也可供从事控制工程的技术人员参考。

版权所有，侵权必究。举报：010-62782989，beiqinquan@tup.tsinghua.edu.cn。

图书在版编目(CIP)数据

自动控制原理（双语教材）/摆玉龙编著. —2版. —北京：清华大学出版社，2018（2024.8重印）
(21世纪高等学校规划教材·电子信息)
ISBN 978-7-302-49563-5

Ⅰ. ①自… Ⅱ. ①摆… Ⅲ. ①自动控制理论－双语教学－高等学校－教材 Ⅳ. ①TP13

中国版本图书馆CIP数据核字(2018)第027574号

责任编辑：郑寅堃　薛　阳
封面设计：傅瑞学
责任校对：李建庄
责任印制：刘海龙

出版发行：清华大学出版社
　　网　　址：https://www.tup.com.cn, https://www.wqxuetang.com
　　地　　址：北京清华大学学研大厦A座　　邮　　编：100084
　　社 总 机：010-83470000　　邮　　购：010-62786544
　　投稿与读者服务：010-62776969, c-service@tup.tsinghua.edu.cn
　　质量反馈：010-62772015, zhiliang@tup.tsinghua.edu.cn
　　课件下载：https://www.tup.com.cn, 010-83470236

印 装 者：三河市人民印务有限公司
经　　销：全国新华书店
开　　本：185mm×260mm　　印　张：24.25　　字　数：586千字
版　　次：2013年8月第1版　　2018年7月第2版　　印　次：2024年8月第9次印刷
印　　数：5201～5700
定　　价：69.00元

产品编号：074741-01

出版说明

随着我国改革开放的进一步深化,高等教育也得到了快速发展,各地高校紧密结合地方经济建设发展需要,科学运用市场调节机制,加大了使用信息科学等现代科学技术提升、改造传统学科专业的投入力度,通过教育改革合理调整和配置了教育资源,优化了传统学科专业,积极为地方经济建设输送人才,为我国经济社会的快速、健康和可持续发展以及高等教育自身的改革发展做出了巨大贡献。但是,高等教育质量还需要进一步提高以适应经济社会发展的需要,不少高校的专业设置和结构不尽合理,教师队伍整体素质亟待提高,人才培养模式、教学内容和方法需要进一步转变,学生的实践能力和创新精神亟待加强。

教育部一直十分重视高等教育质量工作。2007年1月,教育部下发了《关于实施高等学校本科教学质量与教学改革工程的意见》,计划实施"高等学校本科教学质量与教学改革工程"(简称"质量工程"),通过专业结构调整、课程教材建设、实践教学改革、教学团队建设等多项内容,进一步深化高等学校教学改革,提高人才培养的能力和水平,更好地满足经济社会发展对高素质人才的需要。在贯彻和落实教育部"质量工程"的过程中,各地高校发挥师资力量强、办学经验丰富、教学资源充裕等优势,对其特色专业及特色课程(群)加以规划、整理和总结,更新教学内容、改革课程体系,建设了一大批内容新、体系新、方法新、手段新的特色课程。在此基础上,经教育部相关教学指导委员会专家的指导和建议,清华大学出版社在多个领域精选各高校的特色课程,分别规划出版系列教材,以配合"质量工程"的实施,满足各高校教学质量和教学改革的需要。

为了深入贯彻落实教育部《关于加强高等学校本科教学工作,提高教学质量的若干意见》精神,紧密配合教育部已经启动的"高等学校教学质量与教学改革工程精品课程建设工作",在有关专家、教授的倡议和有关部门的大力支持下,我们组织并成立了"清华大学出版社教材编审委员会"(以下简称"编委会"),旨在配合教育部制定精品课程教材的出版规划,讨论并实施精品课程教材的编写与出版工作。"编委会"成员皆来自全国各类高等学校教学与科研第一线的骨干教师,其中许多教师为各校相关院、系主管教学的院长或系主任。

按照教育部的要求,"编委会"一致认为,精品课程的建设工作从开始就要坚持高标准、严要求,处于一个比较高的起点上。精品课程教材应该能够反映各高校教学改革与课程建设的需要,要有特色风格、有创新性(新体系、新内容、新手段、新思路,教材的内容体系有较高的科学创新、技术创新和理念创新的含量)、先进性(对原有的学科体系有实质性的改革和发展,顺应并符合21世纪教学发展的规律,代表并引领课程发展的趋势和方向)、示范性(教材所体现的课程体系具有较广泛的辐射性和示范性)和一定的前瞻性。教材由个人申报或各校推荐(通过所在高校的"编委会"成员推荐),经"编委会"认真评审,最后由清华大学出版社审定出版。

目前，针对计算机类和电子信息类相关专业成立了两个"编委会"，即"清华大学出版社计算机教材编审委员会"和"清华大学出版社电子信息教材编审委员会"。推出的特色精品教材包括：

(1) 21世纪高等学校规划教材·计算机应用——高等学校各类专业，特别是非计算机专业的计算机应用类教材。

(2) 21世纪高等学校规划教材·计算机科学与技术——高等学校计算机相关专业的教材。

(3) 21世纪高等学校规划教材·电子信息——高等学校电子信息相关专业的教材。

(4) 21世纪高等学校规划教材·软件工程——高等学校软件工程相关专业的教材。

(5) 21世纪高等学校规划教材·信息管理与信息系统。

(6) 21世纪高等学校规划教材·财经管理与应用。

(7) 21世纪高等学校规划教材·电子商务。

(8) 21世纪高等学校规划教材·物联网。

清华大学出版社经过三十多年的努力，在教材尤其是计算机和电子信息类专业教材出版方面树立了权威品牌，为我国的高等教育事业做出了重要贡献。清华版教材形成了技术准确、内容严谨的独特风格，这种风格将延续并反映在特色精品教材的建设中。

清华大学出版社教材编审委员会
联系人：魏江江
E-mail: weijj@tup.tsinghua.edu.cn

第2版前言

"他山之石,可以攻玉"。随着中国工程教育改革的发展,借鉴国外先进的工科教学理念,在教学方法、实践教学和实验素材等方面实施教学改革,是当前工程教育改革的趋势所在。双语教学示范课程建设是提升我国高等教育质量工程的重要举措之一。从某种意义上讲,实施双语教学,不仅仅是用外语授课、外语交流,更重要的是要借鉴西方发达国家优秀的原版教材,采用国际上最新的优秀教学模式、教学思想和体系,为学生提供一个国际化的教育大环境,使学生有机会直接接受具有国际化先进水平的科学教育的影响。

《自动控制原理(双语教材)》自从2013年出版以来,受到读者的厚爱和关怀,作者对此表示衷心的感谢。第一版教材考虑到教学内容的难度、学生的英语水平等因素,主要采取中文论述为主,英文为辅的形式,编写了大量的自动控制相关英文阅读材料,同时有针对性地设计了英文MATLAB应用实验,以提高本书作为双语教材的实用性。在综合型双语教学中,教师可以采用母语讲授新内容和新概念,采用双语教学进行概念解析、拓展阅读和复习总结。同时,原汁原味的MATLAB应用实验不仅使得学生需要清楚地了解实验目的,更要耐心地阅读英文,完成实验。设计的实验内容包括MATLAB基础应用、控制系统时域分析、根轨迹法设计、控制系统频域分析等主要内容。

本书是在第一版的基础上加以修改和增订而成的。这次修订主要做了如下工作:①增加了现代控制理论章节,以状态空间模型为基础的现代控制理论,是解决多输入多输出、非线性和时变系统的理论基础,更是最优控制理论、自适应控制、动态系统辨识以及智能控制等分支学科的基础,有必要在自动控制原理课程中作简要介绍;②以附录的形式增加了控制理论术语中英文对照、自动控制理论中的概念解析等内容,有助于学生复习检索相关内容;③修订了全书的习题和文字叙述中的不确切之处,旨在进一步增加本书的可读性。

本书第二版由摆玉龙教授统稿审定,潘强强编写了第8章,摆玉龙修订了全书内容。马真东、陈春梅、陈雪雷、王一朝、王娟、郭鹏飞、魏强、苏坤、段济开和常明恒等做了大量的打印和校对工作。

在编写中编者参考了很多优秀教材和著作。编者向本书中的相关编著者和收录于参考文献中的各位作者表示真诚的谢意。感谢国家自然科学基金项目(编号:41461078)、兰州市科技计划项目(2015-3-34)和西北师范大学研究生培养与课程改革项目对本书出版的资助。

由于作者水平有限,加之时间仓促,错误与不妥之处在所难免,恳请读者批评、指正。

编 者
2018年4月

前　言

随着教育部"高等学校本科教学改革与教学质量工程"的深入开展,双语教学示范课程建设已成为提升我国高等教育质量工程的一项重要举措,因此专业性双语教材的需要日益凸显。作者在多年来从事"自动控制原理"教学工作的基础上,结合在荷兰、日本和澳大利亚访学期间所学到的知识和经验,编著了介于国内中文教材与国外英文教材之间的综合型双语教材。在教材编写上,考虑到教学内容的难度、学生的英语水平等因素,主要采取中文论述为主,英文为辅的形式,编著了大量的自动控制相关英文阅读材料,同时针对性地设计了英文 MATLAB 应用实验,以提高本书作为双语教材的实用性。

"自动控制原理"是自动化类、机电类专业的专业基础课。本书内容包括经典控制理论,计算机控制系统,MATLAB 在控制系统分析、设计和仿真中的应用等。

本书共分 7 章:第 1 章介绍自动控制的一般概念,控制系统的基本组成和分类等;针对 MATLAB 在控制系统分析和设计中的重要性,以英文材料的形式介绍 MATLAB 的特点和编程环境,重点介绍控制工具箱的组成和功用,选择典型案例分析控制系统的特点。第 2 章介绍控制系统数学模型的建立和等效变换,主要包括微分方程模型、传递函数模型和结构图等,实验设计中强调 MATLAB 中变量、矩阵和多项式等数据类型的使用,为控制系统的分析和设计打好基础。第 3 章介绍控制系统的时域分析法,以一阶系统、二阶系统为例,详细介绍各种典型输入下的时域响应,MATLAB 实验设计从模型建立入手,介绍 MATLAB 中各种时域响应的特点。第 4 章介绍根轨迹分析法,在详述根轨迹绘制规则的基础上,以英文材料的形式设计了 MATLAB 实验。第 5 章介绍线性系统的频域分析法,包括基本概念、典型环节的频域特性绘制和频域稳定判据等。第 6 章介绍线性系统的校正方法,主要包括 PID 控制和串联校正等。第 7 章简述计算机控制系统、过程控制系统和机电一体化系统。

本书的最大特点是:借鉴国内外同类教材的优点,以英文阅读材料、概念解析和实验设计的方式,采用英语素材介绍自动控制原理的基本知识。在中文讲述自控原理基本内容的基础上,设计全英文的 MATLAB 设计实例,并且配合了一定数量的典型例题和习题,以培养学生的实验能力、思维能力、信息获取能力与分析设计能力。因此,本书不仅适合电子信息类相关专业"自动控制原理"课程的本科教学需要,同时满足双语教学要求。大量的阅读材料和概念解析,不仅有助于教师的双语教学,而且能让学生通过对比学习,深入理解自动控制的基本原理、基本分析方法和基本设计技术,对后续课程的学习打下坚实的基础。同时激发学生对自动化类专业浓厚的兴趣,锻炼和提高运用所学理论解决实际问题的能力。

本书由摆玉龙教授统稿审定,其中摆玉龙编写了第 1、2 章和第 5 章,邵宇编写第 3、4 章和第 6 章,马永杰、王丽丽编写第 7 章。摆玉龙编写了每章中的 MATLAB 实验设计等英文材料。王丽丽、巨芳子、柴乾隆、高海沙、高燕、李博、黄智慧等绘制了本书的全部插图,并做了大量的打印和校对工作。

  在编写中编者参考了很多优秀教材和著作。编者向收录于参考文献中的各位作者表示真诚的谢意。摆玉龙感谢国家自然科学基金项目（编号：41061038）和甘肃省科技支撑计划（编号：1204GKCA067）对本书出版的资助。

  由于作者水平有限，加之时间仓促，错误与不妥之处在所难免，恳请读者批评、指正。

编 者

2013 年 5 月

目 录

第1章 自动控制系统概述 ... 1

- 1.1 引言 ... 1
- 1.2 自动控制系统的基本概念 ... 2
- 1.3 自动控制理论的发展 ... 3
- 1.4 自动控制系统的基本原理及组成 ... 5
  - 1.4.1 开环控制 ... 6
  - 1.4.2 闭环控制 ... 7
  - 1.4.3 自动控制系统的基本组成 ... 9
  - 1.4.4 自动控制系统的实例分析 ... 11
- 1.5 自动控制系统的分类 ... 16
  - 1.5.1 恒值控制系统和随动控制系统 ... 16
  - 1.5.2 连续系统和离散系统 ... 16
  - 1.5.3 单输入单输出系统和多输入多输出系统 ... 16
  - 1.5.4 线性系统和非线性系统 ... 16
  - 1.5.5 定常系统和时变系统 ... 17
  - 1.5.6 其他类型 ... 17
- 1.6 控制系统的基本要求 ... 18
- 1.7 控制系统的设计概述 ... 20
- 1.8 MATLAB在本章中的应用 ... 22
- 本章小结 ... 26
- 习题 ... 27

第2章 控制系统的数学模型 ... 29

- 2.1 引言 ... 29
- 2.2 控制系统的时域数学模型 ... 31
  - 2.2.1 电气系统 ... 31
  - 2.2.2 机械系统 ... 32
- 2.3 线性系统的传递函数 ... 36
  - 2.3.1 传递函数的定义 ... 36
  - 2.3.2 典型环节及其传递函数 ... 39
  - 2.3.3 电气网络的运算阻抗与传递函数 ... 43
- 2.4 控制系统的结构图及其等效变换 ... 46

2.4.1　结构图的组成 …………………………………………………………… 47
　　　2.4.2　控制系统结构图的建立 ………………………………………………… 47
　　　2.4.3　结构图的等效变换 ……………………………………………………… 49
　　　2.4.4　信号相加点和分支点的移动和互换 …………………………………… 51
　　　2.4.5　结构图简化示例 ………………………………………………………… 53
　2.5　信号流图和梅逊公式 ……………………………………………………………… 55
　　　2.5.1　信号流图 ………………………………………………………………… 55
　　　2.5.2　梅逊公式 ………………………………………………………………… 56
　2.6　闭环传递函数的定义 ……………………………………………………………… 59
　　　2.6.1　闭环系统概述 …………………………………………………………… 59
　　　2.6.2　闭环系统的传递函数 …………………………………………………… 59
　2.7　非线性系统模型概述 ……………………………………………………………… 62
　2.8　MATLAB 在本章中的应用 ……………………………………………………… 69
本章小结 …………………………………………………………………………………… 76
习题 ………………………………………………………………………………………… 77

第 3 章　线性系统的时域分析 ……………………………………………………………… 82

　3.1　引言 ………………………………………………………………………………… 82
　3.2　系统时间响应的性能指标 ………………………………………………………… 83
　3.3　一阶系统的时域分析 ……………………………………………………………… 90
　　　3.3.1　一阶系统的数学模型和结构图 ………………………………………… 90
　　　3.3.2　一阶系统的单位阶跃响应 ……………………………………………… 91
　　　3.3.3　一阶系统的单位脉冲响应 ……………………………………………… 91
　　　3.3.4　一阶系统的单位斜坡响应 ……………………………………………… 92
　　　3.3.5　一阶系统的单位加速度响应 …………………………………………… 92
　3.4　二阶系统的时域分析 ……………………………………………………………… 93
　　　3.4.1　二阶系统的单位阶跃响应 ……………………………………………… 94
　　　3.4.2　欠阻尼二阶系统的动态性能分析 ……………………………………… 96
　3.5　高阶系统时域分析法概述 ………………………………………………………… 100
　3.6　控制系统的稳定性分析 …………………………………………………………… 102
　　　3.6.1　稳定的基本概念和稳定的充分必要条件 ……………………………… 103
　　　3.6.2　代数稳定判据 …………………………………………………………… 104
　3.7　控制系统稳态误差的分析及计算 ………………………………………………… 111
　　　3.7.1　稳态误差的定义 ………………………………………………………… 112
　　　3.7.2　系统类型与稳态误差 …………………………………………………… 113
　　　3.7.3　给定输入信号下的稳态误差计算 ……………………………………… 114
　　　3.7.4　扰动作用下的稳态误差计算 …………………………………………… 116
　　　3.7.5　减少稳态误差的方法 …………………………………………………… 117
　3.8　MATLAB 在本章中的应用 ……………………………………………………… 118

本章小结 ··· 127
习题 ··· 129

第4章 根轨迹法 ·· 134

4.1 引言 ··· 134
4.2 根轨迹的基本概念 ··· 135
4.3 根轨迹绘制的基本规则 ··· 138
4.4 广义根轨迹 ··· 149
4.5 系统性能的根轨迹法分析 ·· 152
    4.5.1 根轨迹分析法概述 ·· 152
    4.5.2 增加开环极点对控制系统的影响 ······································· 154
    4.5.3 增加开环零点对控制系统的影响 ······································· 155
    4.5.4 结论 ·· 155
4.6 MATLAB 在本章中的应用 ·· 156
本章小结 ··· 164
习题 ··· 165

第5章 线性系统的频域分析法 ··· 168

5.1 引言 ··· 168
5.2 频率特性的基本概念 ··· 169
    5.2.1 频率特性的定义 ··· 169
    5.2.2 频率特性的表示方法 ·· 172
5.3 典型环节的频率特性及特性曲线的绘制 ····································· 175
5.4 频域稳定判据及稳定裕量 ··· 187
    5.4.1 奈奎斯特稳定判据 ·· 188
    5.4.2 奈奎斯特稳定判据的应用 ·· 191
    5.4.3 对数稳定判据 ·· 194
    5.4.4 稳定裕量 ·· 195
5.5 频率特性与控制系统性能的关系 ·· 200
    5.5.1 控制系统性能指标 ·· 200
    5.5.2 开环对数幅频特性与性能指标间的关系 ···························· 203
5.6 MATLAB 在本章中的应用 ·· 206
本章小结 ··· 213
习题 ··· 215

第6章 线性系统的校正方法 ·· 221

6.1 引言 ··· 221
6.2 系统校正的基本概念 ··· 222
    6.2.1 受控对象 ·· 222

		6.2.2	性能指标概述	222
		6.2.3	系统校正连接方式	223
		6.2.4	基本控制规律	225
	6.3	系统校正装置		230
		6.3.1	超前校正装置	230
		6.3.2	滞后校正装置	234
		6.3.3	滞后-超前校正装置	236
		6.3.4	超前校正、滞后校正和滞后-超前校正的比较	239
	6.4	反馈校正		239
		6.4.1	反馈校正的特点	239
		6.4.2	反馈校正系统的设计	240
		6.4.3	串联校正与反馈校正比较	240
	6.5	MATLAB 在本章中的应用		242
	本章小结			258
	习题			259

第 7 章 计算机控制系统概述 262

	7.1	引言		262
	7.2	计算机控制系统概述		263
	7.3	A/D 转换采样过程与采样定理		264
		7.3.1	采样过程	264
		7.3.2	采样定理	265
		7.3.3	采样周期在工程应用中的选择方法	267
	7.4	采样信号的复现		268
	7.5	离散控制系统的数学模型		269
		7.5.1	脉冲传递函数的定义	270
		7.5.2	开环采样系统的脉冲传递函数	271
		7.5.3	闭环采样系统的脉冲传递函数	273
	7.6	采样控制系统的稳定性分析		275
	7.7	其他控制系统简介		279
		7.7.1	过程控制系统简介	279
		7.7.2	机电一体化系统简介	281
	7.8	MATLAB 在本章中的应用		283
	本章小结			287
	习题			289

第 8 章 现代控制理论基础 292

	8.1	线性系统的状态空间描述		292
		8.1.1	状态空间描述的基本概念	292

8.1.2　线性定常连续系统状态空间表达式的建立 …………………… 296
　8.2　线性定常系统状态方程的解 ………………………………………………… 307
　　　8.2.1　线性定常齐次状态方程的解 ………………………………… 307
　　　8.2.2　线性定常系统的状态转移矩阵 ……………………………… 309
　　　8.2.3　非齐次状态方程的解 ………………………………………… 310
　　　8.2.4　线性离散系统的解 …………………………………………… 312
　8.3　线性定常系统的可控性和可观测性 ………………………………………… 312
　　　8.3.1　可控性和可观测性的定义 …………………………………… 313
　　　8.3.2　线性定常连续系统的可控性判别 …………………………… 315
　　　8.3.3　线性定常系统可观测性判别 ………………………………… 317
　　　8.3.4　可控性、可观测性与传递函数矩阵的关系 ………………… 319
　8.4　线性定常系统的状态反馈和状态观测器 …………………………………… 320
　　　8.4.1　线性定常系统常用反馈结构 ………………………………… 320
　　　8.4.2　状态反馈与极点配置 ………………………………………… 322
　　　8.4.3　状态观测器 …………………………………………………… 323
　8.5　MATLAB 在本章中的应用 …………………………………………………… 326
　　　8.5.1　State-space equations ………………………………………… 326
　　　8.5.2　Control design using pole placement ……………………… 328
　　　8.5.3　Introducing the reference input …………………………… 330
　　　8.5.4　Observer design ……………………………………………… 332
　本章小结 ……………………………………………………………………………… 335
　习题 …………………………………………………………………………………… 336

附录 A　拉普拉斯变换及反变换 ………………………………………………………… 337

附录 B　z 变换定义及对照表 …………………………………………………………… 339

附录 C　常用校正网络 …………………………………………………………………… 341

附录 D　Bode 图的绘制规则 …………………………………………………………… 343

附录 E　常用 MATLAB 命令 …………………………………………………………… 345

附录 F　自动控制理论中的概念解析 …………………………………………………… 346

附录 G　控制理论术语中英文对照 ……………………………………………………… 352

参考文献 …………………………………………………………………………………… 370

第1章 自动控制系统概述

1.1 引言

随着科学技术的发展,自动控制系统已经普遍出现在人类生产、生活和探索新技术的各个领域中。所谓自动控制就是应用控制器自动地、有目的地操纵被控对象使之具有预定的工作状态。自动控制理论(Automatic Control Theory)是研究自动控制共同规律的技术科学。自动控制技术的广泛应用,不仅将人们从繁重的体力劳动和大量重复性的操作中解放出来,同时也极大地提高了劳动生产率和产品质量。在科学技术的发展历史上,自动控制技术始终起着非常重要的作用。因此,对于工程技术人员和科学工作者来说,掌握一定的控制技术是十分必要的。

下面通过阅读英文材料的形式,介绍自动控制的基本概念、自动控制原理课程的重点内容等。

Reading Material

In our daily life we encounter and experience a variety of dynamic phenomena. The environment we live in is constantly changing. We use many machines and processes that have been developed by humans over the years. To maximize the benefits, some aspects of the machines and processes and the environment we live in will have to be controlled in some desired way. Feedback control is an effective way of achieving the desired control.

Control systems are what make machines and processes function as intended. They are most often based on the principle of feedback whereby the signal to be controlled is compared to a desired reference signal and the discrepancy used to compute corrective control action. These signal processing tasks are implemented through the use of appropriate hardware called the controller.

Every engineering discipline has its own language—concepts, symbols, and words—that engineers in that field use to express ideas. This chapter introduces you to the fundamental language of control system technology, illustrates the key principle of feedback control, and describes the general approach to designing and building feedback control systems.

This course emphasizes the centrality of information processing and control in automatic control systems, introduces the various facets of control system technology, and covers basic system analysis and design methodology as applied to feedback control systems.

本章主要内容：
- 自动控制系统的基本概念
- 自动控制理论的发展
- 自动控制系统的基本原理及组成
- 自动控制系统的分类
- 控制系统的基本要求
- 控制系统的设计概述
- MATLAB在本章中的应用

1.2 自动控制系统的基本概念

我们的现代生活在很大程度上都依赖于自动运行的系统。今天，在人们的日常生活中几乎处处可见自动控制系统的存在，如自动洗衣机、自动售货机、自动电梯等。它们都在一定程度上改变了人类的生活方式，提高了生活质量。所谓自动运行的系统，就是指它的运行不需要人为的干预。自然界有很多这样的例子。如果将人体作为一个例子来考虑，这个系统的持续自动控制是我们生存的基本要求。例如我们的体温保持在37℃的自动温控系统、心跳控制系统、眼球聚焦系统。从肾脏、肺和肝脏的功能来看，它们也可以称为自动系统。这些系统和其他许多人体内的系统一样都是在我们没有任何有意识干预的情况下自动运行的。实际上，在我们周围还有许多自动运行的人造系统。例如，在一个现代化的居室内，温度由温度调节装置自动控制。导航控制系统使汽车自动保持在设定车速，刹车防抱死系统自动防止汽车在湿滑的路面上打滑。在大型办公楼或旅馆，电梯调度系统自动发送车辆搭载乘客。

如今，自动控制技术的应用几乎无处不在，从最初的机械转速、位移的控制到工业过程中温度、压力和流量的控制，从远洋巨轮到深水假艇的控制，从电动假肢到机器人的控制，从电气、机械、航空、化工、核反应、经济管理到生物工程的控制，自动控制理论和技术已经介入许多学科，渗透到各个领域。特别是近年来蓬勃发展的机电一体化（Mechatronics）技术，更是结合了机械、电子和计算机等各方面的优势，研发出数码相机、计算机硬盘等一系列的民用、军用产品。下面通过英文材料的形式，介绍自动控制领域的应用与发展。

Reading Material

Indeed, automatic control system plays an important role in our life. For example, our body is an automatic control system as it always can adjust itself to hold the balance. In fact, we also can find man-made automatic control systems in our daily life. For example, you can find the temperature and humidity in a modern bedroom could be controlled automatically. Navigation control system enables cars to keep a predetermined speed automatically and anti-lock brake system automatically prevents the car from skidding in slippery road. In large office buildings and hotels, the elevator scheduling system automatically sends vehicles carrying passengers.

Industrial engineers, accountants, managers, generals, and many others use the word "control". Although these people come from different walks of life, they are all involved in exercising control in their various areas and their notions of control have an underlying similarity.

Control engineers also use the word "control"; their work is concerned with the control of engineering systems: chart recorders, photo copying machines, robots, diesel engines, satellite positioning, power stations, lifts in high rise buildings, paper-making machines, bread and biscuit making machines, oil refineries, guided missiles and many other machines and processes.

In recent years, mechatronic has surfaced as a philosophy of design based on the integration of mechanical, electrical and software engineering. Many of today's machines and processes are the result of this novel approach and owe their enhanced performance, flexibility and proper operation to the application of some form of electronic or computer control.

In fact, we may consider mechatronic concepts as a natural evolution of control systems technology. Control systems consist of different types of components, and traditionally control engineers have had to deal with the problem of analysis and design of systems having a mix of electrical, electronic, mechanical, liquid flow, gas flow, and thermal components.

所谓的自动化控制(Automatic Control)就是在没有人为操作的前提下,代替人类重复繁重的脑力劳动,对操作对象进行控制。从理论上定义一个自动控制系统首先要从系统的含义说起,系统(System)是指按照某些规律结合在一起的物体(元部件)的组合,它们相互作用,相互依存并能完成一定的任务。由此,我们很容易得出,自动控制系统便是能够实现自动化控制的系统。

Terms and Concepts

From this context, let us note that:

A system is a collection of interconnected components, working together towards some common objective;

A control system is a system whose components have been deliberately configured to collectively achieve a desired objective;

An automatic control system is an interconnection of components forming a system configuration that will provide a desired system response automatically.

1.3 自动控制理论的发展

自动控制理论是研究自动控制共同规律的技术科学,既是一门古老的、已臻成熟的学科,又是一门正在发展的、具有强大生命力的新兴学科。从 1868 年马克斯威尔(J. C.

Maxwell)提出低阶系统稳定性判据至今的一百多年里,自动控制理论的发展可分为四个主要阶段。

第一阶段:经典控制理论(或古典控制理论)的产生、发展和成熟。
第二阶段:现代控制理论的兴起和发展。
第三阶段:大系统控制的兴起和发展。
第四阶段:智能控制的发展阶段。

1. 经典控制理论

经典控制理论(Classical Control Theory)产生并发展于20世纪40—60年代。美国麻省理工学院教授诺伯特•维纳(Norbert Wiener)于1945年发表的论文《控制论:或关于在动物和机器中控制和通信的科学》被认为是控制论的创立之作。控制理论的发展初期,是以反馈理论为基础的自动调节原理,主要用于工业控制。第二次世界大战期间(1938—1945),为了设计和制造飞机及船舶用自动驾驶仪、火炮定位系统、雷达跟踪系统等基于反馈原理的军用设备,进一步促进和完善了自动控制理论的发展。1868年马克斯威尔提出了低阶系统的稳定性判据;1895年数学家劳斯(Routh)和赫尔威茨(Hurwitz)分别独立提出了高斯系统的稳定性判据,即Routh Hurwitz判据;第二次世界大战期间奈奎斯特(H. Nyquist)提出了频率响应理论;1948年伊万斯(W. R. Evans)提出了根轨迹法。至此,控制理论发展的第一个阶段基本完成,形成了以频率法和根轨迹法为主要方法的经典控制理论。

经典控制理论的基本特征:①主要用于线性定常系统的研究,即用于常系数线性微分方程描述的系统的分析和研究;②只用于单输入单输出的反馈控制系统;③只讨论系统输入与输出之间的关系,而忽视系统内部的状态,是一种对系统的外部描述方法。同时,应该指出的是,反馈控制是一种最基本最重要的控制方式,引入反馈信号后,系统对来自内部和外部的干扰响应变得十分迟钝,从而提高了系统的抗干扰能力和控制精度。与此同时,反馈作用又带来了系统稳定性的问题,正是这个曾一度困扰人们的系统稳定性问题,激发了人们对反馈控制系统进行深入研究的热情,推动了自动控制理论的发展和完善。因此从这个意义上来讲,经典控制理论是伴随反馈控制技术的产生和发展而逐渐完善和成熟起来的。

2. 现代控制理论

现代控制理论(Modern Control Theory)于20世纪60年代中期发展成熟。由于经典控制理论只适用于单输入单输出的线性定常系统,只注重系统的外部描述而忽视了系统的内部状态。因而在实际应用中有很大的局限性。

随着航天事业和计算机的发展,20世纪60年代初,在经典控制理论的基础上,以线性代数理论和状态空间分析法为基础的现代控制理论迅速地发展起来。1954年贝尔曼(R. Bellman)提出了动态规划理论;1956年庞特里雅金(L. S. Pontryagin)提出了极大值原理;1960年卡尔曼(R. K. Kalman)提出了多变量最优控制和最优滤波理论。这些理论在数学工具、理论基础和研究方法上不仅提供了系统的外部信息(输出量和输入量),而且还能提供系统内部状态变量的信息。它无论对线性系统或非线性系统,定常系统或时变系统,单变量系统或多变量系统,都是一种有效的分析方法。

3. 大系统理论

20世纪70年代,现代控制理论继续向深度和广度发展,出现了一些新兴的控制方法和理论。例如:①现代频域方法,以传递函数矩阵为数学模型,研究线性定常多变量系统;②自适应控制理论和方法,以系统辨识和参数估计为基础,在实时辨识基础上在线确定最优控制规律;③鲁棒控制法,在保证系统稳定性和其他性能的基础上,设计不变的鲁棒控制器,以处理数学模型的不确定性。

随着控制理论应用范围的不断扩大,从个别小系统的控制,发展到若干相互关联的子系统组成的大系统进行整体控制,从传统的工程控制领域,推广到包括经济管理、生物工程、能源、运输、环境等大型系统。

大系统(Largescale System)理论是过程控制和信息处理相结合的系统工程理论,具有规模庞大、结构复杂、功能综合、目标多样、因素众多等特点。它是一个多输入、多输出、多干扰、多变量的系统。大系统理论目前仍处于发展和开创性阶段。

4. 智能控制理论

智能控制理论(Intelligent Control Theory)是20世纪70年代后,控制理论向深度和广度发展的结果,是人工智能在控制中的应用。智能控制的概念和原理主要是针对被控对象、环境、控制目标或任务的复杂性提出来的,它的指导思想是依据人的思维方式和处理问题的技巧,解决那些目前需要人类智能才能解决的复杂控制问题。被控对象的复杂性体现为模型的不确定性、高度非线性、分布式的传感器和执行器、动态突变、多时间标度、复杂的信息模式、庞大的数据量以及严格的特性指标等。智能控制是驱动智能机器自主地实现其目标的过程,对自主机器人的控制就是一个典型的例子,而环境的复杂性则表现为变化的不确定性和难以辨识。

智能控制是从"仿人"的概念出发的。一般认为,其方法包括学习控制、模糊控制、神经网络控制和专家控制等。

1.4 自动控制系统的基本原理及组成

为了实现各种复杂的控制任务,首先要将被控对象和控制装置按照一定的方式连接起来组成一个有机的整体。在此系统中,被控对象的输出(即被控量)是要求严格加以控制的物理量,可保持为某一恒值,例如温度、压力、液位等,也可以按照某个给定规律运行,例如飞行轨迹、记录曲线等;而控制装置则认为是被控对象施加控制作用的机构总体。理论上讲,可采用不同的原理和方式对被控对象进行控制,而最基本的两种控制形式是开环控制和闭环控制。

Terms and Concepts

Usually, an automatic control system consists of the controlled object and its controller. Based on how the control action is generated, i. e., whether the generation of control action is depended on the actual output, the control systems may be classified as open-loop control systems and closed-loop control systems.

1.4.1 开环控制

开环控制系统(Open-loop Control System)是指系统输出量对系统的调节作用没有影响的系统,即控制设备和控制对象在信号关系上没有形成闭合回路。

Terms and Concepts

An open-loop system utilizes an actuating device to control the process directly without using feedback. Fig. 1-4-1 shows a typical open-loop control system. It is clear that control is based on prior knowledge of the input-output relationship. Current status of the controlled variable is not taken into consideration in making the control decision. Therefore, an open-loop control system is a simple and economical system. It is only possible if the controlled variable and the manipulated variable have a strong, one-to-one relationship. Precise control may not be possible as external factors (disturbances) and changes in system parameters affect the outcome.

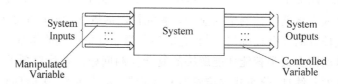

Fig. 1-4-1　A typical open-loop control system

开环系统是按照给定值进行操纵的,信号由给定值至输出量是单向传递的,一定的给定值对应一定的输出量。因此,开环系统的控制精度取决于系统事先的调整精度,对于工作过程中受到的扰动或特性参数的变化无法自动补偿。由于开环控制系统结构简单、成本低廉,使其广泛地应用于系统结构参数稳定和扰动信号较弱的场合,如自动售货机、自动报警器、自动流水线等。

图 1-4-2 给出了一个开环控制系统的结构示意图。从图中可以看出,开环系统只有信号的前向传递,控制器直接作用于被控对象,被控量直接作为整个自动控制系统的输出,整个系统没有信号的反馈。

图 1-4-2　开环控制系统

图 1-4-3 是一个由步进电动机驱动的数控加工机床,其本质也是一个没有反馈环节的开环控制系统。加工程序指令通过运算控制器控制步进电动机精密传动,从而控制台上的工件得以正确加工。

按照自动控制系统的特点,以控制器、执行元件和控制对象等框图形式整理。图 1-4-4 为数控加工机床开环控制框图。此系统的输入量为加工程序指令,输出量为机床工作台的

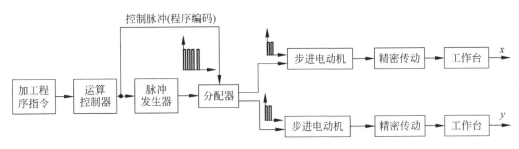

图 1-4-3　数控加工机床示意图

位移,系统的控制对象为工作台,执行机构为步进电动机和传动机构。由图 1-4-4 可见,系统无反馈环节,输出量并不返回来影响控制部分,因此是开环控制。

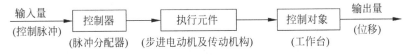

图 1-4-4　数控加工机床开环控制框图

综上所述,开环控制的主要优点是系统结构比较简单,调试方便;但是当工作环境和系统本身的元部件性能参数发生变化时,开环系统的被控变量会受到很大的影响,即抗干扰能力差。一般来说,高精度的开环控制系统要求所有的元部件都有较高的精度和稳定的性能。所以,开环控制对环境和元件有着比较严格的要求。

1.4.2　闭环控制

闭环控制系统(Closed-loop Control System)又常称为反馈控制或按偏差控制,是将控制的结果反馈回来影响当前控制的系统。其中,控制系统的输出对输入有影响,控制器既有控制作用又有调节作用。

Terms and Concepts

A closed-loop control system uses a measurement of the output and feedback of this signal to compare it with the desired output(reference or command). Control systems are most often based on the principle of feedback, whereby the signal to be controlled is compared with a desired reference signal and the discrepancy used to compute corrective control action. These signal processing tasks are implemented through the use of appropriate hardware called the controller. For example, we can take a temperature control of home heating system as a typical example of closed-loop control system.

反馈原理是闭环控制系统的核心机理。反馈又称回馈,是控制论最基本的概念,指将系统的输出返回到输入端并以某种方式改变输入,进而影响系统功能的过程。反馈可分为负反馈和正反馈。在自动控制原理课程中,我们所讲述的反馈系统如果无特殊说明,一般都指负反馈。负反馈使输出起到与输入相反的作用,使系统输出与系统目标的误差减小,系统趋于稳定;正反馈使输出起到与输入相似的作用,使系统偏差不断增大,使系统振荡,可以放大控制作用。对负反馈的研究是控制论的核心问题。

其实,人的一切活动都体现出反馈控制的原理,人本身就是一个具有高度复杂控制能力的反馈控制系统。下面通过解剖从桌上取书的简单动作过程,分析一下它所包含的反馈控制机理。在这里,书的位置是手运动的指令信息,一般称为输入信号。取书时,首先人要用眼睛连续目测手相对于书的位置,并将这个信息送入大脑(称为位置反馈信息);然后由大脑判断手与书之间的距离,产生偏差信号,并根据其大小发出控制手臂移动的命令(称为控制作用或操纵量),逐渐使手与书之间的距离(即偏差)减小。显然,只要这个偏差存在,上述过程就要反复进行,直到误差减小为零,手取到了书。可以看出,大脑控制手取书的过程是一个利用偏差(手与书之间的距离)产生控制作用,并不断使偏差减小直至消除的运动过程;同时,为了取得偏差信号,必须要有手位置的反馈信息,两者结合起来,就构成了反馈控制。显然,反馈控制实际上是一个按照偏差进行控制的过程,因此它也称为按偏差的控制。

人取物视为一个反馈控制系统时,手是被控对象,手的位置是被控量(即系统的输出量),产生控制作用的机构是眼睛、大脑和手臂,统称为控制装置。可以用图 1-4-5 的系统框图来展示这个反馈控制系统的基本组成及工作原理。

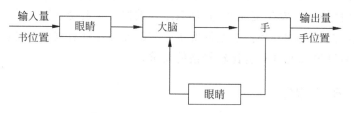

图 1-4-5　人取书的反馈控制系统框图

和开环控制系统相比较,闭环控制系统增加了用于测量输出量的反馈元件,从而组成了反馈网络。图 1-4-6 是一个基本的闭环控制结构示意图。

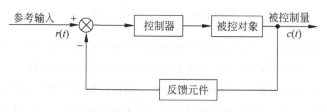

图 1-4-6　闭环控制系统

从生活中的实例来看,电炉箱恒温自动控制系统是一个典型的闭环控制系统。图 1-4-7 给出了电炉箱恒温自动控制系统的示意图。从图中可以看出,电阻丝通过调压变压器主电路加热,炉温期望值用给定电位器预先设置,炉温实际值由热电偶检测,所设的炉温期望值与实际的温度值比较后产生偏差信号,并将其转换成电压,经放大、滤波后,根据预定的控制算法计算出相应的控制量,通过直流伺服电动机和减速器来调节调压变压器的输出电压,从而改变电阻丝中电流的大小,达到控制炉温的目的。

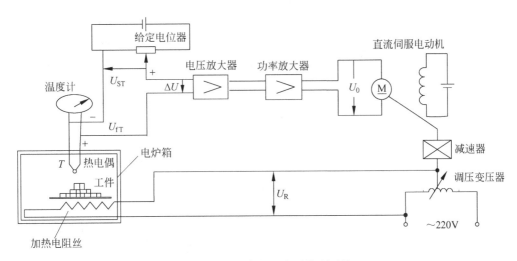

图 1-4-7　电炉箱恒温自动控制系统

利用自动控制的语言,该系统的控制框图如图 1-4-8 所示。图中方框表示了电炉箱恒温自动控制系统的每一个主要组成部分,箭头表示了信号的流向。采用简洁的自动控制语言,可以有效地显示工作流程和每个部件的核心作用。

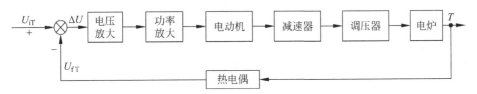

图 1-4-8　电炉箱自动控制框图

由上可见,以电炉箱恒温自动控制系统为代表的闭环控制,其主要优点是自动调节控制量,即任何原因导致被控制量偏离给定值时,就应有相应的控制作用产生,而这种控制作用将减小或消除偏差,使被控量趋于恢复到要求值。因此系统有自动修正输出量的能力,采用反馈控制提高了系统的精度。但是,闭环系统不仅有使用元件多、结构复杂的缺点,而且由于反馈作用,如果系统参数配合不当时,调节过程可能变得很差,甚至出现发散或等幅振荡等不稳定的情况。因此,稳定性成为闭环控制系统的核心问题。

1.4.3　自动控制系统的基本组成

如上所述,自动控制原理适用于电气、机械、航空、化工,甚至经济管理和生物工程等广泛的领域。为了统一各种背景下的被控对象和控制量,在自动控制领域形成了一套自己的语言和表示符号。任何一个自动控制系统都是由被控对象和控制器等基本部件构成的。自动控制系统根据被控对象和具体用途不同,可以有各种不同的结构形式。图 1-4-9 是一个典型自动控制系统的功能框图。图中的每一个方框,代表一个具有特定功能的元件。除被控对象外,控制装置通常是由测量元件、比较元件、放大元件、执行机构、校正元件以及给定元件组成的。这些功能元件分别承担相应的职能,共同完成控制任务。

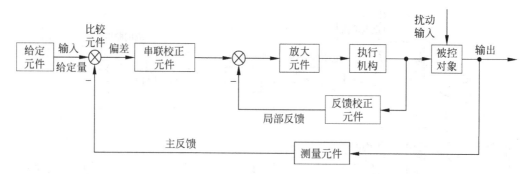

图 1-4-9 典型自动控制系统的功能框图

被控对象(Controlled Object)：一般是指生产过程中需要进行控制的工作机械、装置或生产过程。描述被控对象工作状态的、需要进行控制的物理量就是被控量。

给定元件(Given Element)：主要用于产生给定信号或控制输入信号。

测量元件(Measuring Element)：用于检测被控量或输出量，产生反馈信号。如果测出的物理量属于非电量，一般要转换成电量以便处理，例如图 1-4-7 电炉箱恒温自动控制系统中的热电偶。

比较元件(Components for Comparison)：用于比较输入信号和反馈信号之间的偏差。它可以是一个差动电路，也可以是一个物理元件(如电桥电路、差动放大器、自整角机等)。

放大元件(Amplification Device)：用于放大偏差信号的幅值和功率，使之能够推动执行机构调节被控对象，例如功率放大器、电液伺服阀等。

执行机构(Executive Body)：用于直接对被控对象进行操作，调节被控量，例如阀门、伺服电动机等。

校正元件(Calibration Device)：用于改善或提高系统的性能。常用串联或反馈的方式连接在系统中，例如 RC 网络、测速发电机等。

Terms and Concepts

Fig. 1-4-10 shows a block diagram representation of a dynamic system. Inputs are physical variables external to the system that have some effect(influence) on the state of the system. Outputs are some physical variables of the system; they are useful measures of the state of the system.

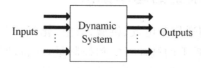

Fig. 1-4-10 Block diagram representation of a dynamic system

From Fig. 1-4-11, we can see a closed-loop control system in action. Controller compares the reference with the measurement and generates the error signal. Then it computes and generates the correction signal as a function of the error signal.

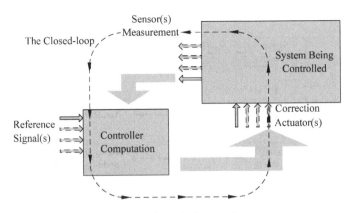

Fig. 1-4-11 A typical control system

A general structure of a closed-Loop control system is given in Fig. 1-4-12. It is clear that the control systems can be divided into information domain and power domain. Each part of the control systems can be analogue or digital devices. In this book, we only concentrate on analogue control systems.

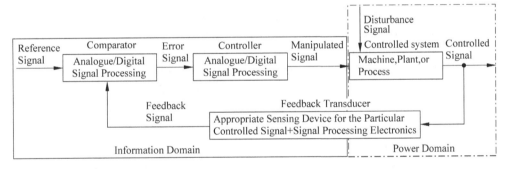

Fig. 1-4-12 A general structure of a closed-loop control system

1.4.4 自动控制系统的实例分析

在工程及实践中有形形色色不同类型的自动控制系统，下面首先通过一段英文阅读材料，介绍自动控制系统的基本组成和基本术语。

1. 自动控制系统概述

Reading Material

Fig. 1-4-13 shows a block diagram of a generic control system. Beginning on the right-hand side of the figure, we see the process that is to be controlled. This is also called the system or the plant. This is usually what we begin with in the control design process. There is an assumed output from the process. This signal Y, is what we want to control to behave in a desired way.

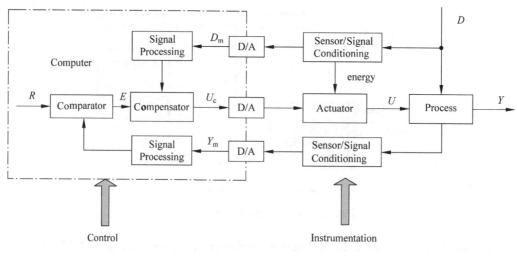

Elements	Signals
Comparator	R: reference, set point
Compensator	E: error; often $E = R - Y_m$
Software Signal Processing	U_c: commanded actuator signal
A/D and D/A Signal Conversions	U: plant input
Hardware Signal Conditioning	D: disturbance
Sensor/Transducer	Y: plant output
Actuator(Final Control Element)	Y_m: measured output
Process, Plant, or System	D_m: measured disturbance

Fig. 1-4-13　A block diagram of a generic control system

The process shown in Fig. 1-4-13 is acted on by two inputs, U and D. The input U comes from an actuator, which is a device that "acts" on the process to affect the system behavior. Usually an actuator is an energy conversion device. The actuator signal is also called the manipulated variable and the actuator may be called the final control element. The actuator signal is what we adjust or manipulate to force changes in the process behavior. The other input, D, is called a disturbance. It is understood to be independent energy source that affects the process behavior, but which we cannot manipulate. Moving to the left of the process we see two measurement blocks labeled sensor/signal conditioning. Like the actuator block, these denote physical hardware. The measurement blocks will typically include some type of transducer that responds to a physical variable as well as an energy conversion. These are collectively called sensors. Often the sensor signal is also acted on by some type of filtering or other signal conditioning to make the signal suitable for processing at the next stage. The figure indicates two sensor/signal conditioning blocks. While we always assume we have some type of measurement of the system behavior(Y), many times we do not have a measurement of the disturbance D. In industrial environments, the term instrumentation is often used to describe the physical actuator and measurement hardware.

Almost all control systems today use some type of digital computations to implement their algorithms. Thus, it is necessary to convert between analog and digital signals in the overall system. In Fig. 1-4-13 these conversions are indicated by the blocks labeled A/D and D/A, representing analog-to-digital and digital-to-analog, respectively. These blocks define the interface between the instrumentation (actuator and measurement devices) and the decision making and control logic executed in a computer or other digital device.

Moving inside the control block in Fig. 1-4-13, we see two blocks labeled signal processing. These blocks may carry out various filtering, scaling, or biasing operations on the incoming signal, often for converting the signals to engineering units. Some of these operations, such as anti-aliasing filtering may also be integrated into the A/D blocks. Finally, the measured, conditioned, and processed value of the measured process output, Y_m, is passed to the comparator block, where it is compared to the reference R, to form an error signal, E. Most often the comparator block simply performs a subtraction, so that the error is just $E = R - Y_m$. The final block to discuss is the compensator. The compensator is the heart of the control system. The compensator uses the error signal as well as any disturbance measurements, D_m, that may be available and then decides on appropriate adjustments to the commanded actuator signal, U_c. These decisions are usually embodied in algorithms called the control law. The combination of the comparator, the compensator, and any associated signal processing is called the controller or simply the control. Finally, the system is said to be closed-loop if the controller is connected to the plant via the actuator and measurement blocks. If these blocks are disconnected, then we say the system is operating in an open-loop mode.

Referring again to Fig. 1-4-13, we would point out several features. First, in general all the signals shown could be vectors. If they are scalars we refer to the system as Single-Input, Single-Output (SISO). If they are vectors, then the system is said to be multivariable or Multi-Input, Multi-Output (MIMO). Also, not all the blocks or the signals shown in the figure are necessarily present in every application. In particular, in many discussions we often assume that the disturbance is not present, the actuator and measurement blocks are lumped in with the plant, the A/D and D/A conversions are transparent, and the comparator is simply a subtraction. Thus, Fig. 1-4-13 effectively reduces to Fig. 1-4-14.

This is a simple unity feedback configuration and is the standard block diagram used to illustrate the basic concepts of control theory in undergraduate courses in feedback control. This figure allows us to illustrate the distinction between two different, but related problems:

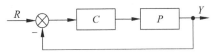

Fig. 1-4-14 Simplified control configuration

(1) Control system engineering: Given a process, determine all the other blocks in Fig. 1-4-13 so that the closed-loop system has desired properties.

(2) Control system design: Given P, determine C in Fig. 1-4-14 to that the closed-

loop system has desired properties.

Although problem(1) is what must be solved in applications, problem(2) is what is typically covered in most introductory course control systems.

We conclude this section by referring again to Fig. 1-4-14. We may think of the control system as fundamentally consisting of three parts. First, the plant produces an output in response to its input. This output is the behavior to be controlled. Second, this behavior is compared to the reference input, which we can think of as the desired behavior or system objective. Third, the error between the actual behavior and the desired behavior is then used by the compensator, which produces a signal to the system that acts to correct the behavior. Notice again that this is a generic description. In engineering we are primarily concerned with the control of physical systems, in which the outputs to be controlled are physical quantities such as voltage, velocity, chemical concentration, etc. However, in general, these ideas are applicable to a wide variety of systems, such as economic systems, natural resource management, decision support systems, etc. In these more abstract settings the inputs and outputs may be described as objectives and performance, respectively. Thus, control system engineering is a true interdisciplinary field, although applications have been in the engineering arena.

2. 电压调节系统

图 1-4-15 给出了一个电压调节系统工作过程。控制目标要求：系统在运行过程中，不论负载如何变化，发电机能够提供由给定电位器设定的规定电压值。当负载恒定，发电机输出规定电压的情况下，偏差电压 $\Delta u = u_r - u = 0$，放大器输出为零，电动机不动，励磁电位器的滑臂保持在原来的位置上，发电机的励磁电流不变，发电机在原动机带动下维持恒定的输出电压。当负载增加使发电机输出电压低于规定电压时，输出电压在反馈处与给定电压经比较后所得的偏差电压 $\Delta u = u_r - u > 0$，放大器输出电压 u_1 便驱动电动机带动励磁电位器的滑臂沿着顺时针旋转，使励磁电流增加，发电机输出电压 u 上升。直到 u 达到规定电压 u_r 时，电动机停止转动，发电机在新的平衡状态下运行，输出满足要求的电压。

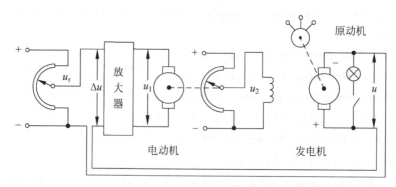

图 1-4-15 电压调节系统原理图

系统中,发电机是被控对象;发电机的输出电压是被控量;给定量是给定电位器设定的电压 u_r。系统框图如图 1-4-16 所示。

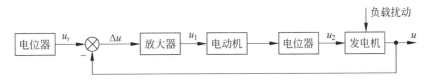

图 1-4-16 电压调节系统框图

3. 高射炮方位角控制系统

采用自整角机作为角度测量元件的高射炮方位角控制系统如图 1-4-17 所示。图中的自整角机工作在变压器状态,自整角发送机 BD 的转子与输入轴连接,转子绕组通入单相交流电;自整角接收机 BS 的转子则与输出轴(炮架的方位角轴)相连接。

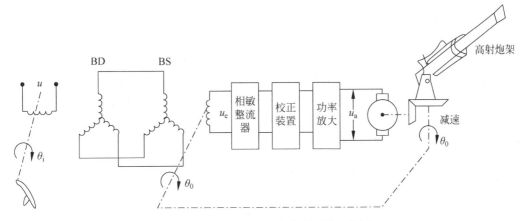

图 1-4-17 高射炮方位角控制系统示意图

当转动瞄准器输入一个角度 θ_i 的瞬间,由于高射炮方位角 $\theta_0 \neq \theta_i$,会出现角位置偏差 θ_e。这时,自整角接收机 BS 的转子输出一个相应的交流调制信号电压 u_e,其幅值与 θ_e 的大小成正比,相位则取决于 θ_e 的极性。当偏差角 $\theta_e > 0$ 时,交流调制信号呈正相位;当 $\theta_e < 0$ 时,交流调制信号呈反相位。该调制信号经相敏整流器解调后,变成一个与 θ_e 的大小和极性对应的直流电压,经校正装置、放大器处理后成为 u_a,驱动电动机带动炮架转动,同时带动自整角接收机的转子将高射炮方位角反馈到输入端。显然,电动机的旋转方向必须是朝着减小或消除偏差 θ_e 的方向转动,直到 $\theta_0 = \theta_i$ 为止。这样,高射炮就指向了手柄给定的方位角上。

系统中,高射炮是被控对象;高射炮方位角 θ_0 是被控量;给定量是由手柄给定的方位角 θ_i。系统框图如图 1-4-18 所示。

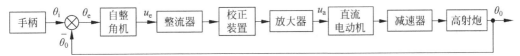

图 1-4-18 高射炮方位角控制系统框图

1.5 自动控制系统的分类

由于自动控制理论研究对象的广泛性,使得自动控制系统的形式是多种多样的。按照不同的标准划分,自动控制系统可以划分为不同的类别,常见的有如下几种。

1.5.1 恒值控制系统和随动控制系统

按给定信号的形式不同,可将系统划分为恒值控制系统(Constant Control System)和随动控制系统(Servo Control System)。

1. 恒值控制系统

恒值控制系统(也称为镇定系统、调节系统或恒值系统)的特点是,给定输入一经设定就维持不变,希望输出量维持在某一特定值上。例如,冶金部门的恒温系统,石油部门的恒压系统等。

2. 随动控制系统

在随动控制系统中,若给定信号的变化规律是事先不能确定的随时间变化的信号(例如函数记录仪、自动火炮系统和飞机自动驾驶仪系统等),则称该系统为自动跟踪系统;若给定输入是预先设定的、按预定规律变化的信号(例如数控机床的输入信号),则称相应的系统为程序控制系统。上述两种系统统称为随动控制系统。

1.5.2 连续系统和离散系统

按照系统内信号的传递形式不同,可将系统划分为连续系统(Continuous System)和离散系统(Discrete System)。

若系统中所有信号都是连续信号,则称为连续系统。如果系统中有一处或几处的信号是离散信号(脉冲序列或数字编码),则称为离散系统(包括采样系统和数字系统)。

随着计算机应用技术的迅猛发展,大量自动控制系统都采用数字计算机作为控制手段。在计算机引入控制系统之后,控制系统就成为离散系统了。

1.5.3 单输入单输出系统和多输入多输出系统

按照输入信号和输出信号的数目,可将系统分为单输入单输出系统[Single Input Single Output(SISO)System]和多输入多输出系统[Multiple Input Multiple Output(MIMO)System]。

单输入单输出系统通常称为单变量系统,这种系统只有一个输入(不包括扰动输入)和一个输出。多输入多输出系统通常称为多变量系统,有多个输入和多个输出。单变量系统可以视为多变量系统的特例。本书的研究对象大部分属于单输入单输出系统。

1.5.4 线性系统和非线性系统

按照组成系统的元件特性分类,可将系统分为线性系统(Linear System)和非线性系统

(Nonlinear System)。

若一个元件的输入与输出的关系曲线为直线,则称该元件为线性元件,否则称为非线性元件。若一个系统中所有的元件均为线性元件,则该系统为线性系统。线性系统有两个重要的特性,即比例性和叠加性。所谓比例性是指当系统的输入量增加 K 倍时,输出量也相应地增加 K 倍。所谓叠加性是指当多个输入量作用于系统时,系统总响应等于各个输入量分别作用于系统的响应之和。系统中只要有一个非线性元件,则该系统称为非线性系统。这时,要用非线性微分(或差分)方程描述其特性。非线性方程的特点是系数与变量有关,或者方程中含有变量及其导数的高次幂或乘积项。例如

$$y^{(2)} + y^2(t) + y(t) = r(t) \tag{1-5-1}$$

$$y(t) = r(t)\sin\omega t \tag{1-5-2}$$

严格地说,实际的物理系统中都含有程度不同的非线性元件,例如放大器和电磁元件的饱和特性,运动部件的死区、间隙和摩擦特性等。由于非线性方程在数学上处理比较困难,目前对不同类型的非线性控制系统的研究还没有统一的方法。但对非线性程度不是太严重的元件,可以用在一定的范围内线性化的方法,从而将非线性控制系统近似为线性控制系统。

Terms and Concepts

As for linear systems, we consider components and systems whose input and output are related by linear differential equation. We can apply Laplace Transform Methods for such systems. In the case of some non-linear systems, we may be able to obtain linear approximations that hold well within narrow ranges of the input and output signals.

1.5.5 定常系统和时变系统

按系统参数是否随时间变化,可以将系统分为定常系统(Time-invariant System)和时变系统(Time-varying System)。

如果控制系统的参数在系统运行过程中不随时间变化,则称为定常系统或者时不变系统,否则称为时变系统。实际系统中的零漂、温度变化、元件老化等影响均属时变因素。严格的定常系统是不存在的,在所考察的时间间隔内,若系统参数的变化相对于系统的运动缓慢得多,则可将其近似作为定常系统来处理。

1.5.6 其他类型

如果被控对象数学模型的结构和参数都是确定的,系统的全部输入信号又均为时间的确定函数,那么系统的输出响应也是确定的,这种系统就称为确定性系统。如果被控对象是确定的,但是系统的信号中含有随机量,如负载的随机变化,电源的随机变动,模型的测量噪声的影响等,就称这种系统为随机系统。因为随机系统及其响应只具有数学统计特性,所以对于随机系统要应用概率统计理论加以研究。如果被控对象本身是不确定的,那么就需要不断地提取这种随机系统在运行过程中的输入输出信息,从中识别对象的模型参数,并不断地修改控制器的参数,以适应系统的随机特性并维持系统的最佳运行状态,这就是现代控制

理论中的模型辨识和现代控制工程中的自适应控制系统。

上述系统参数一般都被认为是集中参数。有些系统的参数不能用集中参数来表示,如电力传输系统的线路阻抗就必须按分布参数来处理。因此,根据参数的分布特性,系统又可分为集中参数系统和分布参数系统。

一般而言,除非特殊说明,本书的研究对象以线性连续定常系统为主,这也是经典控制理论中的基本研究对象。

1.6 控制系统的基本要求

在1.5节中,自动控制系统呈现出研究对象的多样性和复杂性,研究的问题也多种多样,而且不同的问题得出的结论也不尽相同。为了实现自动、有效和精确的控制,必须对控制系统提出一定的要求。根据系统稳态输出和起始过程的特性,对闭环控制系统的基本要求可归纳为三个方面:稳定性、准确性(稳态精度)、快速性与平稳性(动态性能)。三者反映了不同时间区间上被研究对象的特点,这三个方面保证了控制系统的可存在性并在一定条件下实施所希望的控制。

1. 稳定性

稳定性(Stability)是自动控制系统首要考虑的性能要求。从最长的时间区间上讲,稳定性是指系统在 t 于 $(+\infty,-\infty)$ 区间内不发生破坏或不可恢复性运动状态的性质,是系统存在的条件之一。一个稳定的控制系统,其被控制量偏离期望值的初始偏差应随时间的增长而逐渐减小或趋于零。从稳定系统的类型上讲,可以分为以下两类。

(1) 稳定的恒值控制系统:经过系统的扰动,被控制量偏离原来的期望值,经过一定的过渡时间,被控制量应恢复到原来的期望值,此期望值一般为恒定值。

(2) 稳定的随动系统:在自动控制系统的运行中,被控制量应能始终跟踪参考量的变化,参考量的变化一般表现出一定的随机性。

就一般的线性控制系统而言,其稳定性是由系统结构所决定的,与外界因素无关。控制系统中一般含有储能元件或者惯性元件(如绕组的电感、电枢的转动惯量、电炉热容量、物体质量),并且储能元件的能量不可能突变。因此,在有扰动或有输入量时,控制过程不会立即发生,而是有一定的延缓。使被控制量恢复到期望值或者跟踪参考量需要一个时间过程,这一过程称为过渡时间。

例如,反馈控制系统由于被控制对象的惯性,会使控制动作不能及时纠正被控制量的偏差,控制装置的惯性则会使偏差信号不能及时地转换为控制动作。当被控制量回到期望值而使偏差为零时,执行机构惯性向前,致使被控制量超过期望值又产生符号相反的偏差,导致执行机构向相反方向动作,以减小新的偏差。另外,当控制动作已经到位时,又由于被控制对象的惯性,偏差并未减小到零,执行机构继续向前,使被控制量在期望值附近来回摆动,过渡过程呈现振荡形式。若此振荡过程逐渐减弱,系统最后可以达到平衡状态,控制的目的得以实现,称为稳定系统。

从系统运行曲线的角度,稳定性可以这样来表述:系统受到外作用后,其动态过程有振荡的倾向,同时系统具有恢复平衡的能力。如果系统受外力作用后,经过一段时间,其被控量可以达到某一稳定状态,则称系统是稳定的,如图 1-6-1 所示,否则称为不稳定的,如图 1-6-2 所示。其中图 1-6-2(a)为在给定信号作用下,被控制量振荡发散的情况;图 1-6-2(b)为受扰动 $d(t)$ 作用后,被控制量不能恢复平衡的情况。另外,若系统出现等幅振荡,即处于临界稳定的状态,这种情况也视为不稳定。

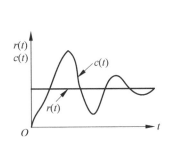

图 1-6-1 稳定系统的动态过程

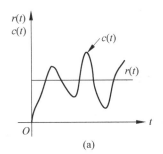

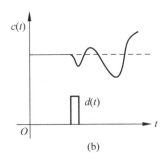

图 1-6-2 不稳定系统的动态过程

显然,不稳定的系统是无法正常工作的。一个能在生产实际中应用的系统,不仅应该是稳定的,而且在动态过程中的振荡也不能太大,否则不能满足生产实际的需要,甚至会导致系统部件的松动和被破坏。

2. 准确性

准确性(Accuracy)就是要求被控变量与设定值之间的误差达到所要求的精度范围。要求被控变量在任何时刻、任何情况下都不超出规定的误差范围,对于高精度控制系统来说,实现起来是困难的。控制的准确性总是用稳态精度来度量。对于稳定的系统,时间足够长就达到了稳态,此时的精度就是稳态精度。稳态精度属于系统的稳态性能。

3. 快速性与平稳性

被控变量由一个值改变到另一个值总是需要一段时间,总是有一个变化过程,这个过程就称为过渡过程,此时系统表现出的特性称为动态性能。人们自然希望过渡过程既快速又平稳,所以快速性和平稳性(Rapidity and Stationary)就是动态性能包含的主要内容。

如果要求一个系统中的被控变量 $c(t)$ 由 0 变到 1,加入对应的输入信号后,输出信号 $c(t)$ 的典型变化曲线如图 1-6-3 所示。图中曲线①和②表示稳定系统的响应,③和④是不稳定系统的响应。

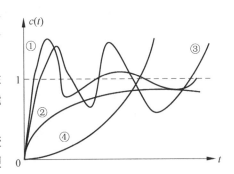

图 1-6-3 系统的典型响应曲线

Terms and Concepts

In control system studies we are mainly concerned with transient and steady state responses of a system which is initially dormant and subjected to step, impulse, or ramp inputs. In chapter 5 of this book, we will investigate the steady state response of the system to a sinusoidal input.

The performance specifications of a control system are: ① disturbance rejection (negligible response to undesirable input signals); ② good transient response (fast, damped); ③ minimum steady state error; ④ low sensitivities (to plant parameter variations); ⑤ robustness (insensitivity to effects not considered in analysis and design phase).

1.7 控制系统的设计概述

自动控制系统的分析与设计是本书研究的重点内容。下面通过一段英文阅读材料,介绍一种通用的自动控制系统设计思想。

Reading Material

Control System Design: The "MAD" Control Theorist

We continue by considering problem (2) defined in 1.4.4 section. Our particular perspective is that there are three essential activities required to design a control system. Referring to Fig. 1-4-14, given a physical system P that we want to control, along with a desired behavior or performance for the controlled system, we determine a control law, C, that will cause the closed-loop system to exhibit the desired behavior by:

(1) Modeling (mathematically) the system based on measurement of essential system characteristics.

(2) Analysis of the model to determine the properties of the system.

(3) Design of the controller which, when coupled with the model of the system, produces the desired closed-loop behavior. This will involve development of:

① control law algorithms.

② measurement and testing techniques for the specific physical system.

③ signal processing and signal conditioning algorithms necessary for interfacing the sensor and controller to the physical system and to each other.

④ simulation studies of the individual components of the control system as well as simulation of the closed-loop system in which all the components are interconnected. Simulation studies are an essential part of the design and development process and are highly dependent on the models obtained from the measurement process.

Note that in addition to these three activities (modeling, analysis, and design), there are two other key activities in the controller development process, although these are associated more with problem (1) (the control engineering processing) than with problem (2). These are:

(1) development of performance specification that define the objective of the control design.

(2) implementation of the controller through software and hardware realizations of the control law, including complete specification of the sensor, signal processing, control elements, final assembly, testing and validation, delivery, and operation of the control system.

The five activities described above are summarized in Fig. 1-7-1, which shows an overall conceptual flowchart of the control system design process. As shown in the figure, starting with a system we wish to control(defined as including the plant, sensors, and actuators), we proceed with two tasks in parallel: defining the required performance specifications and developing a model of the process. The modeling activity will often include some form of measurement to determine key system properties. Note that mathematical modeling is a particularly important part of the process of control system development. By having a framework for describing the system in a precise way, it is possible to develop rigorous techniques for analyzing and designing systems. Once a math model is available and we have decided the goal of the design, it is possible to proceed with

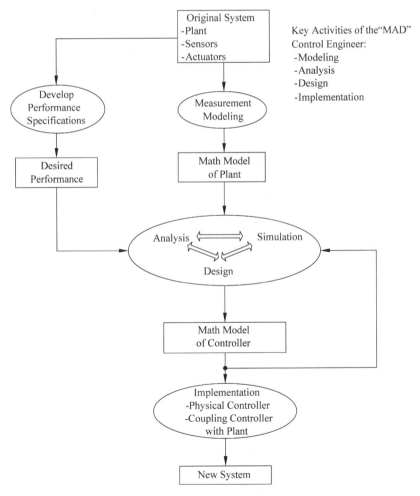

Fig. 1-7-1　Flowchart of control system design process

the analysis of the model and design of the control law. As shown in the figure, simulation cannot be separated from analysis and design, and the process of arriving at a math model of the controller is itself a feedback process. Once a controller model is defined it is necessary to evaluate its effectiveness in combination with the math model of the process (via simulation of the complete control system) before proceeding to implementation.

A laboratory instruction manual will be given in advance. This contains the following: ①learning outcomes; ②introduction to the laboratory equipment; ③experimental procedures; and ④useful hints for the analysis, presentation and discussion of experimental results.

1.8 MATLAB 在本章中的应用

MATLAB 是由美国 MathWorks 公司发布的主要面对科学计算、可视化以及交互式程序设计的高科技计算环境。它将数值分析、矩阵计算、科学数据可视化以及非线性动态系统的建模和仿真等诸多强大功能集成在一个易于使用的视窗环境中,为科学研究、工程设计以及必须进行有效数值计算的众多科学领域提供了一种全面的解决方案。在 MATLAB 的工具箱中,专门设置了针对自动控制系统的多个工具箱,例如控制系统工具箱(Control System Toolbox)、神经网络工具箱(Neural Network Toolbox)、鲁棒控制工具箱(Robust Control Toolbox)等。所有 MATLAB 主包文件和各种工具包都是可读可修改的文件,用户可通过对源程序的修改或加入自己编写程序构造新的专用工具包。本课程的学习将主要利用控制系统工具箱中的部分功能,实现自动控制系统的分析和设计。下面将通过英文材料的形式,复习并掌握 MATLAB 的工作环境,了解 MATLAB 控制系统工具箱中不同命令的功能和实现,演示 MATLAB 自带的示例程序,了解不同结构下控制系统的工作过程及控制效果。

Reading Material

1. Overview of the MATLAB Environment

The MATLAB is a high-performance language for technical computing integrates computation, visualization, and programming in an easy-to-use environment where problems and solutions are expressed in familiar mathematical notation. Typical uses include.

- ☐ math and computation
- ☐ algorithm development
- ☐ data acquisition
- ☐ modeling, simulation, and prototyping
- ☐ data analysis, exploration, and visualization
- ☐ scientific and engineering graphics
- ☐ application development, including graphical user interface building

MATLAB is an interactive system whose basic data element is an array that does not require dimensioning. It allows you to solve many technical computing problems, especially those with matrix and vector formulations, in a fraction of the time it would take to write a program in a scalar non-interactive language such as C or Fortran.

MATLAB features a family of add-on application-specific solutions called toolboxes. Very important to most users of MATLAB, toolboxes allow you to learn and apply specialized technology. Toolboxes are comprehensive collections of MATLAB functions (M-files) that extend the MATLAB environment to solve particular classes of problems. You can add on toolboxes for signal processing, control systems, neural networks, fuzzy logic, wavelets, simulation, and many other areas.

The MATLAB system consists of these main parts:

1) Desktop Tools and Development Environment

This part of MATLAB is the set of tools and facilities that help you use and become more productive with MATLAB functions and files. Many of these tools are graphical user interfaces. It includes: the MATLAB desktop and Command Window, an editor and debugger, a code analyzer, browsers for viewing help, the workspace, files, and other tools.

2) Mathematical Function Library

This library is a vast collection of computational algorithms ranging from elementary functions, like sum, sine, cosine, and complex arithmetic, to more sophisticated functions like matrix inverse, matrix eigenvalues, Bessel functions, and fast Fourier transforms. Each function is a block of code that accomplishes a specific task. To determine the usage of any function, type help [function name] at the MATLAB command window. MATLAB even allows you to write your own functions with the function command.

3) The Language

The MATLAB language is a high-level matrix/array language with control flow statements, functions, data structures, input/output, and object-oriented programming features. It allows both "programming in the small" to rapidly create quick programs, you do not intend to reuse. You can also do "programming in the large" to create complex application programs intended for reuse.

4) Graphics

MATLAB has extensive facilities for displaying vectors and matrices as graphs, as well as annotating and printing these graphs. It includes high-level functions for two-dimensional and three-dimensional data visualization, image processing, animation, and presentation graphics. It also includes low-level functions that allow you to fully customize the appearance of graphics as well as to build complete graphical user interfaces on your MATLAB applications.

5) External Interfaces

The external interfaces library allows you to write C and FORTRAN programs that interact with MATLAB. It includes facilities for calling routines from MATLAB(dynamic linking), for calling MATLAB as a computational engine, and for reading and writing MAT-files.

2. Control System Toolbox

1) Overview of the Control System Toolbox Software

MATLAB technical computing software has a rich collection of functions immediately useful to the control engineer or system theorist. Complex arithmetic, eigenvalues, root-finding, matrix inversion, and fast fourier transforms are just a few examples of important numerical tools found in MATLAB. More generally, the MATLAB linear algebra, matrix computation, and numerical analysis capabilities provide a reliable foundation for control system engineering as well as many other disciplines.

The Control System Toolbox product builds on the foundations of the MATLAB software to provide functions designed for control engineering. This product is a collection of algorithms, written mostly as M-files, that implements common control system design, analysis, and modeling techniques. Convenient Graphical User Interfaces(GUIs) simplify typical control engineering tasks.

Control systems can be modeled as transfer functions, in zero-pole-gain or state-space form, allowing you to use both classical and modern control techniques. You can manipulate both continuous-time and discrete-time systems. Conversions between various model representations are provided. Time responses, frequency responses, and root loci can be computed and graphed. Other functions allow pole placement, optimal control, and estimation. Finally, this product is open and extensible. You can create custom M-files to suit your particular application.

"C:\Program Files\MATLAB\R2009a\toolbox\control"

2) Demos

Control System Toolbox demonstration files show you how to use the toolbox to perform control design tasks in various settings. To run these demos, type demo toolbox control at the MATLAB prompt. This opens the Demos pane in the Help browser showing the Control System Toolbox demos. Alternatively, if you have the Help browser open, you can select the Demos pane directly and follow the same procedure.

3) Building Models

- ☐ Linear(LTI) Models
- ☐ MIMO Models
- ☐ Arrays of Linear Models
- ☐ Model Characteristics
- ☐ Interconnecting Linear Models
- ☐ Converting Between Continuous-and Discrete-Time Systems
- ☐ Reducing Model Order

4) Analyzing Models
- [ ] Quick Start for Performing Linear Analysis Using the LTI Viewer
- [ ] LTI Viewer
- [ ] Simulating Models with Arbitrary Inputs and Initial Conditions
- [ ] Functions for Time and Frequency Response

5) Designing Compensators
- [ ] Quick Start for Tuning Compensator Parameters Using the SISO Design Tool
- [ ] SISO Design Tool
- [ ] Bode Diagram Design
- [ ] Root Locus Design
- [ ] Nichols Plot Design
- [ ] Automated Tuning Design
- [ ] Multi-Loop Compensator Design
- [ ] Functions for Compensator Design

3. Simulink Control Design

The Simulink Control Design software provides tools for linearization and compensator design for control systems and models. Linearized models often simplify compensator design and system analysis. This is useful in many industries and applications, including.

- [ ] aerospace: flight control, guidance, navigation.
- [ ] automotive: cruise control, emissions control, transmission.
- [ ] equipment manufacturing: motors, disk drives, servos.

The Simulink Control Design software works with the Simulink linearization engine and the Control System Toolbox SISO Design Tool. Use it to.

- [ ] compute operating points of models using specifications or simulation.
- [ ] extract linear models from models.
- [ ] tune compensator blocks in models with either single or multi-loop configurations.

The Simulink Control Design software provides a Graphical User Interface for performing linearization and compensator design for Simulink models. With the graphical environment, you can easily inspect and analyze operating points and results of linearization. In addition, you can save and restore settings as well as export results to the MATLAB workspace.

4. Getting Help in MATLAB

MATLAB has a fairly good on-line help; type

`help command name`

for more information on any given command. You do need to know the name of the command that you are looking for; a list of the all the ones used in these tutorials is given in the command listing; a link to this page can be found at the bottom of every tutorial and example page.

在本章中，我们可以通过安装 MATLAB 软件，熟悉 MATLAB 的各个模块。在 MATLAB 控制系统箱中，通过演示示例程序，了解各种控制系统的基本组成和控制效果。其中在热交换控制（Heat Exchange Control）系统中的演示如图 1-8-1 所示。

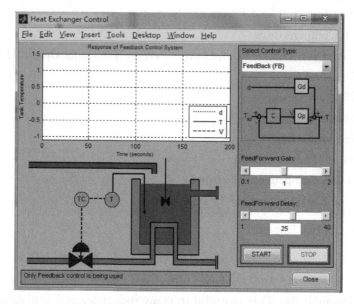

图 1-8-1　热交换控制系统演示图

被控对象是化工反应炉中的液体温度，控制器是热水阀门。系统提供了 4 种可供选择的控制方式，例如开环控制、闭环控制和前馈控制中，精心设计的动画充分展示了各种控制效果，仿真曲线也说明了每一种控制结构的优缺点。

本章小结

（1）自动控制的基本概念是在没有人直接参与的情况下，利用控制装置操纵受控对象，使被控量等于设定的目标值。

（2）自动控制的基本控制结构有开环控制和闭环控制两种。开环控制结构简单，抗扰动能力较差，控制精度一般不高。闭环控制方式的主要特点是抗干扰能力强，控制精度高，但存在能否稳定工作的问题。自动控制原理中主要讨论的是闭环控制方式。

（3）根据不同的需要，可以将自动控制系统按照不同的分类方法进行归类。

（4）一般地，可从稳定（能否正常工作）、准（控制精度）、快（快速响应能力）等方面的性能来评价自动控制系统。而这些性能要求之间往往是相互制约的，因而需要根据不同的工作任务来分析和设计自控系统，使其在满足主要性能要求的同时，兼顾其他性能。

（5）自动控制理论是分析和设计自动控制系统的基础，其发展大致可分为经典控制理论和现代控制理论两大阶段，本书主要介绍经典控制理论。随着生产技术的不断更新和发展，自动控制理论也在不断地发展，了解这方面的知识，对于学习和掌握自动控制技术也是十分必要的。

Summary and Outcome Checklist

Control systems are what make machines and processes function as intended. They are most often based on the principle of feedback whereby the signal to be controlled is compared to a desired reference signal and the discrepancy used to compute corrective control action. These signal processing tasks are implemented through the use of appropriate hardware called the controller.

Tick the box for each statement with which you agree.

- ☐ I am able to explain the meaning of control in the context of machines and processes.
- ☐ I can differentiate between open-loop control and closed-loop control and give an example of each.
- ☐ I am able to sketch the block diagram of a closed-loop control system and name each component and each signal in the diagram.
- ☐ I can identify the function of each component and signal in the above block diagram.
- ☐ I am able to discuss load and load changes in a control system.
- ☐ I am able to list three criteria of good control.
- ☐ I am able to list the benefits of feedback control.
- ☐ I can describe the general approach to designing feedback control systems.

习题

1-1　分析比较开环控制与闭环控制的特征、优缺点和应用场合的不同。

1-2　如题 1-2 图所示为液体自动控制系统原理示意图。在任何情况下,希望液面高度 h 维持不变,试说明系统工作原理,并画出系统原理的框图。

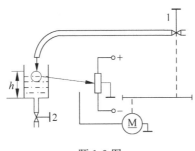

题 1-2 图

1-3　如题 1-3 图所示为水温控制系统。冷水在热交换器中由通入的蒸汽加热,从而得到一定温度的水。冷水流量的变化可用流量计测得。要求:

(1) 说明系统为了保持热水温度为给定值,系统是如何工作的。

(2) 指出系统的被控对象、给定输入量和输出量是什么,并说明冷水流量计的作用。

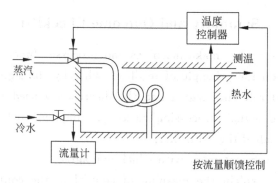

题 1-3 图

(3) 绘制系统的原理结构图。

(4) 指出系统属于何种类型。

1-4 判定下列方程描述的系统是线性定常系统、线性时变系统还是非线性系统。式中 $r(t)$ 是输入信号，$c(t)$ 是输出信号。

(1) $c(t) = 3r(t) + 6\dfrac{\mathrm{d}r(t)}{\mathrm{d}t} + 5\displaystyle\int_0^t r(t)\mathrm{d}t$

(2) $c(t) = 2r(t) + t\dfrac{\mathrm{d}^2 r(t)}{\mathrm{d}t^2}$

(3) $c(t) = [r(t)]^2$

(4) $c(t) = 5 + r(t)\cos\omega t$

(5) $\dfrac{\mathrm{d}^3 c(t)}{\mathrm{d}t^3} + 3\dfrac{\mathrm{d}^2 c(t)}{\mathrm{d}t^2} + 6\dfrac{\mathrm{d}c(t)}{\mathrm{d}t} + c(t) = r(t)$

(6) $t\dfrac{\mathrm{d}c(t)}{\mathrm{d}t} + c(t) = r(t) + 3\dfrac{\mathrm{d}r(t)}{\mathrm{d}t}$

1-5 如题 1-5 图所示为工作台位置液压控制系统。图中，1 为控制电位器，2 为反馈电位器，3 为工作台。该系统可让工作台按照控制电位器给定的信号运动。要求：

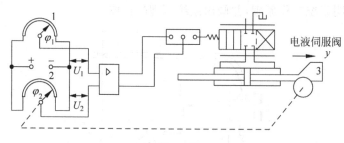

题 1-5 图

(1) 指明系统的输入量、输出量和控制对象。

(2) 绘制系统原理框图。

(3) 说明系统属于何种类型。

第 2 章 控制系统的数学模型

2.1 引言

在第 1 章中,对控制系统的基本概念和主要特点作了概括性的介绍。如果需要对控制系统内部特性及其外部特性进行更深一步的研究,则必须从系统的数学模型入手。所谓控制系统的数学模型就是描述系统在运动过程中各变量之间相互关系的数学表达式。它是分析和设计系统的依据。一个控制系统设计的成败,往往取决于对被控对象动态特性估计的正确程度。因此,控制系统数学模型的建立是控制理论跨学科纵深发展中亟待解决的首要问题。

一个系统数学模型的建立,首先从所研究的物理量谈起。一般而言,系统中变量的关系分为静态关系和动态关系两种。如果系统中各变量随时间变化缓慢,对时间的导数可以忽略不计,就称系统处于静态。当系统的输入量已知时,即可确定系统的输出量及其他变量,这时把描述其变量之间关系的代数方程称为静态数学模型。当系统中的变量对事件的导数不可忽略时,称系统处于动态。而对动态系统而言,为了确定输出量和其他变量,仅仅知道输入量是不够的,还必须要知道一组变量的初始值。因此,把描述其各阶导数之间关系的微分方程称为动态数学模型。

一般来说,建立控制系统数学模型的方法有解析法(又称理论建模法)和实验法(又称系统辨识法)。解析法是对系统的各部分运动和机理进行分析,根据它们所依据的客观规律分别列写相应的运动方程。实验法是人为地给系统施加某种测试信号,记录其输出响应,并用适当的数学模型去逼近,这种方法也称为系统辨识。近年来,系统辨识已发展成为一门独立的学科分支,而本章的重点是研究用解析法建立系统数学模型的方法。

数学模型是描述系统动态特性的数学表达式,从数学的角度可以有多种形式。在经典控制理论中,常用的数学模型是微(差)分方程、传递函数、结构图、信号流图等;在现代控制理论中,采用的是状态空间表达式。结构图、信号流图、状态图是数学模型的图形表达形式。这些模型各有特点和适用范围。

本章主要讨论的是线性定常系统。可以对描述的线性定常微分方程进行积分变换,得出传递函数、框图、信号流图、频率特性等数学描述。线性系统实际上是忽略了系统中某些次要因素,对数学模型进行近似而得到的。以后各章所讨论的系统,一般均指线性化的系统。

下面将通过英文阅读材料的形式,给出分析和设计自动控制系统的一般流程,从中可以看出数学模型在整个过程中的核心地位。

Reading Material

A quantitative mathematical model can be used to analyse a dynamic system and to understand its behavior. The mathematical model describes the relationship between the input and the output of the system. In the present study, attention is confined to those systems for which the input-output relationship is described by ordinary linear differential equations with constant coefficients. Fig. 2-1-1 shows a typical procedure to analysis a control system using a mathematical model.

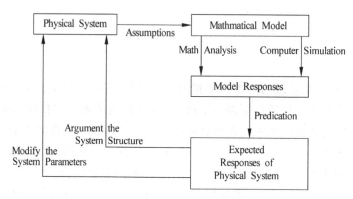

Fig. 2-1-1　Analysis and design using a system model

Mathematical models of dynamic systems (including feedback control systems) are obtained by considering each component/subsystem and developing its transfer function. The individual block diagrams are then arranged in the correct sequence (output of one becomes the input of the next) to obtain the system block diagram. The system block diagram is systematically reduced to obtain the system transfer function. MATLAB may be used to analyse the transfer function models.

The initial step is to obtain the input-output relationship for each component or subsystem in the form of a transfer function. The transfer function blocks are then organised into a block diagram to graphically depict the interconnections in the total system. The block diagram is then systematically reduced to obtain the system transfer function.

The knowledge and experience you gain in creating and using mathematical models of control systems and your understanding of practical issues of implementation equip you with the skills to work in teams engaged in ①designing engineering systems, ②improving existing systems, or ③setting up and commissioning new automation systems.

本章的主要内容：
- □　控制系统的时域数学模型
- □　线性系统的传递函数
- □　控制系统的结构图及其等效变换
- □　信号流图和梅逊公式

☐ 闭环传递函数的定义
☐ 非线性系统模型概述
☐ MATLAB在本章中的应用

2.2 控制系统的时域数学模型

微分方程(Differential Equation)是系统最基本的数学模型，这也是描述系统输入量和输出量之间关系最直接的数学方法。当系统的输入量和输出量都是时间 t 的函数时，其微分方程可以确切地描述系统的运动过程。如果系统的物理参数不随时间变化，则称为定常系统，系统的物理参数不随空间位置变化的系统称为集总参数系统。因此，本节着重研究描述线性、定常、集总参数系统的微分方程。

微分方程是自动控制系统数学模型最基本的形式。传递函数、动态结构图都可由它演化而来。用解析法列写微分方程的一般步骤如下。

(1) 根据控制系统或元件的工作原理(物理、化学和力学等)，确定系统和各元件的输入输出变量。

(2) 从输入端开始按照信号的传递顺序，依据各变量所遵循的各种规律(物理、化学和力学等)，按技术要求忽略一些次要因素，并考虑相邻元件的彼此影响，列出微分方程，一般为微分方程组。

(3) 消去中间变量，求得描述输入量和输出量关系的微分方程。

(4) 标准化。将与输入变量有关的各项放在等号的右侧，与输出变量有关的各项放在等号的左侧，并按降幂排列。最后将系数化简整合为具有一定物理意义的形式。

由此可见，列写微分方程的关键是要了解元件或系统所属学科的有关规律，而不是数学本身。当需要研究的自动控制系统是读者所不熟悉的领域时(例如液压、热力甚至经济系统)，该领域的控制对象背景知识是读者所要首先了解的内容。当建立了相应系统的微分方程等数学模型后，该问题就回归到自动控制系统研究的范畴。本书中所讲述的基本原理即可用于该控制系统研究。

下面以电气系统和机械系统为例，说明如何列写系统或元件的微分方程。这里所举的例子都属于简单系统，而实际系统往往是很复杂的。本书后续章节将逐步介绍如何建立复杂系统的数学模型。

2.2.1 电气系统

电气系统(又称电气网络)是最常见的自动控制系统，其装置是由电阻、电容、电感、运算放大器等元件组成的电路。无源器件是指电阻、电感、电容这类本身不含有电源的器件，像运算放大器这种本身包含电源的器件称为有源器件。仅由无源器件组成的电器网络称为无源网络。如果电器网络中包含有源器件或电源，则称为有源网络。

列写电器网络的微分方程式时都要用到基尔霍夫电流定律和基尔霍夫电压定律。

基尔霍夫电流定律： $$\sum i = 0$$

基尔霍夫电压定律： $$\sum u = 0$$

列写方程时还经常用到理想电阻、电感、电容两端电压、电流与元件参数的关系，它们分别用下面各式表示：

$$u = Ri$$

$$u = L\frac{di}{dt}$$

$$i = C\frac{du}{dt}$$

下面以常见的电路出发，详细介绍电路系统的建模过程。

【例 2-2-1】 图 2-2-1 是由电阻 R、电感 L 和电容 C 组成的无源网络，试列写出以 $u_i(t)$ 为输入量，以 $u_o(t)$ 为输出量的网络微分方程。

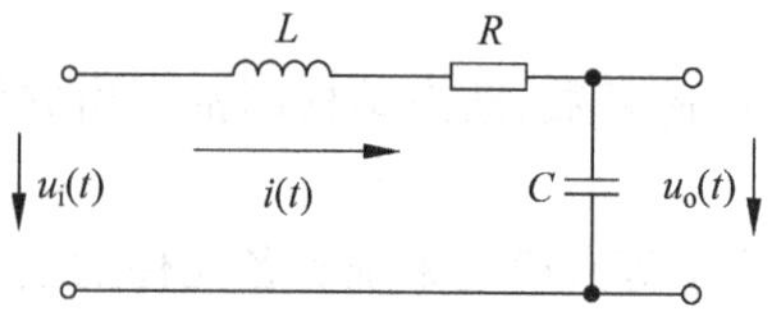

解：设回路电流为 $i(t)$，由基尔霍夫定律可得出回路方程为

$$L\frac{di(t)}{dt} + \frac{1}{C}\int i(t)d(t) + Ri(t) = u_i(t) \tag{2-2-1}$$

图 2-2-1　RLC 无源网络

$$u_o(t) = \frac{1}{C}\int i(t)dt, \quad i(t) = C\frac{du_o(t)}{dt} \tag{2-2-2}$$

消去中间变量 $i(t)$，便可得到描述网络输入输出关系的微分方程为

$$LC\frac{d^2u_o(t)}{dt^2} + RC\frac{du_o(t)}{dt} + u_o(t) = u_i(t) \tag{2-2-3}$$

显然，这是一个典型的二阶线性常系数微分方程，也是图 2-2-1 无源网络的时域数学模型。

【例 2-2-2】 由理想运算放大器组成的电路图如图 2-2-2 所示，电压 $u_i(t)$ 为输入量，电压 $u_o(t)$ 为输出量，求它的微分方程。

解：理想运算放大器正、反相输入端的电位相同，且输出电流为零。根据基尔霍夫电流定律有

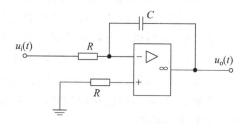

图 2-2-2　电容负反馈电路

$$\frac{u_i(t)}{R} + C\frac{du_o(t)}{dt} = 0 \tag{2-2-4}$$

整理后可得

$$RC\frac{du_o(t)}{dt} = -u_i(t) \tag{2-2-5}$$

式(2-2-5)就是该系统的微分方程，为一阶系统。

2.2.2　机械系统

机械系统是指存在机械运动的装置，一般遵循物理学的力学定律。机械运动包括平移运动(相应的位移称为线位移)和转动(相应的位移称为角位移)两种。

作平移运动的物体要遵循的基本力学定律是牛顿第二定律,即

$$\sum F = m \frac{\mathrm{d}^2 x}{\mathrm{d} t^2}$$

式中,F 为物体所受的力;m 为物体质量;x 为线位移;t 为时间。

转动的物体要遵循如下的牛顿定律:

$$\sum T = J \frac{\mathrm{d}^2 \theta}{\mathrm{d} t^2}$$

式中,T 为物体所受的力矩;J 为物体的转动惯量;θ 为角位移。

运动着的物体,一般都要受到摩擦力的作用,摩擦力 F_C 可表示

$$F_\mathrm{C} = F_\mathrm{B} + F_\mathrm{f} = f \frac{\mathrm{d} x}{\mathrm{d} t} + F_\mathrm{f}$$

式中,x 为位移;$F_\mathrm{B} = f \dfrac{\mathrm{d} x}{\mathrm{d} t}$ 称为黏性摩擦力,它与运动速度成正比,而 f 称为黏性阻尼系数;F_f 表示恒值摩擦力,又称库仑摩擦力。

对于转动的物体,摩擦力的作用体现为如下的摩擦力矩 T_C

$$T_\mathrm{C} = T_\mathrm{B} + T_\mathrm{f} = K_\mathrm{C} \frac{\mathrm{d} \theta}{\mathrm{d} t} + T_\mathrm{f}$$

式中,$T_\mathrm{B} = K_\mathrm{C} \dfrac{\mathrm{d} \theta}{\mathrm{d} t}$ 是黏性摩擦力矩;K_C 称为黏性阻尼系数;T_f 为恒值摩擦力矩。

机械运动的建模过程就是基于上述基本定律的实际运用。一般构成系统的基本器件包括质量块、弹簧和阻尼器等,构成系统的方式有这些器件的串联和并联应用等。下面以例题的形式研究几类常见的机械系统。对于像机器人系统之类的复杂控制系统的建模,都是建立在这些简单模型基础之上的。

【**例 2-2-3**】 图 2-2-3 是一个由弹簧-质量块-阻尼器组成的机械位移系统。m 为物体质量;k 为弹簧系数;f 为阻尼系数;外力 $F(t)$ 为输入量;位移 $x(t)$ 为输出量。试写出系统的运动方程。

解:设质量块 m 相对于初始状态的位移、速度、加速度分别为 $x(t), \dfrac{\mathrm{d} x(t)}{\mathrm{d} t}, \dfrac{\mathrm{d}^2 x(t)}{\mathrm{d} t^2}$,平衡态时弹簧伸长量为 x_o,且有 $mg = k x_\mathrm{o}$,由牛顿运动规律有

$$m \frac{\mathrm{d}^2 x(t)}{\mathrm{d} t^2} = F(t) - F_1(t) - F_2(t) \qquad (2\text{-}2\text{-}6)$$

式中,$F_1(t) = f \dfrac{\mathrm{d} x(t)}{\mathrm{d} t}$ 是阻尼器的阻尼力,其方向与运动方向相反,其大小与运动速度成比例;$F_2(t) = k x(t)$ 是弹簧的弹力,其方向与运动方向相反,其大小与位移成比例。将 $F_1(t)$ 与 $F_2(t)$ 代入式(2-2-6)中,经整理后得到该系统的微分方程为

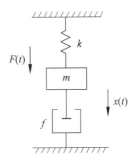

图 2-2-3 机械位移系统

$$m \frac{\mathrm{d}^2 x(t)}{\mathrm{d} t^2} + f \frac{\mathrm{d} x(t)}{\mathrm{d} t} + k x = F(t) \qquad (2\text{-}2\text{-}7)$$

它是一个二阶常系数微分方程。

【**例 2-2-4**】 图 2-2-4 所示的机械转动系统包括一个惯性负载和一个黏性摩擦阻尼器,J 为转动惯量;f 为黏性摩擦系数;ω、θ 为角速度和角位移;T_fz 为作用在该轴上的负载转

阻矩；T 为作用在该轴上的主动外力矩。以 T 为输入量，分别以 ω 和 θ 为输出量，写出对应的运动方程。

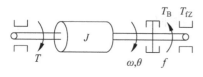

图 2-2-4 机械转动系统

解：根据牛顿转动定律有

$$J\frac{d\omega}{dt}=T-T_B-T_{fZ} \qquad (2-2-8)$$

式中，T_B 为黏性摩擦力矩，且 $T_B=f\omega$，将其代入式(2-2-8)可得

$$J\frac{d\omega}{dt}+f\omega=T-T_{fZ} \qquad (2-2-9)$$

将 $\omega=\dfrac{d\theta}{dt}$ 代入上式可得

$$J\frac{d^2\theta}{dt^2}+f\frac{d\theta}{dt}=T-T_{fZ} \qquad (2-2-10)$$

式(2-2-9)和式(2-2-10)是分别以 ω 和 θ 为输出量的运动方程式。该系统实际上有两个输入量，分别是 T 和 T_{fZ}。

【**例 2-2-5**】 图 2-2-5 所示为恒定磁场他激直流电动机。其中 u 为电枢电压，作为控制输入；m 为作用在电动机轴上的总负载转矩，作为扰动输入；θ 为电动机的转角，作为输出量。

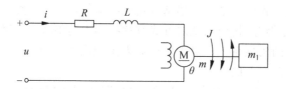

图 2-2-5 恒定磁场他激直流电动机示意图

解：假设电枢反应可忽略不计，电机轴上总转动惯量 J 是常数，各种机械转矩全部归并到负载转矩中，转动轴可认为是刚性轴，电动机电枢回路的电阻、电感全部归并到总电阻 R、电感 L 中。

根据基尔霍夫定律、牛顿定律、直流电机特性，有

$$L\frac{di}{dt}+Ri+e=u \qquad (2-2-11)$$

$$e=C_e\frac{d\theta}{dt} \qquad (2-2-12)$$

$$J\frac{d^2\theta}{dt^2}=m-m_1 \qquad (2-2-13)$$

$$m=C_m i \qquad (2-2-14)$$

式中，R,L——电枢回路总电阻和总电感，Ω，H；

i——电枢电流，A；

e——电动机反电势，V；

u——电枢电压，V；

C_e——电势系数，V·s/rad；

J——电动机轴上的总转动惯量,kg·m²;

m,m_1——电磁转矩、负载转矩,N·m;

C_m——转矩系数,N·m/A。

联立式(2-2-11)~式(2-2-14),消去中间变量 i,e,m,经整理得

$$T_m T_1 \frac{d^3\theta}{dt^3} + T_m \frac{d^2\theta}{dt^2} + \frac{d\theta}{dt} = K_e u - K_m \left(T_1 \frac{dm_1}{dt} + m_1 \right) \quad (2\text{-}2\text{-}15)$$

式中,$T_m = \dfrac{RJ}{C_e C_m}$——电动机的机电时间常数,s;

$T_1 = \dfrac{L}{R}$——电动机的电磁时间常数,s;

$K_e = \dfrac{1}{C_e}$——电枢电压作用系数,rad/(V·s);

$K_m = \dfrac{R}{C_e C_m}$——负载转矩作用系数,rad/(V·m·s)。

若系统的输出量为转速 $n(\text{r/min})$,可将 $\dfrac{d\theta}{dt} = \dfrac{2\pi n}{60}$ 代入式(2-2-15),得

$$T_m T_1 \frac{d^2 n}{dt^2} + T_m \frac{dn}{dt} + n = K_e' u - K_m' \left(T_1 \frac{dm_1}{dt} + m_1 \right) \quad (2\text{-}2\text{-}16)$$

式中,$K_e' = \dfrac{60}{2\pi} K_e$,$K_m' = \dfrac{60}{2\pi} K_m$。

通过以上几个例子可以得出,不同物理系统可以有不同的数学模型;而同一个系统如果所选的输入量、输出量不同时,数学模型也会不同。相似系统是自动控制领域中的一个重要的概念。在物理世界中,实际的系统可能是机械系统,也可能是电气系统或其他类型的系统。如果描述这些系统的微分方程或传递函数具有相同的表现形式,那么无论是动态还是静态,它们的输出响应特性都具有相似性。相似系统的理论无论是对于理论研究,还是工程实际应用,都有着重要的意义。

Reading Material

To understand and control complex systems, one must obtain quantitative mathematical models of these systems. It is necessary therefore to analyze the relationship among the system variables and to obtain a mathematical model. Because the systems under consideration are dynamic in nature, the descriptive equations are usually differential equations. Furthermore, if these equations can be linearized, then the Laplace transform can be utilized to simplify the method of solution. In summary, the approach to dynamic system problems can be listed as follows.

(1) Define the system and its components.

(2) Formulate the mathematical model and list the necessary assumptions.

(3) Write the differential equations describing the model.

(4) Solve the equations for the desired output variables.

(5) Examine the solutions and the assumptions.

(6) If necessary, reanalyze or redesign the system.

2.3 线性系统的传递函数

在电路理论中,为研究电路对输入信号的响应,采用拉普拉斯(Laplace)变换法求解微分方程。通过拉普拉斯变换,将变量从实数域(称为时域)映射到复数 s 域(称为复频域),因此传递函数(Transfer Function)是复频域的模型。它不仅可以表征系统的动态特性,而且可以借其研究系统结构或参数变化对系统性能的影响。传递函数的定义是基于数学上的拉普拉斯变换,需要应用拉普拉斯变换中的叠加原理、导数定理等知识,具体内容可参照本书附录。

Reading Material

Laplace transform method is an efficient method for solving linear differential equations with constant coefficients. The input-output relationship of such systems in the s-domain takes the form of a ratio of two polynomials in s. This ratio is referred to as the transfer function of the system. Many of the characteristics of the system can be ascertained from the transfer function.

2.3.1 传递函数的定义

线性定常系统在零初始条件下,输出量的拉普拉斯变换与输入量的拉普拉斯变换之比,称为该系统的传递函数。

设系统或元件的微分方程一般形式为

$$a_n c^{(n)}(t) + a_{n-1} c^{(n-1)}(t) + \cdots + a_1 \dot{c}(t) + a_0 c(t)$$
$$= b_m r^{(m)}(t) + b_{m-1} r^{(m-1)}(t) + \cdots + b_1 \dot{r}(t) + b_0 r(t) \quad (2\text{-}3\text{-}1)$$

式中,$c(t)$ 为系统输出;$r(t)$ 为系统输入;$a_0, a_1, \cdots, a_n$ 及 $b_0, b_1, \cdots, b_m$ 是由系统结构和参数决定的常数。

设初始条件为零,则有

$$c(0) = \dot{c}(0) = c^{(2)}(0) = \cdots = c^{(n-1)}(0) = 0 \quad (2\text{-}3\text{-}2)$$
$$r(0) = \dot{r}(0) = r^{(2)}(0) = \cdots = r^{(m-1)}(0) = 0 \quad (2\text{-}3\text{-}3)$$

对式(2-3-1)进行拉普拉斯变换,则有

$$(a_n s^n + a_{n-1} s^{n-1} + \cdots + a_1 s + a_0) C(s) = (b_m s^m + b_{m-1} s^{m-1} + \cdots + b_1 s + b_0) R(s) \quad (2\text{-}3\text{-}4)$$

式中,$R(s)$、$C(s)$ 为系统输入 $r(t)$、系统输出 $c(t)$ 的拉普拉斯变换。

由式(2-3-4)可得系统或元件的传递函数为

$$G(s) = \frac{C(s)}{R(s)} = \frac{b_m s^m + b_{m-1} s^{m-1} + \cdots + b_1 s + b_0}{a_n s^n + a_{n-1} s^{n-1} + \cdots + a_1 s + a_0} \quad (2\text{-}3\text{-}5)$$

传递函数的分子多项式(Nominator)$N(s)$、分母多项式(Denominator)$D(s)$分别为

$$N(s) = b_m s^m + b_{m-1} s^{m-1} + \cdots + b_1 s + b_0 \quad (2\text{-}3\text{-}6)$$
$$D(s) = a_n s^n + a_{n-1} s^{n-1} + \cdots + a_1 s + a_0 \quad (2\text{-}3\text{-}7)$$

一般而言,在微分方程模型和传递函数模型之间,用 s^i 代替微分方程式(2-3-1)中的微

分算子 D^i，所得等式右边对应 $N(s)$，左边对应 $D(s)$，即可得传递函数。传递函数与微分方程之间有一一对应关系，所以传递函数也是系统的一种数学模型。在实际工程中，分子多项式 $N(s)$ 和分母多项式 $D(s)$ 是由系统的结构和参数决定的，因此传递函数反映了系统本身的特性。$D(s)$ 称为系统的特征多项式，$D(s)$ 的阶次称为系统的阶次。由式(2-3-5)有

$$C(s)=G(s)R(s) \tag{2-3-8}$$

式(2-3-8)表示输入信号 $R(s)$ 经过 $G(s)$ 的传递后转化成了输出信号 $C(s)$，如图 2-3-1 所示。

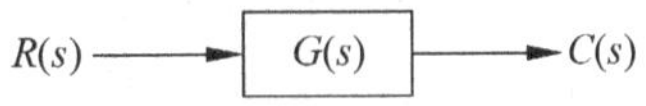

图 2-3-1　传递函数的框图

Terms and Concepts

The transfer function of a linear system is defined as the ratio of Laplace transform of the output variable to the Laplace transform of the input variable, with all initial conditions assumed to be zero. The transfer function of a system(or element)represents the relationship describing the dynamic of the system under consideration. The block diagram is shown in Fig. 2-3-2.

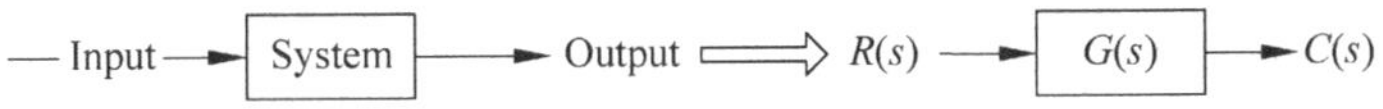

Fig. 2-3-2　The block diagram of transfer function

The input-output relationship of in the Laplace domain is called the transfer function (or G Gain).

以下是关于传递函数的几点说明。

(1) 传递函数是经拉普拉斯变换导出的，拉普拉斯变换是一种线性积分运算，因此传递函数的概念只适用于线性定常系统，它与线性常系数微分方程一一对应。

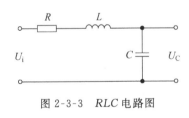

图 2-3-3　RLC 电路图

(2) 传递函数是在零初始条件下定义的，即在零时刻之前，系统处于相对静止状态。因此，传递函数原则上不能反映系统在非零初始条件下的全部运动规律，它是对内部无任何能量存储条件的系统的描述。如果系统的初始条件不为零，如图 2-3-3 所示的电路，说明了这时传递函数的特点。

若有 $Y(s)=G(s)U(s)+V(s)$，其中 $V(s)$ 为非零初始状态下的叠加项。则

$$G(s)=\frac{U_C(s)}{U_i(s)}=\frac{1}{LCs^2+RCs+1} \tag{2-3-9}$$

若 $U_C(0) \neq 0, \dot{U}_C(0) \neq 0$，则有

$$U_C(s)=\frac{1}{LCs^2+RCs+1}U_i(S)+\frac{LCU_C(0)s+[LC\dot{U}_C(0)+RCU_C(0)]}{LCs^2+RCs+1} \tag{2-3-10}$$

其中，

$$U_C(s)=G(s)U_i(s)+V(s); V(s)=\frac{LCU_C(0)s+[LC\dot{U}_C(0)+RCU_C(0)]}{LCs^2+RCs+1}$$

(3) 传递函数只取决于系统本身的结构和元件的参数，是系统的动态数学模型，与输入

信号的具体形式和大小无关。但是同一个系统若选择不同的变量做输入信号和输出信号，所得到的传递函数可能不同，所以谈到传递函数必须指明输入量和输出量。传递函数的概念主要适用于单输入单输出的情况。若线性系统有多个输入信号，在求传递函数时，除了一个有关的输入量以外，其他输入量（包括常值输入量）一概视为零。

(4) 传递函数表示端口关系，不明显表示内部信息。

传递函数不能反映系统或元件的学科属性和物理性质。学科属性和物理性质截然不同的系统可能具有完全相同的传递函数。对于同一物理系统，不同传递函数表示不同端口关系；对于不同物理系统，相同传递函数表示不同端口关系。例如，例 2-2-1 所示的 RLC 电路，当取不同的端口关系时，传递函数式表示出不同的形式。

$$RLC: \frac{U_C(s)}{U_i(s)} = \frac{1}{LCs^2 + RCs + 1} \quad (2\text{-}3\text{-}11)$$

$$RLC: \frac{I_C(s)}{U_i(s)} = \frac{Cs}{LCs^2 + RCs + 1} \quad (2\text{-}3\text{-}12)$$

(5) 传递函数是描述线性定常系统的参数模型（零极点模型）。

式(2-3-5)是传递函数最基本的形式，一般我们将其称为多项式降幂形式。从数学上讲，它也可以改写成零极点形式，表示为

$$G(s) = \frac{C(s)}{R(s)} = K \frac{(s-z_1)(s-z_2)\cdots(s-z_m)}{(s-p_1)(s-p_2)\cdots(s-p_n)} \quad (2\text{-}3\text{-}13)$$

其中，K 为常数；$z_1, z_2, \cdots, z_m$ 称为传递函数的零点；$p_1, p_2, \cdots, p_n$ 为分母多项式方程的根，称为传递函数的极点或特征根。

z_i 和 p_j 可以是实数也可以是复数。若为复数，必须共轭成对出现。在复平面上，一般用"×"表示极点，用"○"表示零点。在自动控制原理中，零极点的位置会决定系统的本质特点（如稳定性和快速性等）。在本书的后续章节中，将详细讨论极点和零点在控制系统中的重要作用。

由此可知，对于实际的物理元件和系统而言，线性定常系统必须满足以下条件。

① 控制系统传递函数的输入量与它所引起的响应（输出量）之间的传递函数需要满足分子多项式 $N(s)$ 的最高阶次 m 总是小于分母多项式 $D(s)$ 的阶次 n，即 $m < n$。这是由于系统中总是含有惯性元件以及受到系统能源的限制。

② 零点与极点或为实数，或为共轭复数。

(6) 传递函数在形式上确定了系统的固有特性，不仅可以传递输入与输出之间的信息关系，而且能够体现系统与外界的联系方式。在图 2-3-4 所示的模型中，图 2-3-4(a)与图 2-3-4(b)中的机械系统的组成元件和相互之间的连接方式是一样的。区别在于图 2-3-4(a)中 $F(t)$ 直接作用于物体 m；图 2-3-4(b)中位移量 $x(t)$ 通过 f_1 和 k 间接作用于物体 m。

由图 2-3-4(a)可知

$$m\ddot{Y}(t) + (f_1 + f_2)\dot{Y}(t) + kY(t) = F(t) \quad (2\text{-}3\text{-}14)$$

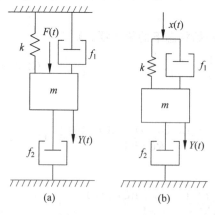

图 2-3-4 机械系统

即
$$\frac{Y(s)}{F(s)} = \frac{1}{ms^2 + (f_1 + f_2)s + k} \tag{2-3-15}$$

由图 2-3-4(b)可知

$$\frac{Y(s)}{X(s)} = \frac{f_1 s + k}{ms^2 + (f_1 + f_2)s + k} \tag{2-3-16}$$

由式(2-3-15)和式(2-3-16)可知,传递函数的分母相同,即它们的固有特性相同;传递函数的分子不同,即它们与外界的联系不同。

Terms and Concepts

By definition, the transfer function of a component or system is the ratio of the transformed output to the transformed input. A transfer function may be defined only for a linear, stationary (constant parameter) system. A nonstationary system, often called a time-varying system, has one or more time-varying parameters, and the Laplace transformation may not be utilized. Furthermore, a transfer function is an input-output description of the behavior of a system. Thus, the transfer function description does not include any information concerning the internal structure of the system and its behavior. The following part gives the main characteristic of the transfer function.

(1) It is based on the linear and stationary system.

(2) The zero initial condition is the basic requirement for deriving the transfer function.

(3) It is a ratio between the input and the output.

(4) $N(s)$ will be lower order than $D(s)$.

(5) It shows the input-output relationship.

(6) The denominator $D(s)$ of transfer function is called the characteristic function since it contains all the physical characteristic of the systems.

(7) The numerator polynomial $N(s)$ is a function how the input enters the systems.

2.3.2 典型环节及其传递函数

不同的控制系统,它们的组成、所用的元部件及其功能都是不同的。但从控制理论出发,只要数学模型相同,其动态性能相同。因此不论控制系统的物理功能有何不同,都认为它们由几种典型环节组成。下面介绍自动控制系统中最常见的典型基本环节。在以后章节的讨论中,这些基本环节在组成各类复杂系统中起到重要的作用。

在以下叙述中,设 $r(t)$ 为环节的输入信号,$c(t)$ 为输出信号,$G(s)$ 为传递函数。其中,每个典型环节中的阶跃响应,将在第 3 章中详细介绍。

1. 放大环节(比例环节)

放大环节(Proportional Component),也称比例环节,其动态方程是

$$c(t) = Kr(t) \tag{2-3-17}$$

由式(2-3-17)可求得放大环节的传递函数

$$G(s) = \frac{C(s)}{R(s)} = K \tag{2-3-18}$$

(a) 功能框图

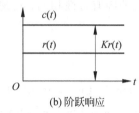

(b) 阶跃响应

图 2-3-5 放大环节

式中，K 为常数，称为放大系数。放大环节又称为比例环节，它的输出量与输入量成比例，它的传递函数是一个常数。放大比例环节的框图如图 2-3-5(a)所示。

当 $r(t) = 1(t)$ 时，$c(t) = K1(t)$，比例环节能立即成比例地响应输入量的变化，比例环节的阶跃响应如图 2-3-5(b)所示。比例环节是自动控制系统中遇到最多的一种，例如电子放大器、齿轮减速器、杠杆机构、弹簧、电位器等。

2. 惯性环节

惯性环节(Inertial Element)的微分方程是

$$T\frac{\mathrm{d}c(t)}{\mathrm{d}t} + c(t) = r(t) \tag{2-3-19}$$

由式(2-3-19)求得惯性环节的传递函数

$$G(s) = \frac{C(s)}{R(s)} = \frac{1}{Ts+1} \tag{2-3-20}$$

式中，T 称为惯性环节的时间常数。当 $T = 0$ 时就变成了放大环节。惯性环节的框图如图 2-3-6(a)所示。

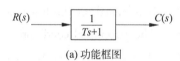

(a) 功能框图

当 $r(t) = 1(t)$ 时，$R(s) = \frac{1}{s}$ 这时可得

$$C(s) = G(s)R(s) = \frac{1}{Ts+1} \cdot \frac{1}{s} = \left(\frac{1}{s} - \frac{1}{s + \frac{1}{T}}\right) \tag{2-3-21}$$

转换为时域表达式，有

$$c(t) = 1 - \mathrm{e}^{-\frac{t}{T}} \tag{2-3-22}$$

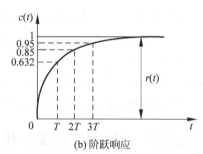

(b) 阶跃响应

惯性环节的阶跃响应曲线如图 2-3-6(b)所示。由图可见，当输入量发生突变时，输出量不会突变，只能按指数规律逐渐变化，这表明该环节具有惯性。在电子系统中，如图 2-3-6(c)所示的一阶惯性调节器可以作为一个很好的例子。由于运算放大器的开环增益很大、输入阻抗很高，所以有

$$\begin{cases} i_0(t) = -i_\mathrm{f}(t) = \dfrac{u_\mathrm{i}(t)}{R_0} \\ i_\mathrm{f}(t) = \dfrac{u_\mathrm{o}(t)}{R_1} + C_1 \dfrac{\mathrm{d}u_\mathrm{o}(t)}{\mathrm{d}t} \end{cases} \tag{2-3-23}$$

于是可得

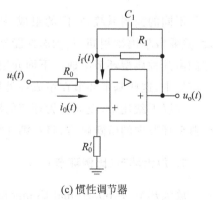

(c) 惯性调节器

图 2-3-6 惯性环节

$$\frac{u_{\mathrm{i}}(t)}{R_0} = -\left[\frac{u_{\mathrm{o}}(t)}{R_1} + C_1 \frac{\mathrm{d}u_{\mathrm{o}}(t)}{\mathrm{d}t}\right] \tag{2-3-24}$$

对式(2-3-24)进行拉普拉斯变换,并整理后可得

$$\frac{U_{\mathrm{o}}(s)}{U_{\mathrm{i}}(s)} = \frac{K}{Ts+1} \left(T = R_1 C_1, K = -\frac{R_1}{R_0}\right) \tag{2-3-25}$$

3. 积分环节

积分环节(Integrating Element)的动态方程是

$$c(t) = \int r(t) \mathrm{d}t \tag{2-3-26}$$

由式(2-3-26)可求得积分环节的传递函数

$$G(s) = \frac{C(s)}{R(s)} = \frac{1}{s} \tag{2-3-27}$$

积分环节的输出量等于输入量的积分,其框图如图 2-3-7(a)所示。

当 $r(t) = 1(t)$ 时,$R(s) = \frac{1}{s}$,则可得到

$$C(s) = G(s)R(s) = \frac{1}{Ts} \cdot \frac{1}{s} = \frac{1}{Ts^2} \tag{2-3-28}$$

转换为时域形式,有

$$c(t) = \frac{1}{T}t \tag{2-3-29}$$

其阶跃响应曲线如图 2-3-7(b)所示。由图可见,输出量随时间的增长而不断增加,增长的斜率为 $m = \frac{1}{T}$。积分环节的特点是它的输出量为输入量对时间的积累。因此,凡是输出量对输入量有存储和积累特点的

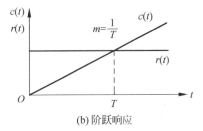

图 2-3-7 积分环节

元件一般都含有积分环节。例如水箱的水位与流量,电炉箱的温度与热流量(或功率),机械运动中的转速与转矩,位移与速度,速度与加速度,电容的电量与电流等。积分环节也是自动控制系统中经常遇到的环节之一。特别是在后续章节中,积分环节的多少将影响系统的整体跟踪性能,将是提高系统稳态精度的主要手段之一。

4. 振荡环节

振荡环节(Oscillating Element)的微分方程是

$$T^2 \frac{\mathrm{d}^2 c(t)}{\mathrm{d}t^2} + 2\xi T \frac{\mathrm{d}c(t)}{\mathrm{d}t} + c(t) = r(t) \quad (0 \leqslant \xi \leqslant 1) \tag{2-3-30}$$

振荡环节的传递函数是

$$G(s) = \frac{C(s)}{R(s)} = \frac{1}{T^2 s^2 + 2\xi T s + 1} = \frac{\omega_{\mathrm{n}}^2}{s^2 + 2\xi \omega_{\mathrm{n}} s + \omega_{\mathrm{n}}^2} \quad (0 \leqslant \xi \leqslant 1) \tag{2-3-31}$$

式中,T、ξ、ω_{n} 皆为常数,且 $\omega_{\mathrm{n}} = 1/T$。$T$ 为该环节的常数;ω_{n} 为无阻尼振荡频率;ξ 为阻尼比。上述传递函数为二阶环节,当 $0 \leqslant \xi \leqslant 1$ 时,该环节称为振荡环节,因为这时的输出函数具有振荡的形式。振荡环节的框图如图 2-3-8(a)所示。

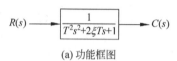

(a) 功能框图

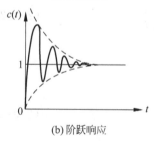

(b) 阶跃响应

图 2-3-8 振荡环节

当 $\xi=0$ 时，$c(t)$ 为等幅自由振荡（又称为无阻尼振荡），其振荡频率为 ω_n。

当 $\xi \geqslant 1$ 时，$c(t)$ 为非周期响应，这时系统已经不是振荡环节了。

当 $0<\xi<1$ 时，$c(t)$ 为减幅振荡（又称为欠阻尼振荡），其振荡频率为 ω_d，ω_d 称为阻尼振荡频率。这时的响应为

$$c(t) = 1 - \frac{e^{-\xi\omega_n t}}{\sqrt{1-\xi^2}}\sin(\omega_d t + \Psi) \quad (2\text{-}3\text{-}32)$$

式中，$\omega_d = \omega_n\sqrt{1-\xi^2}$，$\Psi = \arctan\dfrac{\sqrt{1-\xi^2}}{\xi}$，其阶跃响应曲线如图 2-3-8(b) 所示。在控制系统中，若包含两种不同形式的储能单元，这两种单元的能量又能相互交换，在能量储存和交换过程中，就可能出现振荡而构成振荡环节。例如，由于电感和电容是两种不同的储能元件，电感储存的磁能和电容储存的电能相互交换就有可能形成振荡过程。第3章中将会对二阶系统的动态性能做详细的分析。

5. 微分环节

微分环节（Differentiation Element）在传递函数中有三种类型：纯微分环节、一阶微分环节、二阶微分环节。它们的微分方程分别为

$$C(t) = \frac{dr(t)}{dt} \quad (2\text{-}3\text{-}33)$$

$$C(t) = \tau \frac{dr(t)}{dt} + r(t) \quad (2\text{-}3\text{-}34)$$

$$C(t) = \tau^2 \frac{d^2 r(t)}{dt^2} + 2\xi\tau \frac{dr(t)}{dt} + r(t) \quad (2\text{-}3\text{-}35)$$

相应的传递函数分别为

$$G(s) = s \quad (2\text{-}3\text{-}36)$$

$$G(s) = \tau s + 1 \quad (2\text{-}3\text{-}37)$$

$$G(s) = \tau^2 s^2 + 2\tau\xi s + 1 \quad (2\text{-}3\text{-}38)$$

式中，ξ 和 τ 是常数，称 τ 为该环节的时间常数。图 2-3-9(a) 集中显示了三种微分类典型环节的框图。

微分环节中起主要作用的是零点而不是极点，于是在暂态过程中环节的输出量将含有与输入量微分成比例的分量，环节的名称就由此得来。

当 $r(t)=1(t)$ 时，$R(s)=\dfrac{1}{s}$ 可以得出

$$C(s) = G(s)R(s) = \tau s \frac{1}{s} = \tau \quad (2\text{-}3\text{-}39)$$

$$c(t) = \tau\delta(t) \quad (2\text{-}3\text{-}40)$$

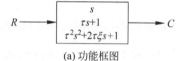

(a) 功能框图

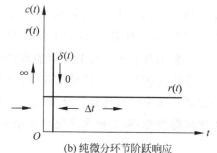

(b) 纯微分环节阶跃响应

(c) 一阶微分环节阶跃响应

图 2-3-9 微分环节

式中，$\delta(t)$为单位脉冲函数，则纯微分环节的阶跃响应曲线如图 2-3-9(b)所示。一阶微分环节的阶跃响应如图 2-3-9(c)所示。

6. 延迟环节

延迟环节(Delay Link)的动态方程是

$$c(t) = r(t-\tau) \quad (2\text{-}3\text{-}41)$$

式中，τ 是常数，称为该环节的延迟时间。由式(2-3-41)可见，延迟环节任意时刻的输出值等于以前的输入值，也就是说，输出信号比输入信号延迟了 τ 个时间单位，从本质上讲是非线性环节，它的传递函数是

$$G(s) = \frac{C(s)}{R(s)} = e^{-\tau s} = \frac{1}{e^{\tau s}} \quad (2\text{-}3\text{-}42)$$

若将 $e^{\tau s}$ 按泰勒(Taylor)级数展开，可得

$$e^{\tau s} = 1 + \tau s + \frac{\tau^2 s^2}{2!} + \frac{\tau^3 s^3}{3!} + \cdots \quad (2\text{-}3\text{-}43)$$

由于 τ 很小，所以可只取前两项，$e^{\tau s} \approx 1+\tau s$，于是由式(2-3-42)有

$$G(s) = \frac{1}{e^{\tau s}} \approx \frac{1}{\tau s+1} \quad (2\text{-}3\text{-}44)$$

式(2-3-44)表明，在延迟时间很小的情况下，延迟环节可用一个小惯性环节来代替。

上述各种典型环节，是从数学模型的角度来划分的，它们是系统传递函数最基本的构成因子。在和实际元件相联系时，应注意以下几点。

(1) 系统的典型环节是按数学模型的共性来划分的，它与系统中使用的元件并非都是一一对应的，一个元件的数学模型可能是若干个典型环节的数学模型的组合，而若干个元件的数学模型的组合也可能就是一个典型的数学模型。

(2) 同一装置(元件)，如果选取的输入输出量不同，它可以成为不同的典型环节。如直流电动机以电枢电压为输入、转速为输出时，它是一个二阶振荡环节。但若以电枢电流为输入、转速为输出时，它却是一个积分环节。

(3) 在分析和设计系统时，将被控对象(或系统)的数学模型进行分解，就可以了解它是由哪些典型环节所组成的。因此，掌握典型环节的动态特性将有助于对系统动态特性的分析研究。

(4) 典型环节的概念只适用于能够用线性定常数学模型描述的系统。

2.3.3　电气网络的运算阻抗与传递函数

求传递函数一般都要列写微分方程。然而对于电气网络，如果采用电路理论中运算阻抗的概念和方法，不列写微分方程也可方便地求出相应的传递函数。这种建模方法在基于电气元件的自动控制系统中非常重要。

就运算阻抗而言，电阻 R 的运算阻抗为其本身 R，电感 L 的运算阻抗为 Ls，电容 C 的运算阻抗为 $\frac{1}{Cs}$，其中 s 是拉式变换的复参量。把普通电路中的电阻 R、电感 L、电容 C 全换成相应的运算阻抗，把电流 $i(t)$ 和电压 $u(t)$ 换成其拉式变换式 $I(s)$ 和 $U(s)$，把运算阻抗当作普通电阻。在零初始条件下，电路中的运算阻抗和电流、电压的拉普拉斯变换式 $I(s)$、$U(s)$

之间的关系满足各种电路定律,利用各种电路定律(如欧姆定律、基尔霍夫电流定律和基尔霍夫电压定律),经过一定的代数运算,就可以求解 $I(s)$、$U(s)$ 及相应的传递函数,同时能够求得相应的运算电路图。

【例 2-3-1】 在图 2-3-10(a)中,电压 $U_1(t)$ 和 $U_2(t)$ 分别是输入量和输出量。求该电路传递函数 $G(s)=\dfrac{U_2(s)}{U_1(s)}$。

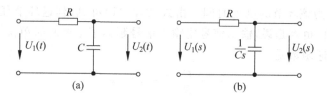

图 2-3-10 RC 电路

解:将电路图 2-3-10(a)变换成运算电路图 2-3-10(b),R 与 $\dfrac{1}{Cs}$ 组成简单的串联电路,于是

$$G(s)=\frac{U_2(s)}{U_1(s)}=\frac{\dfrac{1}{Cs}}{R+\dfrac{1}{Cs}}=\frac{1}{RCs+1} \tag{2-3-45}$$

由该传递函数可知,这是一个惯性环节。

【例 2-3-2】 在图 2-3-11(a)中,电压 $U_1(t)$、$U_2(t)$ 分别为输入量和输出量,求传递函数 $G(s)=\dfrac{U_2(s)}{U_1(s)}$。

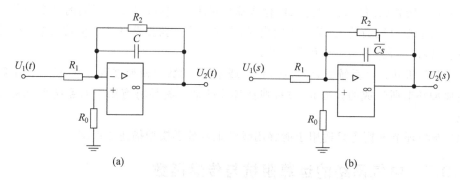

图 2-3-11 运放电路

解:将图 2-3-11(a)变换成运算电路图 2-3-11(b),根据理想运算放大器反相输入时的特性,有

$$G(s)=\frac{U_2(s)}{U_1(s)}=-\frac{R_2\dfrac{1}{Cs}}{R_2+\dfrac{1}{Cs}}\bigg/R_1=-\frac{R_2}{R_1(R_2Cs+1)} \tag{2-3-46}$$

由此传递函数可知,其含有一个放大环节和一个惯性环节。

【例 2-3-3】 在图 2-3-12 中，电压 $U_1(t)$、$U_2(t)$ 分别是输入量和输出量，求传递函数 $G(s) = \dfrac{U_2(s)}{U_1(s)}$。

解：根据理想运算放大器反相输入时的特性，有

$$G(s) = \frac{U_2(s)}{U_1(s)} = -\frac{\dfrac{1}{Cs}}{R} = -\frac{1}{RCs} \qquad (2\text{-}3\text{-}47)$$

由该传递函数可知，其含有一个积分环节。

【例 2-3-4】 在图 2-3-13 中，电压 $U_1(t)$、$U_2(t)$ 分别是输入量和输出量，求传递函数。

解：根据理想运算放大器反相输入时的特性，有

$$G(s) = \frac{U_2(s)}{U_1(s)} = -\frac{R}{\dfrac{1}{Cs}} = -RCs \qquad (2\text{-}3\text{-}48)$$

由该传递函数可知，其含有纯微分环节和放大环节。

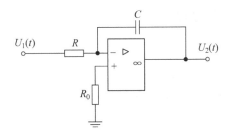 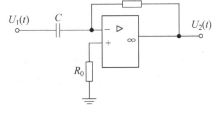

图 2-3-12　积分电路　　　　　　　图 2-3-13　微分电路

【例 2-3-5】 求图 2-3-14(a)所示 RC 网络的传递函数。

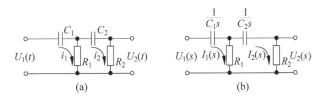

图 2-3-14　RC 级联网络

解：将图 2-3-14(a)变换成运算电路图 2-3-14(b)，同时依据电路知识可写出

$$U_1(s) = \frac{1}{C_1 s} I_1(s) + R_1 [I_1(s) - I_2(s)] \qquad (2\text{-}3\text{-}49)$$

$$R_1 [I_2(s) - I_1(s)] + \frac{1}{C_2 s} I_2(s) + R_2 I_2(s) = 0 \qquad (2\text{-}3\text{-}50)$$

$$U_2(s) = R_2 I_2(s) \qquad (2\text{-}3\text{-}51)$$

消去方程组中的变量 $I_1(s)$、$I_2(s)$，得

$$G(s) = \frac{U_2(s)}{U_1(s)} = \frac{R_1 R_2 C_1 C_2 s^2}{R_1 R_2 C_1 C_2 s^2 + (R_1 C_1 + R_2 C_2 + R_1 C_2) s + 1} \qquad (2\text{-}3\text{-}52)$$

由式(2-3-52)可得出，两级高通滤波器的传递函数由二阶纯微分环节和振荡环节组成。

Reading Material

In chapter 1, you were introduced to the principle of basic control theory. You learnt that all feedback control systems have a common structure irrespective of the system being controlled.

In this chapter, you will study a systematic procedure to obtain quantitative mathematical models of feedback control systems. You will confine your attention to linear systems and obtain the input-output relationship for each component and subsystem in the form of a transfer function. In the following part, the transfer function blocks will be organised into a block diagram to graphically depict the interconnections in the total system. The block diagram will be then systematically reduced to obtain the system transfer function.

2.4 控制系统的结构图及其等效变换

控制系统的传递函数框图又称为动态结构图,简称框图(Block Diagram),它们是以图形表示的数学模型。结构图能够非常清楚地表示输入信号在系统各元件之间的传递过程,利用结构图又可以方便地求出复杂系统的传递函数,因而结构图是分析控制系统的一个简明而又有效的工具。本节介绍如何绘制系统结构图以及结构图的等效变换。

Terms and Concepts

The design of control systems can be divided into two distinct areas. One is concerned with the design of individual components, the other with the design of overall systems by utilizing existing components. The former belongs to the domain of instrumentation engineers; the latter, the domain of control engineers. This is a control text, so we are mainly concerned with utilization of existing components. Consequently, our discussion of control components stresses their functions rather than their structures.

Control components can be mechanical, electrical, hydraulic, pneumatic, or optical devices. Depending on whether signals are modulated or not, electrical devices again are divided into AC (Alternating Current) or DC (Direct Current) devices. Even a cursory introduction of these devices can easily take up a whole text, so this will not be attempted. Instead, we select a number of commonly used control components, discuss their functions, and develop their transfer function. The loading problem will be considered in this development. We then show how these components are connected to form control systems. Block diagrams of these control systems are developed. Finally, we discuss manipulation of block diagrams. Mason's formula is introduced to compute overall transfer functions of block diagrams.

□ The Block diagram representation of the system relationship is prevalent in control systems engineering.

☐ Block diagram consists of unidirectional, operational block that represents the transfer function of the variable of interest.
☐ Block diagram transformations and reduction techniques are derived by considering the algebra of the diagram variables.

2.4.1 结构图的组成

不论一个控制系统多么复杂或是多么简单，它的结构图必须而且只有 4 个基本要素组成。

1. 函数方框（Function Block）

函数方框表示对信号进行数学变换。方框中写入元件或系统的传递函数，输出变量等于方框中的传递函数与输入变量的乘积，如图 2-4-1(a) 所示。

2. 方框的输出信号流线（The Output Signal）

方框的输出信号流线是指带有箭头的有向线段。箭头表示信号的传递方向，线上标明信号的拉普拉斯变换函数，如图 2-4-1(b) 所示。

3. 分支点（测量点、引出点）（A Pickoff Points, Branch Points）

分支点表示信号的引出或测量位置。从同一位置引出的信号，在数值和性质各方面完全相同，如图 2-4-1(c) 所示。

4. 相加点（综合点）（Summing Point）

对两个或是两个以上信号进行加减运算。"＋"表示相加，"－"表示相减，通常"＋"号可以省略不写，如图 2-4-1(d) 所示。

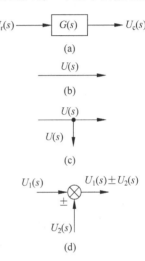

图 2-4-1　结构图基本要素

Terms and Concepts

Any system described by a set of linear differential (and/or linear algebraic) equations can be represented graphically by a block diagram.
☐ Circle is the symbol for a summing operation.
☐ Arrows pointing toward the circle indicate input signals.
☐ An arrow pointing away from the circle indicates the output.
☐ A sign($\pm$) placed near the arrow head indicates whether the particular signal is to be added or subtracted.

2.4.2 控制系统结构图的建立

结构图的建立一般是进行控制系统分析的首要步骤。同微分方程模型的建立类似，需要充分了解控制系统的背景知识，按照控制系统所属领域的规律逐步进行。一般而言，建立控制系统结构图的步骤如下。

（1）列写出系统各元件的微分方程。在建立方程时应分清各元件的输入量、输出量，同时应考虑相邻元部件之间是否有负载效应。

（2）在零初始条件下，对各微分方程进行拉普拉斯变换，并将变换式写成标准形式。

（3）由标准变换式利用结构图的 4 个基本单元，分别画出各元部件的子结构图。

（4）按照系统中信号的传递顺序，依次将各元部件的子结构图连接起来，便可得到系统的结构图。

【例 2-4-1】 在图 2-4-2(a)中，电压 $U_1(t)$、$U_2(t)$ 分别是输入量和输出量，试建立系统的结构图。

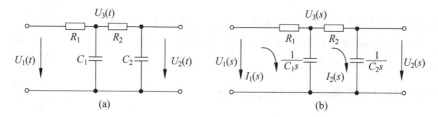

图 2-4-2 RC 滤波电路图

解：图 2-4-2(a)所对应的运算电路图如图 2-4-2(b)所示。设中间变量 $I_1(s)$、$I_2(s)$ 和 $U_3(s)$，其中 $U_3(s)$ 是连接 R_1 和 R_2 之间的电位。从输出量 $U_2(s)$ 写出系统方程式(2-4-1)，其后的方程由上一式右侧的未知量入手，依次写至输入端。

$$U_2(s) = \frac{1}{C_2 s} I_2(s) \tag{2-4-1}$$

$$I_2(s) = \frac{1}{R_2}[U_3(s) - U_2(s)] \tag{2-4-2}$$

$$U_3(s) = \frac{1}{C_1 s}[I_1(s) - I_2(s)] \tag{2-4-3}$$

$$I_1(s) = \frac{1}{R_1}[U_1(s) - U_3(s)] \tag{2-4-4}$$

按照上述方程式的顺序，根据输入量和输出量可依次得到如图 2-4-3(a)～图 2-4-3(d)所示的子方框结构图。每一个子结构图充分体现了所对应的微分方程的输入输出关系。最后，从最终输出量开始顺序绘制系统整体结构图，如图 2-4-4 所示。

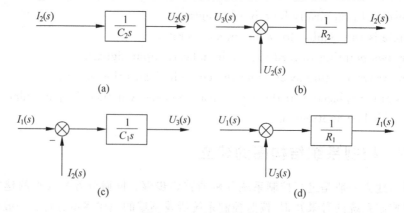

图 2-4-3 子框图的建立

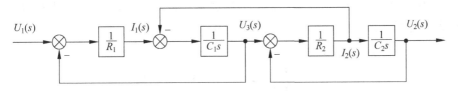

图 2-4-4　RC 滤波电路整体结构图

2.4.3　结构图的等效变换

一般情况下,系统整体传递函数是分析系统动态性能的首要条件。但是,一个复杂的结构图,其方框的连接必然是错综复杂的,为了得到系统的输入量与输出量之间的系统函数,必须对系统的结构图进行简化。结构图的变换相当于在结构图上进行数学方程式的运算,因此必须遵循变换前后数学关系保持不变的原则,通常又称结构图等效变换。

结构图简化(Block Diagram Reduction)的一般方法是移动引出点或比较点,变换比较点,进行方框运算,将串联、并联反馈连接的方框合并。以下是几种常见的框图连接方式。

1. 串联方框的简化(等效)(In Series)

前一环节的输出量是后一环节的输入量的连接称为环节的串联。如图 2-4-5(a)所示,传递函数分别为 $G_1(s)$ 和 $G_2(s)$ 的两个方框,若 $G_1(s)$ 的输出量作为 $G_2(s)$ 的输入量,则称为串联连接。

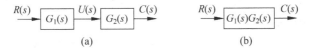

图 2-4-5　方框串联连接及其简化

由图 2-4-5(a)有

$$U(s) = G_1(s)R(s) \tag{2-4-5}$$

$$C(s) = G_2(s)U(s) \tag{2-4-6}$$

由式(2-4-6)消去 $U(s)$,得

$$C(s) = G_1(s)G_2(s)R(s) = G(s)R(s) \tag{2-4-7}$$

式中,$G(s)=G_1(s)G_2(s)$ 是串联方框的等效传递函数,原来的两个方框现在等效成一个方框,可用图 2-4-5(b)的框图表示。由此可知串联连接的等效方框的传递函数等于各个方框传递函数之乘积。这个结论可以推广到 n 个串联方框的情况。

在考虑两环节是否为串联时要注意以下两点。

(1) 环节之间应无负载效应,否则要考虑将它们作为一个整体,而不能分为两个独立的部分。

(2) 串联连接的环节之间应无分支点和综合点,否则它们就不是串联。

2. 并联方框的简化(等效)(In Parallel)

输入量相同,输出量相加或相减的连接称为并联。如图 2-4-6(a)所示,传递函数分别

为 $G_1(s)$ 和 $G_2(s)$ 的两个方框,如果它们有相同的输入量,而输出量等于两方框输出量的代数和,则称为并联连接。

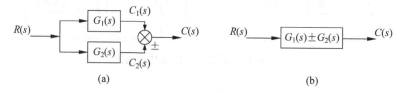

图 2-4-6 框图并联连接及简化

由图 2-4-6(a)有

$$C_1(s) = G_1(s)R(s) \tag{2-4-8}$$

$$C_2(s) = G_2(s)R(s) \tag{2-4-9}$$

$$C(s) = C_1(s) \pm C_2(s) \tag{2-4-10}$$

由式(2-4-8)~式(2-4-10)消去 $C_1(s)$ 和 $C_2(s)$,得

$$C(s) = [G_1(s) \pm G_2(s)]R(s) = G(s)R(s) \tag{2-4-11}$$

式中,$G(s)=[G_1(s)\pm G_2(s)]$ 是并联方框的等效传递函数,可用图 2-4-6(b)的方框表示。由此可知两个方框并联连接的等效传递函数,等于各个方框传递函数的代数和。这个结论可以推广到 n 个方框并联连接的情况。

3. 反馈方框的简化(等效)

如果将系统或环节的输出反馈到输入端与输入信号进行比较,就构成了反馈连接,如图 2-4-7(a)所示,若传递函数分别为 $G(s)$ 和 $H(s)$,反馈信号 $B(s)$ 在相加点前取"+"号表示为正反馈,表示输入信号与反馈信号相加;取"-"号表示相减,是负反馈。负反馈是自动控制系统中经常会碰到的基本结构形式。

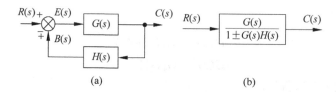

图 2-4-7 框图的反馈连接及简化

由图 2-4-7(a)有

$$C(s) = G(s)E(s) \tag{2-4-12}$$

$$B(s) = H(s)C(s) \tag{2-4-13}$$

$$E(s) = R(s) \mp B(s) \tag{2-4-14}$$

由式(2-4-12)~式(2-4-14)消去中间变量 $E(s)$ 和 $B(s)$,得

$$C(s) = G(s)[R(s) \mp H(s)C(s)] \tag{2-4-15}$$

于是有

$$C(s) = \frac{G(s)}{1 \pm G(s)H(s)} R(s) \tag{2-4-16}$$

则可得到

$$\Phi(s) = \frac{C(s)}{R(s)} = \frac{G(s)}{1 \pm G(s)H(s)} \qquad (2\text{-}4\text{-}17)$$

式中,$G(s)$为前向通道传递函数;$H(s)$为反馈传递函数;$\Phi(s)$为闭环传递函数;$G(s)H(s)$为系统的开环传递函数。式(2-4-17)即为反馈连接的等效传递函数,"+"适用于负反馈系统,"-"用于正反馈系统。式(2-4-16)可用图2-4-7(b)的方框表示。对照图2-4-7(a)和图2-4-7(b)可知,变换前后输入量和输出量之间的关系不变,因此图2-4-7(a)和图2-4-7(b)两图等效。反馈连接的等效传递函数是最重要的结构图化简工具,一般可以有效地减少信号的耦合和交叉,简化系统之间的相互关系。

2.4.4 信号相加点和分支点的移动和互换

1. 信号相加点的移动

信号相加点移动的原则是,原信号不变,移动前后保证相加结果不变。图2-4-8清楚地表示出相加点移动的等效方法。图2-4-8(a)表示相加点往前移;图2-4-8(b)表示相加点往后移。

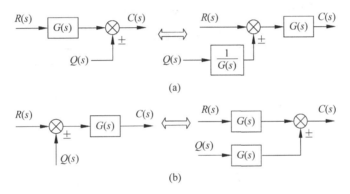

图 2-4-8 相加点移动

2. 信号分支点移动

信号分支点移动的原则是,原各点信号不变,信号分支点移动后要保证该分支信号不变。分支点移动的等效结构图如图2-4-9所示。其中图2-4-9(a)表示分支点前移,图2-4-9(b)表示分支点后移。

3. 信号相加点互换

根据加法交换律,两个或多个相邻相加点位置互换时,互换前后的结果不变。即在简化过程中,可以根据需要交换相邻的相加点的位置。

4. 信号分支点的互换

如果若干个引出点相邻,则表明同一个信号送到多处。因此,相邻点互换位置完全不改变信号的性质,即这种变换不需要作任何传递函数的变换。特别指出,相邻的分支点和相加

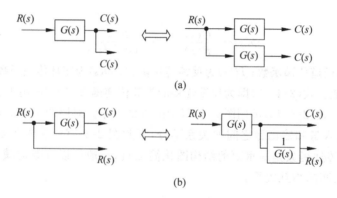

图 2-4-9 分支点的移动

点是不能交换的,否则会引出错误的结果。

分支点前移时,必须在分出支路串入具有相同传递函数的函数方框;分支点后移时,必须在分出支路串入具有相同传递函数倒数的函数方框。相加点前移时,必须在移动的相加支路中串入具有相同传递函数倒数的函数方框;相加点后移时,必须在移动的相加支路中串入具有相同传递函数的函数方框。

框图变换时经常需要用到的变换规则如表 2-4-1 所示。

表 2-4-1 框图变换规则

变 换	原 框 图	等 效 框 图
支点前移	$A \to G \to AG$, AG	$A \to G \to AG$, $G \to AG$
支点后移	$A \to G \to AG$, A	$A \to G \to AG$, $\frac{1}{G} \to A$
加点前移	$A \to G \to AG \xrightarrow{+}_{-} AG-B$, B	$A \xrightarrow{+}_{-} \to G \to AG-B$, $\frac{1}{G} \leftarrow B$
加点后移	$A \xrightarrow{+}_{-} A-B \to G \to AG-BG$, B	$A \to G \to AG \xrightarrow{+}_{-} AG-BG$, $B \to G \to BG$
变单位反馈	$A \xrightarrow{+}_{-} \to G \to B$, H	$A \to \frac{1}{H} \xrightarrow{+}_{-} \to H \to G \to B$

续表

变换	原框图	等效框图
相加点变位	$A \to \otimes \to A-B$; B 输入到 $\otimes$（负）	先 A 与 B 作 $A-B$（上支路），再复制 $A-B$
	$A \to \otimes \to A-B$（上支路输出 A）	等效图：B 先加到前端
	$A \xrightarrow{+} \otimes \xrightarrow{A-B} \otimes \xrightarrow{A-B+C}$，$B$ 负，C 正	$A \xrightarrow{+} \otimes \xrightarrow{A+C} \otimes \xrightarrow{A-B+C}$，$C$ 正，B 负

Terms and Concepts

In general, by systematic reduction of the block diagram, the transfer function relating an input to an output can be easily obtained. The following steps are generally taken during the reduction of a block diagram.

(1) Combine blocks in cascade.

(2) Move a summing point behind(ahead of)a block.

(3) Move a pickoff point behind(ahead of)a block.

(4) Eliminating a feedback loop.

2.4.5 结构图简化示例

利用结构图的变换规则简化系统的结构图时,可根据具体情况采取不同的简化方法。如果结构图只有简单的串、并联和反馈连接时,可先计算简单的串、并联和反馈连接部分,然后再逐步简化整个结构图。如果结构图中存在交叉连接或交叉反馈时,则先做分支点或综合点的移动,消去交叉现象后,再按简单连接方式逐步简化。

【例 2-4-2】 简化图 2-4-10(a)所示的多回路系统,求其闭环传递函数 $\dfrac{C(s)}{R(s)}$ 及 $\dfrac{E(s)}{R(s)}$。

解:该框图有三个反馈回路,由 $H_1(s)$ 组成的回路称为主回路,另外两个回路是副回路。由于存在着由分支点和相加点形成的交叉点 A 和 B,首先要解除交叉。可以将分支点 A 后移到 $G_4(s)$ 的输出端,或将相加点 B 前移到 $G_2(s)$ 的输入端后再交换相邻相加点的位置,或同时移动 A 和 B。这里采用将点 A 后移的方法将图 2-4-10(a)化简为图 2-4-10(b)。化简 G_3、G_4、H_3 副回路后得到图 2-4-10(c)。对于图 2-4-10(c)中的副回路再进行串联和反馈简化得到图 2-4-10(d)。由该图求得

$$\frac{C(s)}{R(s)} = \frac{\dfrac{G_1 G_2 G_3 G_4}{1 + G_2 G_3 H_2 + G_3 G_4 H_3}}{1 + \dfrac{G_1 G_2 G_3 G_4 H_1}{1 + G_2 G_3 H_2 + G_3 G_4 H_3}} = \frac{G_1 G_2 G_3 G_4}{1 + G_2 G_3 H_2 + G_3 G_4 H_3 + G_1 G_2 G_3 G_4 H_1} \quad (2\text{-}4\text{-}18)$$

$$\frac{E(s)}{R(s)} = \frac{1}{1 + \dfrac{G_1 G_2 G_3 G_4 H_1}{1 + G_2 G_3 H_2 + G_3 G_4 H_3}} = \frac{1 + G_2 G_3 H_2 + G_3 G_4 H_3}{1 + G_2 G_3 H_2 + G_3 G_4 H_3 + G_1 G_2 G_3 G_4 H_1} \quad (2\text{-}4\text{-}19)$$

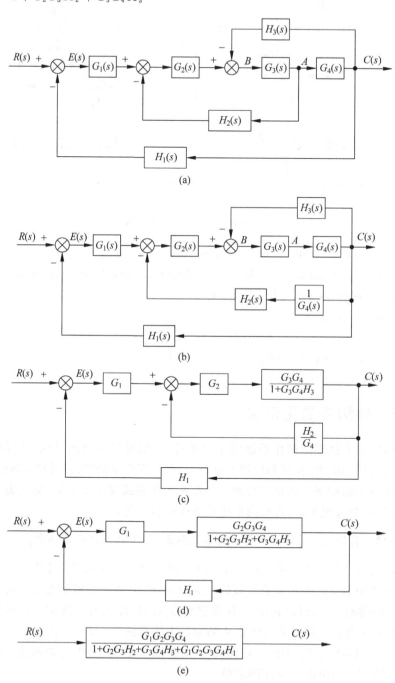

图 2-4-10 多回路框图的化简

2.5 信号流图和梅逊公式

2.5.1 信号流图

信号流图(Signal-Flow Graph)是一种表示线性化代数方程组变量间关系的图示方法。信号流图由节点和支路组成。每一个节点表示系统的一个变量,而每两个节点间的连接支路为这两个变量之间信号的传输关系。信号流向用支路上的箭头表示,传输关系(增益、传递函数)则标注在支路上。

下面以图 2-5-1 所示某个系统的信号流图为例,介绍信号流图的一些术语。

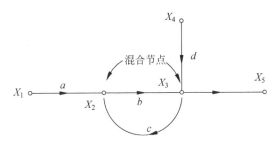

图 2-5-1 典型的信号流图

节点:用以表示变量或信号的点称为节点,用符号"。"表示,如 X_1, X_2, X_3, X_4, X_5。

传输:两个节点之间的增益或传递函数称为传输。

支路:联系两个节点并标有信号流向的定向线段称为支路,如 a, b, c, d。

源点:只有输出支路,没有输入支路的节点称为源点,它对应于系统的输入信号,或称为输入节点,如 X_1, X_4。

阱点:只有输入支路,没有输出支路的节点称为阱点,它对应于系统的输出信号,或称为输出节点,如 X_5。

混合节点:既有输入支路,又有输出支路的节点称为混合节点,如 X_2, X_3。

通路:沿支路箭头方向而穿过各相连支路的途径,称为通路。如果通路与任一节点相交不多于一次,就称为开通路;如果通路的终点就是通路的起点,并且与任何其他节点相交的次数不多于一次,则称为闭通路或回路;如果通路通过某一节点多于一次,那么这个通路既不是开通路,又不是闭通路。

回路增益:回路中各支路传输的乘积,称为回路增益。

在此,将信号流图的基本性质归纳如下,以图 2-5-2 为例。

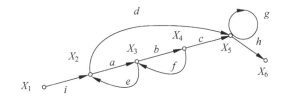

图 2-5-2 典型的信号流图

(1) 节点标志系统的变量。一般节点自左向右顺序设置，每个节点标志的变量是所有流向该节点的信号之代数和，而从同一节点流向各支路的信号均用该节点的变量表示。例如，在图 2-5-2 中，节点 X_3 标志的变量是来自节点 X_2 和节点 X_4 的信号之和，它同时又流向节点 X_2 和节点 X_4。

(2) 支路相当于乘法器，信号流经支路时，被乘以支路增益而变换为另一信号。例如，在图 2-5-2 中，来自节点 X_2 的变量被乘以支路增益 a，来自节点 X_4 的变量被乘以支路增益 f，自节点 X_3 流向节点 X_4 的变量被乘以支路增益 b。

(3) 信号在支路上只能沿箭头单向传递，即只有前因后果的因果关系。

(4) 对于给定的系统，节点变量的设置是任意的，因此信号流程图不是唯一的。

Terms and Concepts

A signal-flow graph is a diagram consisting of nodes that are connected by several directed branches and is a graphical representation of a set of linear relations.

The basic element of a signal-flow graph is a unidirectional path segment called a branch, which relates the dependency of an input and an output variable in a manner equivalent to a block of a block diagram. The input and output points or junctions are called nodes.

A path is a branch or a continuous sequence of branches that can be traversed from one signal to another signal.

A loop is a closed path that originates and terminates on the same node, and along the path no node is met twice. Two touching loops are said to be nontouching if they do not have a common node.

2.5.2　梅逊公式

首先，我们定义两个术语。

前向通路及前向通路传递函数：信号从输入端开始，沿着箭头方向传递到输出端时，每个方块和综合点只经过一次的通路，称为前向通路。前向通路上所有传递函数的乘积，称为前向通路传递函数。

回路及回路传递函数：信号传递的起点就是其终点，而且每个方块和综合点只通过一次的闭合通路，称为回路。回路中所有传递函数的乘积（包括代表回路反馈极性的正、负号），称为回路传递函数。

Terms and Concepts

Mason loop rule: A rule that enables the user to obtain a transfer function by tracing paths and loops within a system.

用梅逊公式，可不经过任何结构变换，一步写出系统的传递函数。

$$\Phi(s) = \frac{\sum_{i=1}^{n} P_i \Delta_i}{\Delta} \qquad (2-5-1)$$

其中，$\Delta = 1 - \sum L_i + \sum L_i L_j - \sum L_i L_j L_k + \cdots$ 称为特征式。

P_i——从输入端到输出端第 i 条前向通路的总传递函数。

Δ_i——在 Δ 中，将与第 i 条前向通路相接触的回路所在项除去后所余下的部分，称为余子式。

$\sum L_i$——所有单回路的"回路传递函数"之和。

$\sum L_i L_j$——两两不接触回路，其"回路传递函数"乘积之和。

$\sum L_i L_j L_k$——所有三个互不接触回路，其"回路传递函数"乘积之和，"回路传递函数"指反馈回路的前向通路和反馈通路的传递函数之积，并且包含表示反馈极性的正负号。

【例 2-5-1】 试应用梅逊公式求图 2-5-3 所示框图的传递函数。

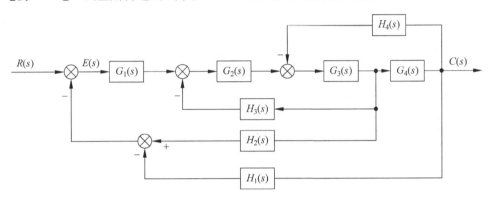

图 2-5-3 系统结构图

解：本题信号流图如图 2-5-4 所示。

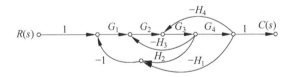

图 2-5-4 信号流图

$$P_1 = G_1 G_2 G_3 G_4 \tag{2-5-2}$$

$$L_1 = -G_2 G_3 H_3 \tag{2-5-3}$$

$$L_2 = -G_1 G_2 G_3 H_2 \tag{2-5-4}$$

$$L_3 = G_1 G_2 G_3 G_4 H_1 \tag{2-5-5}$$

$$L_4 = -G_3 G_4 H_4 \tag{2-5-6}$$

$$\Delta = 1 - (L_1 + L_2 + L_3 + L_4) \tag{2-5-7}$$

$$\Delta_1 = 1 \tag{2-5-8}$$

$$P = \frac{C}{R} = \frac{\sum_{k=1}^{n} P_k \Delta_k}{\Delta} = \frac{G_1 G_2 G_3 G_4}{1 + G_2 G_3 H_3 + G_1 G_2 G_3 H_2 + G_3 G_4 H_4 - G_1 G_2 G_3 G_4 H_1} \tag{2-5-9}$$

注意：应用梅逊公式可以方便地求出系统的传递函数，而不必进行结构图变换。但当结构图较复杂时，容易遗漏前向通路、回路或互不接触回路，因此在使用时应特别注意。

【**例 2-5-2**】 系统的框图如图 2-5-5 所示，用梅逊公式求系统的传递函数。

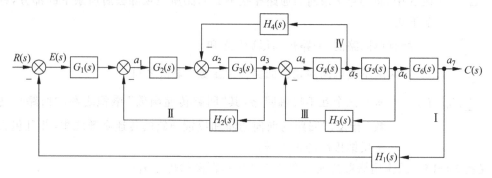

图 2-5-5 系统结构图

解：系统共有 4 个单回路，由图 2-5-5 求得

$$\sum L_i = -G_1 G_2 G_3 G_4 G_5 G_6 H_1 - G_2 G_3 H_2 - G_4 G_5 H_3 - G_3 G_4 H_4 \quad (2\text{-}5\text{-}10)$$

图中只有 Ⅱ、Ⅲ 两个回路不接触，则有

$$\sum L_i L_j = \sum L_2 L_3 = (-G_2 G_3 H_2)(-G_4 G_5 H_3) = G_2 G_3 G_4 G_5 H_2 H_3 \quad (2\text{-}5\text{-}11)$$

$$\sum L_i L_j L_k = 0 \quad (2\text{-}5\text{-}12)$$

$$\Delta = 1 - \sum L_i + \sum L_i L_j = 1 + G_1 G_2 G_3 G_4 G_5 G_6 H_1 + G_2 G_3 H_2$$
$$+ G_4 G_5 H_3 + G_3 G_4 H_4 + G_2 G_3 G_4 G_5 H_2 H_3 \quad (2\text{-}5\text{-}13)$$

图中只有一条前向通路，则有

$$P_1 = G_1 G_2 G_3 G_4 G_5 G_6 \quad (2\text{-}5\text{-}14)$$

且所有回路均与之接触，则有

$$\Delta_1 = 1$$

所以

$$\Phi(s) = \frac{P_1 \Delta_1}{\Delta} = \frac{G_1 G_2 G_3 G_4 G_5 G_6}{1 + G_1 G_2 G_3 G_4 G_5 G_6 H_1 + G_2 G_3 H_2 + G_4 G_5 H_3 + G_3 G_4 H_4 + G_2 G_3 G_4 G_5 H_2 H_3}$$
$$(2\text{-}5\text{-}15)$$

Terms and Concepts

In general, the linear dependence P between the independent variable x_i (often called the input variable) and a dependent variable x_j is given by Mason's signal-flow gain formula:

$$P = \frac{\sum_k P_{ijk} \Delta_{ijk}}{\Delta} \quad (2\text{-}5\text{-}16)$$

where

$P_{ijk} = k$th path from variable x_i to variable x_j,

Δ = determinant of the graph,

Δ_{ijk} = cofactor of the path P_{ijk},

And the summation is taken over all possible k paths from x_i to x_j. The cofactor Δ_{ijk} is the determinant with the loops touching the kth path removed. The determinant Δ is

$$\Delta = 1 - \sum_{n=1}^{N} L_n + \sum_{m=1,q=1}^{M,Q} L_m L_q - \sum L_r L_s L_t + \cdots \tag{2-5-17}$$

where L_q equals the value of the qth loop transmittance.

2.6 闭环传递函数的定义

2.6.1 闭环系统概述

在第 1 章中，详细介绍了闭环控制系统的基本组成部分，包括控制对象、执行机构、检测装置、给定环节和比较环节。因此，闭环控制系统一般有以下几个特点。

（1）系统输出量对控制作用有直接影响。
（2）有反馈环节，能够利用反馈减小误差。
（3）当出现干扰时，可以自动减弱其影响。
（4）系统可能工作不稳定。

在针对闭环控制系统进行分析和设计时，一般采取传递函数作为基本的数学模型。闭环传递函数是依据输入和输出定义的，即输出信号的拉普拉斯变换与输入信号的拉普拉斯变换的比。由于输入变化时，输出会相应变化，所以相除后，传递函数中不含输入信号项，即与输入信号无关。

本节在传递函数基本概念的基础上，引入不同定义下的传递函数概念，例如开环传递函数、误差传递函数等。这些概念主要讨论在不同输入下的同一系统的数学模型，主要目的是观察除了输入信号之外，其他因素（例如扰动）对系统的影响。在这样的背景下，如果系统不变，所得的传递函数分母都相同，分母表明系统的本质，分子体现外界信息的变化。而且在针对多输入系统研究中，线性系统的叠加原理起到了重要的作用。

2.6.2 闭环系统的传递函数

自动控制系统在实际工作中会受到两类信号的作用。一类是有用信号；另一类就是扰动信号。有用信号包括参考输入、控制输入、指令输入及给定值，通常加在系统的输入端，而扰动信号一般作用在受控对象或其他元部件上，甚至夹杂在指令信号中。基于后面章节的需要，下面介绍几个系统传递函数的概念。

图 2-6-1 就是模拟这种实际情况的典型控制系统框图。其中 $R(s)$ 为参考输入信号，$E(s)$ 为偏差信号，$N_1(s)$ 为系统噪声，$N_2(s)$ 为测量噪声，$Y(s)$ 为反馈信号，$C(s)$ 为输出信号。

1. 系统的开环传递函数

在反馈控制系统中，定义前向通路的传递函数与反馈通路的传递函数之积为开环传递

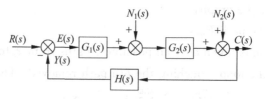

图 2-6-1 典型系统框图

函数。图 2-6-1 中的前向通路中包括两个函数方框和两个相加点,则前向通路的传递函数 $G(s)$ 为

$$G(s) = G_1(s)G_2(s) \tag{2-6-1}$$

显然,在系统框图中,将反馈信号 $Y(s)$ 在相加点前断开后,反馈信号与偏差信号之比 $\dfrac{Y(s)}{E(s)}$ 就是该系统的开环传递函数,即

$$\dfrac{Y(s)}{E(s)} = G_o(s) = G_1(s)G_2(s)H(s) \tag{2-6-2}$$

2. 输出 $C(s)$ 相对于参考输入 $R(s)$ 的闭环传递函数

此时,令 $N_1(s)=0$,$N_2(s)=0$,图 2-6-1 可变为图 2-6-2,称 $\Phi(s)=\dfrac{C(s)}{R(s)}$ 为输出对于参考输入的闭环传递函数。

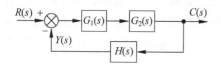

图 2-6-2 $N_1(s)$ 和 $N_2(s)$ 为零时的系统框图

于是有

$$\Phi(s) = \dfrac{C(s)}{R(s)}$$
$$= \dfrac{G_1(s)G_2(s)}{1+G_1(s)G_2(s)H(s)} \tag{2-6-3}$$

当 $H(s)=1$ 时,可得 $\Phi(s) = \dfrac{G_1(s)G_2(s)}{1+G_1(s)G_2(s)}$,称为单位反馈。

由式(2-6-3)中的 $\Phi(s)$ 可得系统在输入 $R(s)$ 的输出量 $C(s)$ 为

$$C(s) = \Phi(s)R(s) = \dfrac{G_1(s)G_2(s)}{1+G_1(s)G_2(s)H(s)}R(s) \tag{2-6-4}$$

对于理想的系统模型则要求 $\Phi(s)=1$,即输出信号能够完美跟踪输入信号。此时若前向通路的传递函数 $G(s)\to\infty$,则式(2-6-3)中的分母也趋近于无穷大,式(2-6-3)可近似为

$$\Phi(s) = \dfrac{C(s)}{R(s)} \approx \dfrac{G_1(s)G_2(s)}{G_1(s)G_2(s)H(s)} = \dfrac{1}{H(s)} \tag{2-6-5}$$

对于理想的模型则需 $H(s)=1$。

3. 偏差信号 $E(s)$ 相对于参考输入 $R(s)$ 的闭环传递函数

在图 2-6-3 中,偏差信号 $E(s)$ 的大小反映误差的大小,所以有必要了解偏差信号与参考输入的关系。令 $N_1(s)=0$,$N_2(s)=0$,则称 $\Phi_e(s)=\dfrac{E(s)}{R(s)}$ 为偏差信号对于参考输入的闭环传递函数。这时图 2-6-1 变为图 2-6-3,输出量是 $E(s)$,输入量是 $R(s)$,前向通路的传递

函数为1。具体做法是：①由输入 $R(s)$ 到输出 $E(s)$，明确前向通路；②原图中的剩余部分，全部作为反馈网络。

则此时有

$$\Phi_e(s) = \frac{E(s)}{R(s)}$$

$$= \frac{1}{1+G_1(s)G_2(s)H(s)} \quad (2\text{-}6\text{-}6)$$

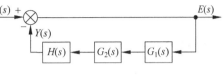

图 2-6-3　$E(s)$ 与 $R(s)$ 的框图

对于理想的系统模型则希望误差为 0，即 $\Phi_e(s)=0$，可知此时式(2-6-6)分母应趋于无穷大，即 $G_1(s)G_2(s)H(s)\to\infty$。可得

$$\Phi_e(s) = \frac{E(s)}{R(s)} \approx \frac{1}{\infty} \approx 0 \quad (2\text{-}6\text{-}7)$$

4. 输出 $C(s)$ 相对于扰动输入 $N_1(s)$ 的闭环传递函数

此时令 $R(s)=0$，$N_2(s)=0$，称 $\Phi_{n_1}(s)=\dfrac{C(s)}{N_1(s)}$ 为输出对于扰动信号 $N_1(s)$ 的闭环传递函数，这时输入量变为 $N_1(s)$，图 2-6-1 变为图 2-6-4。

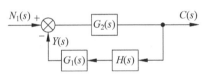

图 2-6-4　$R(s)$ 和 $N_2(s)$ 为零时的系统框图

因而可求得如图 2-6-4 所示系统在干扰作用下的闭环传递函数为

$$\Phi_{n_1}(s) = \frac{C(s)}{N_1(s)}$$

$$= \frac{G_2(s)}{1+G_1(s)G_2(s)H(s)} \quad (2\text{-}6\text{-}8)$$

进一步由 $\Phi_{n_1}(s)$ 可求得在输入信号 $N_1(s)$ 作用下的输出信号为

$$C(s) = \Phi_{n_1}(s)N_1(s) = \frac{G_2(s)}{1+G_1(s)G_2(s)H(s)}N_1(s) \quad (2\text{-}6\text{-}9)$$

对于理想的控制系统，则希望扰动信号对系统的影响为零。从数学上讲，要求 $\Phi_{n_1}(s)=\dfrac{C(s)}{N_1(s)}=0$，此时需要 $G_1(s)H(s)\to\infty$，可得

$$\Phi_{n_1}(s) \approx \frac{G_2(s)}{G_1(s)G_2(s)H(s)} = \frac{1}{G_1(s)H(s)} \approx 0 \quad (2\text{-}6\text{-}10)$$

5. 输出 $C(s)$ 相对于扰动输入 $N_2(s)$ 的闭环传递函数

此时令 $R(s)=0$，$N_1(s)=0$，称 $\Phi_{n_2}(s)=\dfrac{C(s)}{N_2(s)}$ 为输出对于扰动信号 $N_2(s)$ 的闭环传递函数，这时输入量变为 $N_2(s)$，图 2-6-1 变为图 2-6-5 所示。可求得闭环传递函数为

$$\Phi_{n_2}(s) = \frac{C(s)}{N_2(s)}$$

$$= \frac{1}{1+G_1(s)G_2(s)H(s)} \quad (2\text{-}6\text{-}11)$$

对于理想的控制系统模型，我们希望

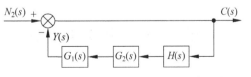

图 2-6-5　$R(s)$ 和 $N_1(s)$ 为零时的系统框图

$$\Phi_{n_2}(s) = \frac{C(s)}{N_2(s)} = 0, 则此时需要 G_1(s)G_2(s)H(s) \to \infty, 可求得$$

$$\Phi_{n_2}(s) = \frac{C(s)}{N_2(s)} = \frac{1}{1+G_1(s)G_2(s)H(s)} \approx \frac{1}{G_1(s)G_2(s)H(s)} \approx 0 \quad (2\text{-}6\text{-}12)$$

6. 结论

通过比较以上几种闭环传递函数可知,对于理想的控制系统,可得出如表 2-6-1 所示的结果。

表 2-6-1 几种理想控制系统的闭环传递函数

输入量	输出量	闭环传递函数	条件
$R(s)$	$C(s)$	$\frac{C(s)}{R(s)}=1$	$G_1(s)G_2(s) \to \infty$
$R(s)$	$E(s)$	$\frac{E(s)}{R(s)}=0$	$G_1(s)G_2(s)H(s) \to \infty$
$N_1(s)$	$C(s)$	$\frac{C(s)}{N_1(s)}=0$	$G_1(s)H(s) \to \infty$
$N_2(s)$	$C(s)$	$\frac{C(s)}{N_2(s)}=0$	$G_1(s)G_2(s)H(s) \to \infty$

由表 2-6-1 可知,当 $G_1(s) \to \infty$ 时,即满足各项条件,可以达到理想的控制效果(完美,误差为零)。然而,由于 $G_1(s)$ 一般是自动控制系统的控制装置,显然控制装置的增益增加时,系统的各项性能会相应改善,但是同时出现的稳定性问题也需要引起足够的重视。在后续章节中,这部分内容还会深入地讨论。

Terms and Concepts

$$\frac{Y(s)}{R(s)} = \frac{G(s)}{1+G(s)H(s)} \quad (2\text{-}6\text{-}13)$$

This closed-loop transfer function is particularly important because it represents many of the existing practical control systems.

The reduction of the block diagram shown as follow in Fig. 2-6-6 to a single block representation is one example of several useful block diagram reductions.

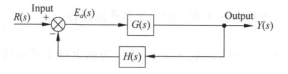

Fig. 2-6-6 Negative feedback control system

2.7 非线性系统模型概述

在工程实际问题中,纯粹理想的线性系统几乎是不存在的,因为组成系统的所有元件在不同程度上都具有非线性特性,即实际元件的输入量和输出量之间都存在不同程度的非线性,所以它们的动态方程应是非线性微分方程。由于本书的定位是经典控制理论,在章节设计上对于非线性自动控制系统不做深入的讨论。在本节中,将首先介绍非线性环节的主要特征,然后针对部分非线性模型,介绍简单的线性化方法。

在控制理论中,按非线性程度不同可以分为两类。第一类是"非本质非线性"特性,即非线性特性在工作点附近不存在饱和、继电、死区、滞环等,如图 2-7-1(a)所示曲线图;第二类是"本质非线性"特性,即非线性特性在工作点附近存在饱和、继电、死区、滞环等,如图 2-7-1(b)所示曲线图。

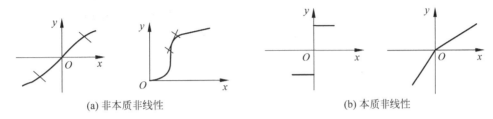

(a) 非本质非线性   (b) 本质非线性

图 2-7-1 两类曲线图

1. 典型非线性环节

为了简化对问题的分析,通常将本质非线性特性用简单的折线来代替,称为典型非线性特性。常见的典型非线性环节主要有以下几种。

1) 饱和特性

饱和特性的数学表达式为

$$y = \begin{cases} M, & x > a \\ kx, & |x| \leqslant a \\ -M, & x < -a \end{cases} \quad (2-7-1)$$

式中,a 为线性区宽度;k 为线性区斜率,如图 2-7-2 所示。

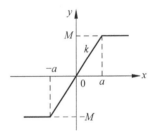

图 2-7-2 饱和特性曲线

饱和特性的特点是,输入信号超过某一范围后,输出不再随输入的变化而变化,而是保持在某一常数上。这在控制系统中是普遍存在的,例如常见的调节器。当输入信号较小而工作在线性区时,可视为线性元件。但当输入信号较大而工作在饱和区时,就必须作为非线性来处理。

2) 死区特性

死区特性的数学表达式为

$$y = \begin{cases} 0, & |x| \leqslant a \\ k[x - a\,\text{sign}(x)], & |x| > a \end{cases} \quad (2-7-2)$$

式中,$\text{sign}(x)$ 为符号函数,如图 2-7-3 所示。死区特性常见于许多控制设备与控制装置中,如各种测量元件的不灵敏区,在死区内虽有输入信号,但其输出为零。

3) 滞环特性

滞环特性又称为间隙特性,其数学表达式为

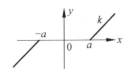

图 2-7-3 死区特性曲线

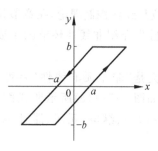

图 2-7-4 滞环特性曲线

$$y = \begin{cases} k[x - a\text{sign}(x)], & \dot{y} \neq 0 \\ b\text{sign}(x), & \dot{y} = 0 \end{cases} \quad (2\text{-}7\text{-}3)$$

如图 2-7-4 所示，滞环特性表示为正向与反向特性不是重叠在一起，而是在输入输出曲线上出现闭合环路。当输入信号小于间隙 a 时，输出为零。只有在 $x > a$ 时，输出才随输入线性变化。当输入反向时，其输出则保持在方向发生变化时的输出值上，直到输出反向变化 $2a$ 后，输出才线性变化。例如，铁磁元件的磁滞、齿轮传动中的齿隙、液压传动中的油隙等均属于这类特性。

4）继电特性

继电特性的数学表达式为

$$y = \begin{cases} 0, & -ma < x < a, \dot{x} > 0 \\ 0, & -a < x < ma, \dot{x} < 0 \\ b\text{sign}(x), & |x| \geq a \\ b, & x \geq ma, \dot{x} < 0 \\ -b, & x \leq -ma, \dot{x} > 0 \end{cases} \quad (2\text{-}7\text{-}4)$$

分别取 $a=0, m=1, m=-1$ 时三种特殊情况下的继电特性曲线，如图 2-7-5 所示。继电特性不仅包含死区特性，而且具有滞环特性。

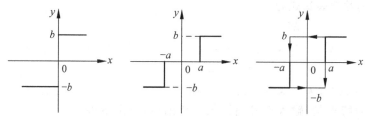

(a) $a=0$ 时继电特性曲线　　(b) $m=1$ 时继电特性曲线　　(c) $m=-1$ 时继电特性曲线

图 2-7-5 继电的特性曲线

2. 线性化过程

在研究控制系统动态过程中，会遇到求解非线性微分方程的问题。然而对于高阶非线性微分方程来说，在数学上不可能求得一般形式的解。这样，对于非线性特性元件和系统的研究在理论上是很困难的，因此在可能的条件下将非线性方程进行线性化。所谓线性化 (Linearization) 就是在工程实践中，控制系统都有一个额定的工作状态和工作点，当变量在工作点附近的小范围内变化时，把非线性特性用线性特性来代替。非线性方程线性化最常用的方法就是小偏差线性化，主要是利用数学分析中的泰勒级数，忽略二阶及二阶以上的高阶各项，得到只包含偏差一次项的线性化方程，以此代替原有的非线性方程。这种线性化的方法称为微偏法。

下面来具体讨论这种线性化的方法。设一个非线性元件的输入为 x、输出为 y，相应的数学表达式为 $y = f(x)$，如图 2-7-6 所示。

在预期点工作点邻域展开成以偏差量 $\Delta x = x - x_0$ 表示的泰勒级数(Taylor series),有

$$y = f(x) = f(x_0) + f'(x_0)(x - x_0) + \frac{1}{2}f''(x_0)(x - x_0)^2 + \cdots \tag{2-7-5}$$

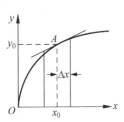

图 2-7-6 非线性特性的线性化

若在工作点(x_0, y_0)附近增量$(x - x_0)$的变化很小,一般都满足微量的要求,则可略去式中$(x - x_0)^2$项及其后面所有的高阶项。这样式(2-7-5)近似表示为

$$y = y_0 + K(x - x_0) \tag{2-7-6}$$

或写成

$$\Delta y = K\Delta x \tag{2-7-7}$$

式中,$y_0 = f(x_0)$,$K = f'(x_0)$,$\Delta y = y - y_0$,$\Delta x = x - x_0$。小偏差线性化的精确度要比忽略非线性因素的简化性处理所得到的线性方程精确得多。式(2-7-6)就是非线性方程的线性化。

如果已知静态时变量间的非线性关系式,采用小偏差线性化的最终目的是求变量间的传递函数。具体非线性模型线性化的步骤如下。

(1) 按系统数学模型的建立方法,列写系统中每一部分的微分方程式。
(2) 确定系统的工作点,并分别求出工作点处各变量的工作状态。
(3) 对存在的非线性函数,检验是否适合线性条件,若符合就进行线性化处理。
(4) 将其余线性方程,按偏量形式处理,其中,对变量直接用偏量形式写出,对常量因其偏量为零,故消去此项。
(5) 量化所有偏量化方程,消去中间变量,最后得到只含系统总输入和总输出偏量的线性化微分方程。

例如,在静态时,函数 y 是变量 x_1 和 x_2 的非线性函数,即

$$y = f(x_1, x_2) \tag{2-7-8}$$

设在工作点(x_{10}, x_{20})附近,$y = f(x_1, x_2)$有连续偏导数,$x_1(t)$和$x_2(t)$有连续导数。根据全导数公式,得

$$\frac{dy}{dt} = \frac{\partial f}{\partial x_1}\frac{dx_1}{dt} + \frac{\partial f}{\partial x_2}\frac{dx_2}{dt} \tag{2-7-9}$$

在(x_{10}, x_{20})附近,$\left.\frac{\partial f}{\partial x_1}\right|_{x_0}$ 和 $\left.\frac{\partial f}{\partial x_2}\right|_{x_0}$ 认为是常数,则式(2-7-9)变成

$$\frac{dy}{dt} = \left(\frac{\partial f}{\partial x_1}\right)_0 \frac{dx_1}{dt} + \left(\frac{\partial f}{\partial x_2}\right)_0 \frac{dx_2}{dt} \tag{2-7-10}$$

式(2-7-10)是一个常系数线性微分方程,在零初始条件下取拉普拉斯变换,得

$$sY(s) = \left(\frac{\partial f}{\partial x_1}\right)_0 sX_1(s) + \left(\frac{\partial f}{\partial x_2}\right)_0 sX_2(s) \tag{2-7-11}$$

方程两边约去 s,得

$$Y(s) = \left(\frac{\partial f}{\partial x_1}\right)_0 X_1(s) + \left(\frac{\partial f}{\partial x_2}\right)_0 X_2(s) \tag{2-7-12}$$

式(2-7-12)就是对本系统进行小偏差线性化后得到的变量拉普拉斯变换式间的关系式。

上述方法可以推广到 n 元以上的函数。函数 y 是变量 $x_1, x_2, \cdots, x_n$ 的非线性函数,即
$$y = f(x_1, x_2, \cdots, x_n)$$
在工作点 $(x_{10}, x_{20}, \cdots, x_{n0})$ 附近有连续偏导数和连续导数。则有
$$Y(s) = \left(\frac{\partial f}{\partial x_1}\right)_0 X_1(s) + \left(\frac{\partial f}{\partial x_2}\right)_0 X_2(s) + \cdots + \left(\frac{\partial f}{\partial x_n}\right)_0 X_n(s) \quad (2\text{-}7\text{-}13)$$

对于线性化问题有如下说明。

(1) 系统工作在一个正常的工作状态,有一个稳定的工作点。

(2) 线性化的基本条件是非本质非线性的,即在预期工作点的邻域内存在关于变量的各阶导数或偏导数。也就是说,非线性函数属于单值、连续、光滑的非本质非线性函数。不符合这个条件的非线性函数不能展开成泰勒级数,因此不能采用小偏差线性化方法,这种非线性特性称为本质非线性。

(3) 在很多情况下,对于不同的预期工作点,线性化后的方程的形式是一样的,但各项系数及常数项可能不同。

(4) 线性化后的微分方程通常是增量方程,在实用上为了简便通常直接采用 y 和 x 来表示增量。

3. 两相伺服电动机

两相伺服电动机属于微型交流异步电动机,使用交流电源。两相伺服电动机最常用的控制方法是幅相控制,又称电容控制。它的接线方法如图 2-7-7 所示。图中 SM2～表示两相伺服电动机,它有两个绕组:激磁绕组和控制绕组。i_f、i_c 为激磁绕组和控制绕组的电流,u_f、u_c 为两绕组电压。电容 C 称为移相电容,它与激磁绕组串联后接到交流电源上。串接电容 C 的目的是使激磁绕组和控制绕组的电压相位差为 $90°$ 左右,以便产生旋转磁场。电容控制是通过改变控制绕组的电压的大小控制电动机的。

两相伺服电动机的电耦合关系复杂,要想从电磁角度出发去分析和推导动态数学模型是困难的。下面根据电机的静态特性曲线、力学原理和小偏差线性化的概念推导两相伺服电动机的传递函数。

当电动机的电流和转速不变时,称电机处于静态。实验表明,静态时两相伺服电动机的转速 ω 是控制绕组的电压 U(有效值)和电磁转矩 T 的函数,即

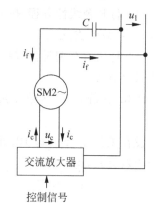

图 2-7-7 两相伺服电容电动机的电容控制

$$\omega = \omega(U, T) \quad (2\text{-}7\text{-}14)$$

以 U 为参变量时,ω 和 T 的关系曲线称为机械特性,如图 2-7-8(a) 所示。以 T 为参变量时,ω 和 U 的关系曲线称为调节特性,如图 2-7-8(b) 所示。

由图可知,两相伺服电动机的机械特性和调节特性具有明显的非线性,在不同的位置有不同的斜率。对式 (2-7-14) 取时间变量 t 的导数,得

$$\frac{d\omega}{dt} = \frac{\partial \omega}{\partial U}\frac{dU}{dt} + \frac{\partial \omega}{\partial T}\frac{dT}{dt} \quad (2\text{-}7\text{-}15)$$

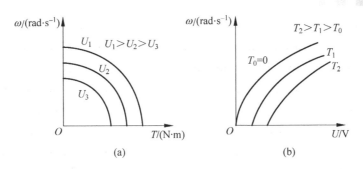

图 2-7-8 静态特性

在工作点附近,$\dfrac{\partial \omega}{\partial U}$ 及 $\dfrac{\partial \omega}{\partial T}$ 视为常数,则在零初始条件下对式(2-7-15)两边取拉普拉斯变换并约去 s 后得

$$\Omega(s) = \frac{\partial \omega}{\partial U} U(s) + \frac{\partial \omega}{\partial T} T(s) \tag{2-7-16}$$

下面进行力学分析以求出 T 和 ω 的关系。把电机轴上的总阻转矩当作是扰动力矩,把它看成系统的另一个输入量。求 $\Omega(s)$ 与 $U(s)$ 之间的传递函数时,令扰动力矩为零,于是有

$$T = J \frac{\mathrm{d}\omega}{\mathrm{d}t} \tag{2-7-17}$$

取拉普拉斯变换后得

$$T(s) = Js\Omega(s) \tag{2-7-18}$$

将式(2-7-18)代入式(2-7-16)得

$$\Omega(s) = \frac{\partial \omega}{\partial U} U(s) + \frac{\partial \omega}{\partial T} Js\Omega(s) \tag{2-7-19}$$

于是可得两相伺服电动机的传递函数

$$G(s) = \frac{\Omega(s)}{U(s)} = \frac{\dfrac{\partial \omega}{\partial U}}{-J\dfrac{\partial \omega}{\partial T}s + 1} = \frac{K}{\tau_{\mathrm{m}} s + 1} \tag{2-7-20}$$

式中,$K = \dfrac{\partial \omega}{\partial U}$,它是调节特性的斜率;$\tau_{\mathrm{m}} = -J\dfrac{\partial \omega}{\partial T}$,其中 $\dfrac{\partial \omega}{\partial T}$ 是机械特性的斜率。因 $\dfrac{\partial \omega}{\partial T} < 0$,故 $\tau_{\mathrm{m}} > 0$。

因静态特性的非线性,当两相伺服电动机在较大转速范围内运行时,K 与 τ_{m} 不是固定的,K 与 τ_{m} 变化约 2~4 倍。

由上可见,此节的学习目的是了解真实物理器件与自动控制原理之间的关系。例如,电动机的传递函数为一个二阶或一阶系统。

Reading Material

Motors are indispensable in many control systems. They are used to turn antennas, telescopes, and ship rudders; to close and open valves; to drive tapes in recorders, and rollers in steel mills; and to feed paper in printers. There are many types of motors: dc, ac, stepper, and hydraulic. The magnetic field of a dc motor can be excited by a circuit connected in series with the armature circuit; it can also be excited by a field circuit that is independent of the armature circuit or by a permanent magnet. We discuss in this text only separately excited dc motors. DC motors used in control systems, also called servomotors, are characterized by large torque to rotor-inertia ratios, small sizes, and better linear characteristics. Certainly, they are more expensive than ordinary dc motors.

A common actuator in control systems is the DC motor. It directly provides rotary motion with wheels or drums can provide transitional motion. The electric circuit of the armature and the free body diagram of the rotor are shown in Fig. 2-7-9.

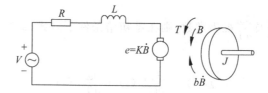

Fig. 2-7-9   The control system of DC motor

For this example, we will assume the following values for the physical parameters. These values were derived by experiment from an actual motor.

- [ ] moment of inertia of the rotor$(J) = 0.01 \text{kg} \cdot \text{m}^2$.
- [ ] damping ratio of the mechanical system$(b) = 0.1 \text{N} \cdot \text{s/m}$.
- [ ] electromotive force constant$(K = K_e = K_t) = 0.01 \text{N} \cdot \text{m/A}$.
- [ ] electric resistance$(R) = 1 \Omega$.
- [ ] electric inductance$(L) = 0.5 \text{H}$.
- [ ] input(V): source voltage.
- [ ] output(theta): position of shaft.
- [ ] the rotor and shaft are assumed to be rigid.

the motor torque, T, is related to the armature current, i, by a constant factor K_t. The back emf, e, is related to the rotational velocity by the following equations:

$$T = K_t i \tag{2-7-21}$$

$$e = K_e \dot{\theta} \tag{2-7-22}$$

In SI units (which we will use), K_t (armature constant) is equal to K_e (motor constant).

From the figure above, we can write the following equations based on Newton's law combined with Kirchhoff's law:

$$J\ddot{\theta} + b\dot{\theta} = Ki \tag{2-7-23}$$

$$L\frac{\mathrm{d}i}{\mathrm{d}t} + Ri = V - K\dot{\theta} \tag{2-7-24}$$

Transfer Function

Using Laplace Transforms, the above modeling equations can be expressed in terms of s.

$$s(Js+b)\theta(s) = KI(s) \qquad (2\text{-}7\text{-}25)$$

$$(Ls+R)I(s) = V - Ks\theta(s) \qquad (2\text{-}7\text{-}26)$$

By eliminating $I(s)$ we can get the following open-loop transfer function, where the rotational speed is the output and the voltage is the input.

$$\frac{\theta}{V} = \frac{K}{(Js+b)(Ls+R)+K^2} \qquad (2\text{-}7\text{-}27)$$

2.8 MATLAB 在本章中的应用

在本章控制系统数学模型的建立中,需要应用多项式、矩阵和数组等多种数学工具。下面将通过英文阅读材料的形式,主要介绍在 MATLAB 环境中,如何进行多项式的输入及输出,如何实现数组及矩阵的输入输出等。这些技术在后续的章节中需要进一步的应用。

Reading Material

1. Vectors

Let's start off by creating something simple, like a vector. Enter each element of the vector(separated by a space)between brackets, and set it equal to a variable. For example, to create the vector a, enter into the MATLAB command window(you can "copy" and "paste" from your browser into MATLAB to make it easy):

```
a = [1 2 3 4 5 6 9 8 7]
```

MATLAB should return:

```
a =
    1 2 3 4 5 6 9 8 7
```

To generate a series that does not use the default of incrementing by 1, specify an additional value with the colon operator(first:step:last). In between the starting and ending value is a step value that tells MATLAB how much to increment(or decrement, if step is negative) between each number it generates. To generate a vector with elements between 0 and 20, incrementing by 2(this method is frequently used to create a time vector), use

```
t = 0:2:20
    t =
        0 2 4 6 8 10 12 14 16 18 20
```

Manipulating vectors is almost as easy as creating them. First, suppose you would like to add 2 to each of the elements in vector 'a'. The equation for that looks like:

```
b = a + 2
b =
    3 4 5 6 7 8 11 10 9
```

Now suppose, you would like to add two vectors together. If the two vectors are the same length, it is easy. Simply add the two as shown below:

```
c = a + b
c =
    4  6  8  10  12  14  20  18  16
```

Subtraction of vectors of the same length works exactly the same way.

MATLAB sometimes stores such a list in a matrix with just one row, and other times in a matrix with just one column. In the first instance, such a 1-row matrix is called a row-vector; in the second instance, such a 1-column matrix is called a column-vector. Either way, these are merely different ways for storing vectors, not different kinds of vectors.

2. Matrices and Arrays

The most basic MATLAB data structure is the matrix: a two-dimensional, rectangular shaped data structure capable of storing multiple elements of data in an easily accessible format. These data elements can be numbers, characters, and logical states of true or false, or even other MATLAB structure types. MATLAB uses these two-dimensional matrices to store single numbers and linear series of numbers as well. In these cases, the dimensions are 1-by-1 and 1-by-n respectively, where n is the length of the numeric series. MATLAB also supports data structures that have more than two dimensions. These data structures are referred to as arrays in the MATLAB documentation.

MATLAB is a matrix-based computing environment. All of the data that you enter into MATLAB is stored in the form of a matrix or a multidimensional array. Even a single numeric value like 100 is stored as a matrix (in this case, a matrix having dimensions 1-by-1):

```
A = 100;
Name      Size              Bytes Class
  A       1x1               8 double array
```

Entering matrices into MATLAB is the same as entering a vector, except each row of elements is separated by a semicolon(;) or a return:

```
B = [1 2 3 4; 5 6 7 8; 9 10 11 12]
B =
    1   2   3   4
    5   6   7   8
    9  10  11  12
```

The square brackets operator constructs two-dimensional matrices only, (including 0-by-0, 1-by-1, and 1-by-n matrices).

Matrices in MATLAB can be manipulated in many ways. For one, you can find the transpose of a matrix using the apostrophe key:

```
C = B'
C =
    1   5   9
```

```
    2    6   10
    3    7   11
    4    8   12
```

It should be noted that if **C** had been complex, the apostrophe would have actually given the complex conjugate transpose. To get the transpose(the two commands are the same if the matrix is not complex).

Now you can multiply the two matrices **B** and **C** together. Remember that order matters when multiplying matrices.

```
D = B * C
    D =
        30    70   110
        70   174   278
       110   278   446
D = C * B
D =
       107   122   137   152
       122   140   158   176
       137   158   179   200
       152   176   200   224
```

Another option for matrix manipulation is that you can multiply the corresponding elements of two matrices using the .* operator(the matrices must be the same size to do this).

```
E = [1 2; 3 4]
F = [2 3; 4 5]
G = E.* F
E =
    1    2
    3    4
F =
    2    3
    4    5
G =
    2    6
   12   20
```

If you have a square matrix, like **E**, you can also multiply it by itself as many times as you like by raising it to a given power.

```
E^3
    ans =
    37    54
    81   118
```

If wanted to cube each element in the matrix, just use the element-by-element cubing.

```
E.^3
    Ans =
```

```
    1    8
   27   64
```

You can also find the inverse of a matrix.

```
X = inv(E)
X =
   -2.0000    1.0000
    1.5000   -0.5000
```

Or its eigenvalues:

```
eig(E)
ans =
   -0.3723
    5.3723
```

There is even a function to find the coefficients of the characteristic polynomial of a matrix. The "poly" function creates a vector that includes the coefficients of the characteristic polynomial.

```
p = poly(E)
p =
    1.0000   -5.0000   -2.0000
```

Remember that the eigenvalues of a matrix are the same as the roots of its characteristic polynomial:

```
roots(p)
ans =
    5.3723
   -0.3723
```

3. Functions

To make life easier, MATLAB includes many standard functions. Each function is a block of code that accomplishes a specific task. MATLAB contains all of the standard functions such as sin, cos, log, exp, sqrt, as well as many others. Commonly used constants such as pi, and i or j for the square root of -1, are also incorporated into MATLAB.

```
sin(pi/4)
ans =
    0.7071
```

To determine the usage of any function, type help [function name] at the MATLAB command window. MATLAB even allows you to write your own functions with the function command.

4. Plotting

It is also easy to create plots in MATLAB. Suppose you wanted to plot a sine wave as a function of time. First make a time vector (the semicolon after each statement tells MATLAB we don't want to see all the values) and then compute the sin value at each time. Fig. 2-8-1 shows the sine wave as a function of time.

```
t = 0:0.25:7;
y = sin(t);
plot(t,y)
```

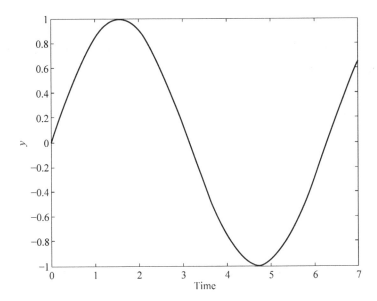

Fig. 2-8-1 Sine wave as a function of time

The plot contains approximately one period of a sine wave. Basic plotting is very easy in MATLAB, and the plot command has extensive add-on capabilities. I would recommend you to learn more about it.

5. Polynomials

In MATLAB, a polynomial is represented by a vector. To create a polynomial in MATLAB, simply enter each coefficient of the polynomial into the vector in descending order. For instance, let's say you have the following polynomial:

$$s^4 + 3s^3 - 15s^2 - 2s + 9 \tag{2-8-1}$$

To enter this into MATLAB, just enter it as a vector in the following manner.

```
x = [1 3 -15 -2 9]
    x =
       1   3   -15   -2   9
```

MATLAB can interpret a vector of length $n+1$ as an nth order polynomial. Thus, if your polynomial is missing any coefficients, you must enter zeros in the appropriate place in the vector. For example

$$s^4 + 1 \tag{2-8-2}$$

would be represented in MATLAB as:

```
y = [1 0 0 0 1]
```

You can find the value of a polynomial using the polyval function. For example, to find the value of the above polynomial at $s=2$

```
z = polyval([1 0 0 0 1],2)
    z =
        17
```

You can also extract the roots of a polynomial. This is useful when you have a high-order polynomial such as

$$s^4 + 3s^3 - 15s^2 - 2s + 9 \qquad (2\text{-}8\text{-}3)$$

Finding the roots would be as easy as entering the following command.

```
roots([1 3 -15 -2 9])
    ans =
        -5.5745
         2.5836
        -0.7951
         0.7860
```

Let's say you want to multiply two polynomials together. The product of two polynomials is found by taking the convolution of their coefficients. MATLAB's functions evident that it will do this for you.

```
x = [1 2];
y = [1 4 8];
z = conv(x,y)
    z =
        1   6   16   16
```

Dividing two polynomials is just as easy. The deconv function will return the remainder as well as the result. Let's divide z by y and see if we get x.

```
[x, R] = deconv(z,y)
    x =
        1   2
    R =
        0   0   0   0
```

As you can see, this is just the polynomial/vector x from before. If y had not gone into z evenly, the remainder vector would have been something other than zero.

If you want to add two polynomials together which have the same order, a simple z= x+y will work(the vectors x and y must have the same length). In the general case, the user-defined function, poly add can be used. To use poly add, copy the function into an m-file, and then use it just as you would any other function in the MATLAB toolbox. Assuming you had the poly add function stored as a m-file, and you wanted to add the two uneven polynomials, x and y, you could accomplish this by entering the command:

```
z = polyadd(x,y)
x =
    1   2
y =
    1   4   8
z =
    1   5   10
```

6. Matrix

MATLAB has a number of functions that create different kinds of matrices. Some create specialized matrices like the Hankel or Vandermonde matrix. The functions shown in the Tab. 2-8-1 below create matrices for more general use.

Tab. 2-8-1 Different Functions of Matrix in MATLAB

Function	Description
ones	Create a matrix or array of all ones
zeros	Create a matrix or array of all zeros
eye	Create a matrix with ones on the diagonal and zeros elsewhere
accumarray	Distribute elements of an input matrix to specified locations in an output matrix, also allowing for accumulation
diag	Create a diagonal matrix from a vector
magic	Create a square matrix with rows, columns, and diagonals that add up to the same number
rand	Create a matrix or array of uniformly distributed random numbers
randn	Create a matrix or array of normally distributed random numbers and arrays
randperm	Create a vector (1-by-n matrix) containing a random permutation of the specified integers

Most of these functions return matrices of type double (double-precision floating point). However, you can easily build basic arrays of any numeric type using the ones, zeros, and eye functions.

```
A = magic(5)
A =
    17  24   1   8  15
    23   5   7  14  16
     4   6  13  20  22
    10  12  19  21   3
    11  18  25   2   9
```

Note that the elements of each row, each column, and each main diagonal add up to the same value: 65.

通过这一部分内容的学习,读者应当充分了解在 MATLAB 环境下的多项式输入、方程求根和矩阵计算等内容。由于数学模型中的传递函数在分析和设计自动控制系统中至关重要，后续章节中需要利用这些知识，在 MATLAB 环境下建立模型，以进行分析和研究。

本章小结

（1）描述控制系统动态特性的数学表达式称为数学模型。在分析系统性能之前必须建立系统的数学模型。

（2）控制系统的数学模型有多种，常见的有微分方程、传递函数、动态结构图以及第 5 章将要介绍的频率特性等。各种数学模型间可以相互转换。微分方程是表述系统动态特性的最基本形式；传递函数是零初始条件下，线性定常系统输出量的拉普拉斯变换与系统输入量的拉普拉斯变换之比。一般在自动控制系统分析和设计中，需要首先求出系统的传递函数模型。

（3）针对结构图模型，一般有两方面的问题：结构图的建立和结构图的化简。结构图的建立需要熟悉所研究的控制系统，按照基本步骤逐步进行。结构图的化简遵循的是等效变换原则，即在保持被变换部分的输入量和输出量之间的数字关系不变的前提下，进行简化。这一过程对应着微分方程消除中间变量的过程。

（4）一个控制系统可以看作由若干个典型环节组成，掌握这些典型环节的特性，将有益于对整个系统的分析。典型环节包括放大环节、积分环节、惯性环节和振荡环节等。本章中详细介绍了每一种环节的微分方程模型、传递函数模型和结构图表示。同时，给出了各种典型环节的时域阶跃信号响应。

（5）系统的传递函数可以分为开环传递函数、闭环传递函数、误差传递函数等。这些传递函数可以从不同的侧面反映系统的性能，在控制系统的分析和设计中需要选择使用。

（6）非线性控制系统包括具有饱和特性、死区特性、滞环特性和继电特性等典型的非线性特征。对于简单的非线性系统，可以采用在工作点附近线性化的简单方法。

（7）本章的 MATLAB 应用中，介绍了在 MATLAB 环境下，基本的变量、多项式和矩阵应用等。这些基础知识将有助于后续章节中利用 MATLAB 进行自动控制系统的分析和设计。

Summary and Outcome Checklist

Mathematical models of dynamic systems (including feedback control systems) are obtained by considering each component/subsystem and developing its transfer function. The individual block diagrams are then arranged in the correct sequence (output of one becomes the input of the next) to obtain the system block diagram. The system block diagram is systematically reduced to obtain the system transfer function.

MATLAB may be used to analyse the transfer function models.

Tick the box for each statement with which you agree:

☐ I am able to derive differential equations of simple electrical, mechanical, and electromechanical devices and systems;

☐ I am able to derive transfer functions of simple electrical, mechanical, and

- electromechanical devices and systems;
- ☐ I am able to develop block diagram models of complex dynamic systems, and by systematic reduction of the block diagram, obtain the system transfer function;
- ☐ Given the transfer function of a dynamic system, I am able to use MATLAB to obtain the step response and the pole-zero map.

2-1 如题 2-1 图所示为一转动物体，J 表示转动惯量，f 表示黏性摩擦系数。若输入为转矩 $M(t)$，输出为轴角位移 $\theta(t)$ 或角速度 $\omega(t)$，求传递函数。

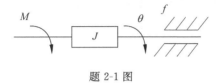

题 2-1 图

2-2 设机械系统如题 2-2 图所示，试求与图示系统具有相同传递函数的电模拟系统。

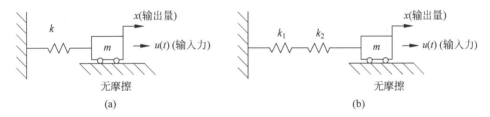

题 2-2 图

2-3 已知电机系统如题 2-3 图所示，试画出系统的结构图，并求传递函数 $G(s) = \dfrac{X(s)}{E(s)}$。（提示：设电磁线圈反电势 $e_b = K_1 \dfrac{\mathrm{d}x}{\mathrm{d}t}$，线圈电流 i_2 对衔铁产生的力 $F_0 = K_2 i_2$。）

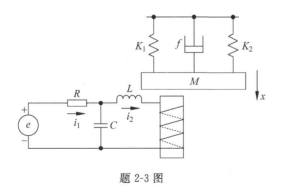

题 2-3 图

2-4 求题 2-4 图所示机械系统的微分方程和传递函数。图中 $F(t)$ 为输入量；位移 $x(t)$ 为输入量；m 为质量；k 为弹簧的弹性系数；f 为黏滞阻尼系数。

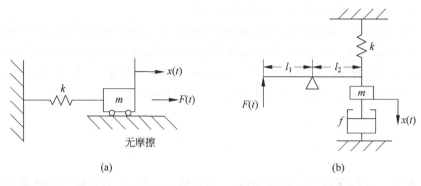

题 2-4 图

2-5 求题 2-5 图所示机械系统的微分方程式和传递函数。图中位移 x_i 为输入量,位移 x_o 为输出量,k_1、k_2 为弹簧的弹性系数,f 为黏滞阻尼系数,图 2-5(a)的重力忽略不计。

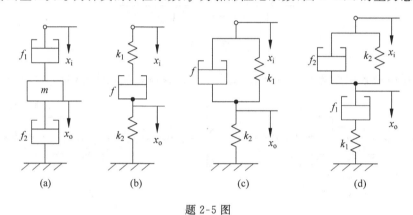

题 2-5 图

2-6 求题 2-6 图所示无源电网络的传递函数,图中电压 $u_1(t)$ 是输入量,电压 $u_2(t)$ 是输入量。

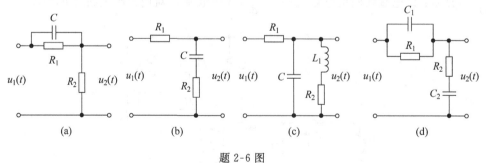

题 2-6 图

2-7 求题 2-7 图所示有源电网络的传递函数,图中电压 $u_1(t)$ 是输入量,电压 $u_2(t)$ 是输入量。

2-8 已知系统传递函数 $\dfrac{C(s)}{R(s)} = \dfrac{2}{s^2+3s+2}$,且初始条件为 $C(0)=-1, \dot{C}(0)=0$。试求系统在输入 $r(t)=1(t)$ 作用下的输出 $C(t)$。

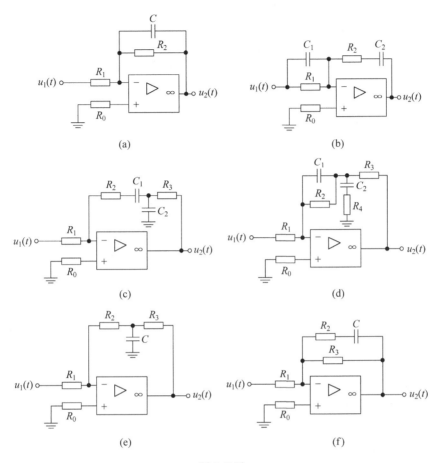

题 2-7 图

2-9 已知给定系统的结构图如题 2-9 图所示,试求出传递函数 $\dfrac{C(s)}{R(s)}$、$\dfrac{E(s)}{R(s)}$。

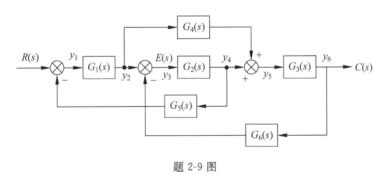

题 2-9 图

2-10 试简化如题 2-10 系统结构图所示,并分别求出传递函数 $\dfrac{C_1(s)}{R_1(s)}$,$\dfrac{C_1(s)}{R_2(s)}$,$\dfrac{C_2(s)}{R_1(s)}$ 及 $\dfrac{C_2(s)}{R_2(s)}$。

2-11 试求题 2-11 图所示系统的传递函数 $\dfrac{C(s)}{R(s)}$。

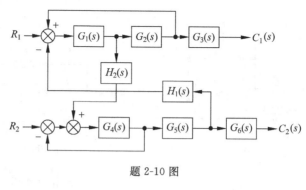

题 2-10 图

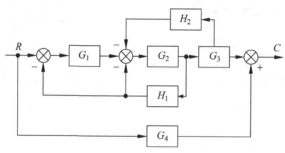

题 2-11 图

2-12 （武汉科技大学硕士研究生入学考试试题）系统结构图如题 2-12 图所示。通过框图化简，试求图示系统的传递函数 $X_o(s)/X_i(s)$。

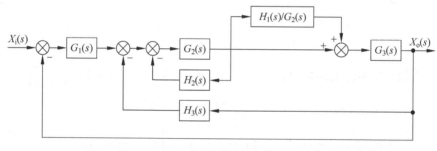

题 2-12 图

2-13 （中国科学院自动化研究所硕士研究生入学考试试题）控制系统框图如题 2-13 图所示。

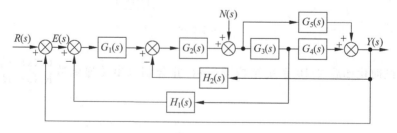

题 2-13 图

(1) 用框图等效变换法,求系统的闭环传递函数 $Y(s)/R(s)$ 和 $E(s)/R(s)$。

(2) 欲使系统的输出 $Y(s)$ 不受扰动 $N(s)$ 的影响,应满足什么条件?

2-14　(华南理工大学硕士研究生入学考试试题)试分别用简化结构图、信号流图方法求题 2-14 图所示系统的传递函数 $\dfrac{C(s)}{R(s)}$。

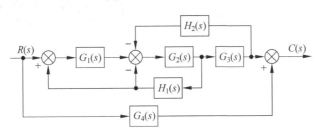

题 2-14 图

2-15　(大连理工大学硕士研究生入学考试试题)给定某系统的传递函数为:
$$\frac{C(s)}{R(s)} = \frac{10}{s(0.1s+1)}$$
如果用电阻、电容、运算放大器等元件构成该系统的模拟装置,试画出该模拟装置的电路原理,并计算出电阻、电容的参数值。

第3章 线性系统的时域分析

3.1 引言

分析和设计控制系统的首要任务是建立系统的数学模型。一旦获得合理的数学模型,就可以采用不同的分析方法来分析系统的动态性能和稳态性能,同时得出改善系统性能的措施。经典控制理论中常用的工程分析方法有三种,即时域分析法(Time Domain Analysis)、根轨迹法(Root Locus Method)和频率响应法(Frequency Response Method)。不同的方法有不同的特点和适用范围,本章主要讨论线性控制系统的时域分析法。

Reading Material

Control systems are required to have satisfactory transient and steady state responses. These are contradictory requirements, and the final design is often a compromise between the two. Quantitative performance measures are required to carry forward the design process and to make comparisons between competing designs. Quantitative performance measures of feedback control systems will enable quantification of design objectives so that design can be carried out logically and systematically (remove fuzziness associated with customer requirements) and enable meaningful comparisons to be made between competing designs. The criteria for selection of performance measures include: ① design objectives translate into simple design rules; ② easy to analyse; ③ experimentally measurable. Design specifications for control systems normally include several time response indices for a specified (test) reference input as well as a desired steady-state accuracy. This topic introduces you to the time domain performance measures that are widely used by control practitioners.

In the following activities, you will explore the relationships between the commonly used quantitative performance measures of feedback control systems and the location of the system transfer function poles and zeros in the s-plane. You will learn that a second order transfer function with damping ratio $\xi \approx 0.7$ has the optimum transient response, while its natural frequency ω_n determines the duration of the transients. It is therefore very desirable to accomplish such a transfer function for feedback control systems.

本章主要内容：
- [ ] 系统时间响应的性能指标
- [ ] 一阶系统的时域分析
- [ ] 二阶系统的时域分析
- [ ] 高阶系统时域分析法概述
- [ ] 控制系统的稳定性分析
- [ ] 控制系统稳态误差的分析及计算
- [ ] MATLAB 在本章中的应用

3.2 系统时间响应的性能指标

时域分析法是指在时间域内研究系统在典型输入信号的作用下，其输出响应随时间变化规律的方法。按照第 1 章中介绍的对自动控制系统的基本要求，根据时域分析法研究系统性能，可以定量描述包括稳定性（Stability）、快速性（Speed）和准确性（Accuracy）等各项性能指标。时域分析法的物理概念清晰，准确度较高，在已知系统结构和参数并建立了系统的微分方程后，使用时域分析法比较方便，从而它也成为控制系统研究中直观而有效的方法之一。

下面通过阅读英文资料，辨析时域与时变系统的基本定义。

Terms and Concepts

Time domain：the mathematical domain that incorporates the time response and the description of a system in terms of time.

Time-varying system：a system for which one or more parameters may vary with time.

1. 典型的输入信号

研究系统的动态特性，就是研究系统在输入信号作用下，考察输出量是如何按输入量的变化而变化的，即系统对输入如何产生响应。控制系统的响应是由系统本身的结构参数、初始状态和输入信号的形式所决定的。为了便于分析和设计，提出两个假定，第一个假定，在输入信号作用于系统的初始状态（$t=0$）时，系统相对静止，即为零初始状态；第二个假定，需要假定一些典型的输入信号，作为系统的实验信号。选取这些实验信号时应考虑以下三个方面。

（1）选取的输入信号的典型性应反映系统工作的大部分情况。

（2）选取的输入信号具有一定代表性，且其数学表达式简单，以便于数学分析和实验研究。

（3）应选取能使系统工作在最不利情况下的输入信号作为典型的实验输入信号。

综上所述，选取常用的典型输入信号有：阶跃信号、斜坡（速度）信号、加速度（抛物线）信号、脉冲信号和正弦信号。

1）阶跃信号

阶跃信号（Step Signal）表示输入中瞬时的变化，如图 3-2-1(a)所示。它的数学表达

式为

$$r(t) = \begin{cases} A, & t \geq 0 \\ 0, & t < 0 \end{cases} \quad (3\text{-}2\text{-}1)$$

式中，A 为一常数。当 $A=1$ 时称为单位阶跃信号，记为 $1(t)$。阶跃信号 $r(t)$ 的拉普拉斯变换为

$$\mathcal{L}[r(t)] = \frac{A}{s} \quad (3\text{-}2\text{-}2)$$

阶跃信号是评价系统动态性能时应用较多的一种典型外作用。在实际运用中，最经常采用的实验信号就是阶跃信号，它可以表示指令的突然转换、电源的突然接通、设备故障和负荷的突变等。由于阶跃信号的频谱具有很宽的频带，通常也作为测试信号，等价于应用无数个频率范围很宽的正弦信号。

2) 斜坡信号

斜坡信号（Ramp Signal）是指由零值开始随时间作线性增长的信号，如图 3-2-1(b) 所示。它的数学表达式为

$$r(t) = \begin{cases} At, & t \geq 0 \\ 0, & t < 0 \end{cases} \quad (3\text{-}2\text{-}3)$$

式中，A 为一常数。当 $A=1$ 时称为单位斜坡信号。斜坡信号 $r(t)$ 的拉普拉斯变换为

$$\mathcal{L}[r(t)] = \frac{A}{s^2} \quad (3\text{-}2\text{-}4)$$

斜坡信号可以看作是阶跃信号的积分，有时又称为速度信号。斜坡信号的变化要比阶跃信号快一个等级，它具有测试系统将如何对随时间变化的信号做出响应的能力。大型船闸的匀速升降、列车的匀速前进、主拖动系统发出的位置信号等都可以看成斜坡信号。

3) 加速度信号

加速度（抛物线）信号（Parabolic Signal）可以看成是数学上的抛物线函数，特点是随时间以等加速度不断增长，如图 3-2-1(c) 所示。它的数学表达式为

$$r(t) = \begin{cases} \frac{1}{2}At^2, & t \geq 0 \\ 0, & t < 0 \end{cases} \quad (3\text{-}2\text{-}5)$$

式中，A 为一常数。当 $A=1$ 时称为单位加速度信号。加速度信号 $r(t)$ 的拉普拉斯变换为

$$\mathcal{L}[r(t)] = \frac{A}{s^3} \quad (3\text{-}2\text{-}6)$$

加速度信号可以看作是斜坡信号的积分，有时又称为抛物线信号。加速度信号的变化要比斜坡快一个等级，在实际运用中很少发现有必要使用变化比加速度信号更快的测试信号。

4) 脉冲信号

脉冲信号（Impulse Signal）可以看成是一个持续时间极短的信号，如图 3-2-1(d) 所示。它的数学表达式为

$$r(t) = \begin{cases} \dfrac{A}{\varepsilon}, & 0 < t < \varepsilon \\ 0, & t < 0 \text{ 或 } t > \varepsilon \end{cases} \quad (3\text{-}2\text{-}7)$$

式中，A 为一常数，ε 为脉冲的宽度。当 $\varepsilon \to 0$ 时称为单位理想脉冲信号，记为 $\delta(t)$，如图 3-2-1(e)所示，即

$$\delta(t) = \begin{cases} \infty, & t = 0 \\ 0, & t \neq 0 \end{cases} \tag{3-2-8}$$

且

$$\int_{-\infty}^{\infty} \delta(t) = 1 \tag{3-2-9}$$

(理想)脉冲信号 $\delta(t)$ 的拉普拉斯变换为

$$\mathcal{L}[\delta(t)] = \int_0^{\infty} \delta(t) \mathrm{e}^{-st} \mathrm{d}t = \lim_{\varepsilon \to 0} \int_0^{\varepsilon} \frac{1}{\varepsilon} \mathrm{e}^{-st} \mathrm{d}t = \lim_{\varepsilon \to 0} \left[\frac{1}{\varepsilon} \frac{-\mathrm{e}^{-st}}{s} \right]_0^{\varepsilon}$$

$$= \lim_{\varepsilon \to 0} \frac{1}{\varepsilon s} \left[1 - \left(1 - \varepsilon s + \frac{1}{2!} \varepsilon^2 s^2 - \cdots \right) \right] = 1 \tag{3-2-10}$$

$\delta(t)$ 所描述的脉冲信号实际上是无法获得的，在现实中不存在，只有数学意义，但是它却是一个重要的数学工具。在工程实践中，当 ε 远小于被控对象的时间常数时，这种窄脉冲信号就可近似地当成 $\delta(t)$ 信号。脉冲电压信号、冲击力等都可以近似为脉冲信号。

5）正弦信号

正弦信号（Sinusoidal Signal）如图 3-2-1(f)所示。它的数学表达式为

$$r(t) = A \sin \omega t \tag{3-2-11}$$

式中，A 为振幅，ω 为角频率。

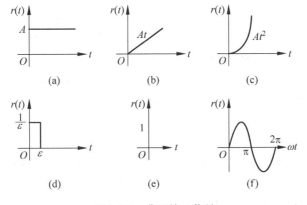

图 3-2-1 典型输入信号

正弦信号可以求得系统对不同频率的正弦函数输入的稳态响应，称为频率响应，将在后续章节阐述。在实际过程中，交流电源、电磁波、机车设备受到的振动、电源和机械振动的噪声等都可以近似为正弦信号。

以上所述输入信号是按时间变化规律来划分的，实际使用时它们可以是任何不同的物理量，如温度、电压、电流、转角、转速、压力等。同时这些测试信号都具有在数学上描述简单和实验室容易实现的特点。究竟采用哪一种典型输入信号分析系统的特性，取决于系统在正常工作情况下最常见的输入信号形式。如果系统的输入信号是一个突变的量，则应选取阶跃信号；如果系统的输入信号是一个瞬时冲击的函数，则选取脉冲信号最为合适；如果系统的输入信号是随时间逐渐变化的函数，则应选取斜坡信号。一般来说，控制系统在实验

信号的基础上设计出来以后,在实际信号的作用下,系统响应特性都能满足要求。表 3-2-1 给出了 5 种典型输入信号的时域表达式和复域表达式。

表 3-2-1　典型输入信号

名　　称	时域表达式	复域表达式
单位阶跃信号	$1(t), t \geq 0$	$\dfrac{1}{s}$
单位斜坡信号	$t, t \geq 0$	$\dfrac{1}{s^2}$
单位加速度信号	$\dfrac{1}{2}t^2, t \geq 0$	$\dfrac{1}{s^3}$
单位脉冲信号	$\delta(t), t = 0$	1
正弦信号	$A\sin\omega t$	$\dfrac{A\omega}{s^2+\omega^2}$

Reading Material

In practice, actual input signal to the control system is quite varied and unpredictable, and/or too complex to analyse. When carrying out mathematical analysis of linear systems, we can consider the input magnitude to be unity without any loss of generality. In experimental investigation of practical systems, however, the magnitude of the input test signal should be carefully selected (depends on the particular system being tested). Otherwise one or more components of the system may be forced to operate outside their linear range. The linear model will not be applicable then, and the output of the system will differ significantly from that predicted by the transfer function analysis.

2. 单位冲激响应

设系统的输入响应 $R(s)$ 与输出信号 $C(s)$ 之间的传递函数是 $G(s)$,则有

$$C(s) = G(s)R(s) \qquad (3\text{-}2\text{-}12)$$

若输入信号是单位冲激函数 $\delta(t)$,即 $r(t)=\delta(t)$,则

$$\begin{cases} R(s) = \mathcal{L}[\delta(t)] = 1 \\ C(s) = G(s) \\ c(t) = \mathcal{L}^{-1}[G(s)] = g(t) \end{cases} \qquad (3\text{-}2\text{-}13)$$

在零初始条件下,当系统的输入信号是单位冲激函数 $\delta(t)$ 时,系统的输出信号称为系统的单位冲激响应。系统的单位冲激响应就是系统传递函数 $G(s)$ 的拉普拉斯反变换 $g(t)$。同传递函数一样,单位冲激函数也是系统的数学模型。

对式(3-2-12)两边取拉普拉斯反变换,并利用拉普拉斯变换的卷积定理可得

$$c(t) = g(t) * r(t) = \int_0^t g(\tau)r(t-\tau)\mathrm{d}\tau = \int_0^t g(t-\tau)r(\tau)\mathrm{d}\tau \qquad (3\text{-}2\text{-}14)$$

可见,输出信号 $c(t)$ 等于单位冲激响应 $g(t)$ 与输入信号 $r(t)$ 的卷积。

3. 系统的时间响应

若系统输出信号的拉普拉斯变换是 $C(s)$，则系统的时间响应 $c(t)$ 是

$$c(t) = \mathcal{L}^{-1}[C(s)] \qquad (3\text{-}2\text{-}15)$$

根据拉普拉斯反变换中的部分分式法可知，有理分式 $C(s)$ 的每一个极点（分母多项式的根）都对应于 $c(t)$ 中一个时间响应项，即运动模态，而 $c(t)$ 就是由 $C(s)$ 的所有极点所对应的时间响应项（运动模态）的线性组合。不同极点所对应的运动模态如表 3-2-2 所示。

表 3-2-2 极点与运动模态

极 点	运 动 模 态
实数单极点 σ	$k e^{\sigma t}$
m 重实数极点 σ	$(k_1 + k_2 t + \cdots + k_m t^{m-1}) e^{\sigma t}$
一对复数极点 $\sigma \pm j\omega$	$k e^{\sigma t} \sin(\omega t + \phi)$
m 重复数极点 $\sigma \pm j\omega$	$e^{\sigma t}[k_1 \sin(\omega t + \phi_1) + k_2 t \sin(\omega t + \phi_2) + \cdots + k_m t^{m-1} \sin(\omega t + \phi_m)]$

若系统的输入信号是 $R(s)$，传递函数是 $G(s)$，则零初始条件下有

$$C(s) = G(s) R(s)$$

可见，输出信号拉普拉斯变换式的极点是由传递函数的极点和输入信号拉普拉斯变换式的极点组成的。通常把传递函数极点所对应的运动模态称为该系统的自由运动模态或振型，或称为该传递函数或微分方程的模态或振型。系统的自由运动模态与输入信号无关，也与输出信号的选择无关。传递函数的零点并不形成运动模态，但它们却影响各模态在响应中所占的比例，因而也影响时间响应及其曲线形状。

系统的时间响应中，与传递函数极点对应的时间响应分量称为瞬（暂）态分量，与输入信号极点对应的时间响应分量称为稳态分量。

根据数学中拉普拉斯变换的微分性质和积分性质可以推导出线性定常系统的下述重要特性：系统对输入信号导数的响应，等于系统对该输入信号响应的导数；系统对输入信号积分的响应，等于系统对该输入信号响应的积分，积分常数由零输出的初始条件确定。可见，一个系统的单位阶跃响应、单位冲激响应和单位斜坡响应中，只要知道一个，就可通过微分或积分运算求出另外两个。

Reading Material

Poles are the set of values of s that make the value of the transfer function infinity. Zeros are the set of values of s that make the value of the transfer function zero. The characteristics of poles and zeros of a transfer function can be described as the following.

(1) Poles are those values of s that make the transfer function infinity.

(2) The equation obtained by equating denominator of the transfer function to zero is referred to as the characteristic equation.

(3) Poles are the roots of the characteristic equation.

(4) Zeros are those values of s that make the transfer function zero.

(5) Zeros are the roots of the equation obtained by equating the numerator of the transfer function to zero.

(6) Poles play much more important role than zeros in determining the system dynamics.

4. 动态响应与稳态响应

系统在输入信号的作用下,其输出随时间变化的过程称为系统的时间响应。任何一个控制系统的时间响应都由瞬态响应和稳态响应两大部分组成。

1) 瞬态过程

瞬态响应(Transient Response)又称为过渡过程或动态响应,是指系统在典型输入信号作用下,系统输出量从初始状态到最终状态的响应过程,由于实际系统具有惯性、摩擦以及其他原因,系统输出量不可能完全复现输入量的变化。根据系统结构和参数选择的情况,动态过程表现为衰减、发散或等幅振荡形式。显然,一个可以实际运行的控制系统,其动态过程必须是衰减的,换句话说,系统必须是稳定的。动态响应除了提供系统稳定性的信息外,还可以提供响应速度及阻尼情况等信息,这些信息用动态性能来描述。

2) 稳态响应

稳态响应(Steady-State Response)又称稳态过程,是指系统在典型输入信号作用下,当时间 t 趋于无穷时,系统输出量的表现方式。它表征系统输出量最终复现输入量的程度,提供系统有关稳态误差的信息。从理论上讲,只有时间趋于无穷大时才进入稳态过程,但这在工程应用中是无法实现的,因此在工程中只讨论典型输入信号加入后一段时间里的动态过程,在这段时间里反映了系统主要的动态性能指标。而在这段时间之后,认为系统进入了稳态过程。

Reading Material

The time response of a control system is usually divided into two parts: transient response and steady-state response. If a time response is denoted by $c(t)$, then the transient response and steady-state response may be denoted by $c_t(t)$ and $c_{ss}(t)$. The steady-state response is simply the fixed response when time reaches infinity, i.e.,

$$c_{ss}(t) = \lim_{t \to \infty} c(t)$$

Therefore, a sine wave is considered as steady-state response because its behavior is fixed as time approaches infinity.

Transient response is defined as the part of the response that goes to zeros as time becomes large. Therefore, $c_t(t)$ has the property of

$$\lim_{t \to \infty} c_t(t) = 0$$

The transient response of a control system is of importance, since it is a part of the dynamic behavior of the system; and the difference between the response and the input or the desired response, before the steady state is reached, must be closely watched. If the steady-state response of the output does not agree with the steady state of the reference exactly, the system is said to have a steady-state error.

5. 瞬态性能指标与稳态性能指标

在许多实际情况中,控制系统所需要的性能指标常以时域量值的形式体现,即通常使用单位阶跃信号作为测试信号,来计算在时间域的瞬态和稳态的性能。因为单位阶跃信号响应比较容易,而且阶跃信号对系统来说是最严峻的工作状态,如果系统在阶跃信号作用下的性能指标能满足要求,那么系统在其他形式的输入信号下,其性能指标一般也可满足要求。

1) 瞬态性能指标

描述稳定的系统在单位阶跃信号作用下瞬态过程随时间的变化状况的指标,称为瞬态性能指标(Transient Performance Specifications)。图 3-2-2 给出了控制系统的单位阶跃响应曲线,根据图示定义如下瞬态性能指标。

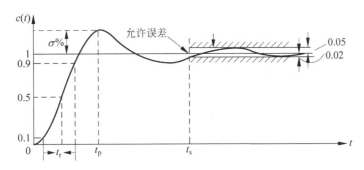

图 3-2-2 控制系统的单位阶跃响应曲线

(1) 上升时间 t_r(Rise Time):指响应曲线从零时刻开始第一次上升到稳态值所需要的时间,也可以指响应曲线从稳态值的 10% 上升到稳态值的 90% 所需要的时间。

(2) 峰值时间 t_p(Peak Time):指响应曲线从零时刻开始到达第一个峰值的时间。

(3) 超调量 $\sigma\%$(Percentage Overshoot):指响应曲线的最大峰值与稳态值之差和相对稳态值之比的百分数,即

$$\sigma\% = \frac{c(t_p) - c(\infty)}{c(\infty)} \times 100\% \tag{3-2-16}$$

若系统输出响应单调变化,则无超调量。超调量也称最大超调量,或百分比超调量。

(4) 调节时间 t_s(Setting Time):指响应曲线从零时刻开始到达并保持在稳态值±5%或±2%范围内所需的最短时间。调节时间又称为过渡过程时间。

(5) 振荡次数 N:在调整时间 t_s 内响应曲线振荡次数。

上述这些指标描述了瞬态响应过程,反映了系统的动态性能。其中,上升时间 t_r、峰值时间 t_p 均表征系统响应初始阶段的快速性;调节时间 t_s 表示系统过渡过程的持续时间,从总体上反映了系统的快速性;最大超调量 $\sigma\%$ 反映了系统动态过程的平稳性,即用超调量表示实际响应与期望响应的接近程度。由控制系统本身的特性决定,这些要素通常是相互矛盾的,因而必须做出折中的选择。

Terms and Concepts

Test input signal: an input signal used as a standard test of a system's ability to respond adequately.

Performance index: a quantitative measure of the performance of a system.

Overshoot: the amount by which the system output response processed beyond the desired response.

Peak time: the time for a system to respond to a step input and rise to a peak response.

Rise time: the time for a system to respond to a step input and attain a response equal to a percentage of the magnitude of the input.

Setting time: the time required for the system output to settle within a certain percentage of the input amplitude.

2) 稳态性能指标

当响应时间大于调节时间时,系统进入稳态过程。稳态误差(Steady-State Error)是描述系统稳态性能的一种性能指标。其定义为

$$e_{ss} = \lim_{t \to \infty}[r(t) - c(t)] = \lim_{t \to \infty} e(t) \tag{3-2-17}$$

稳态误差是控制系统精度和抗干扰能力的一种度量,反映控制系统复现或跟踪输入信号的能力。

3.3 一阶系统的时域分析

用一阶微分方程描述的控制系统称为一阶系统。一阶系统在控制工程中应用广泛,一些控制部件及简单系统(如 RC 网络、发电机、空气加热器、液面控制系统等)都可用一阶系统来描述。有些高阶系统的特征,常可用一阶系统的特征来近似表征。

Terms and Concepts

The order of a system is defined as the degree of its characteristic polynomial. A first-order system is represented by a first-order differential equation.

3.3.1 一阶系统的数学模型和结构图

如图 3-3-1 所示 RC 滤波电路为一阶系统,其微分方程为

$$RC \frac{dc(t)}{dt} + c(t) = r(t) \tag{3-3-1}$$

当该电路的初始条件为零时,一阶系统的传递函数为

$$G(s) = \frac{C(s)}{R(s)} = \frac{1}{Ts + 1} \tag{3-3-2}$$

式中,$T=RC$ 为时间常数,它是表征系统惯性的重要特征参数,反映了系统过渡过程的品质。T 越小,系统响应越快。

一阶系统的结构图及其简化形式如图 3-3-2(a)和图 3-3-2(b)所示。

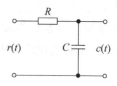

图 3-3-1 RC 滤波电路

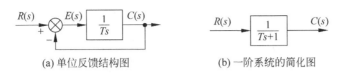

(a) 单位反馈结构图　　(b) 一阶系统的简化图

图 3-3-2　一阶系统结构图

3.3.2　一阶系统的单位阶跃响应

单位阶跃函数的拉普拉斯变换为 $R(s)=1/s$，故输出的拉普拉斯变换式为

$$C(s) = G(s)R(s) = \frac{1}{Ts+1} \cdot \frac{1}{s} = \frac{1}{s} - \frac{T}{Ts+1} \quad (3\text{-}3\text{-}3)$$

取 $C(s)$ 的拉普拉斯逆变换，得

$$c(t) = C_{ss} + C_{tt} = 1 - e^{-\frac{t}{T}} \qquad t \geqslant 0 \quad (3\text{-}3\text{-}4)$$

由式(3-3-4)可以看出：

(1) 输出量的初始值为零，而终值将变为 1。
(2) C_{ss} 称为稳态分量，它的变化规律由输入信号的形式决定；C_{tt} 称为暂态分量。
(3) 该响应曲线的一个重要特征是，当 $t=T$ 时，$c(t)$ 的数值等于 0.632，即响应 $c(t)$ 达到了其总变化的 63.2%。
(4) 该响应曲线时间常数 T 越小，系统的响应越快。

下面分析一阶系统的输入单位阶跃信号的特性指标。

1. 调节时间 t_s

图 3-3-3 为一阶系统的单位响应曲线。当 $t=T$ 时，指数响应曲线将从 0 上升到稳定值的 63.2%；当 $t=2T$ 时，指数响应曲线上升到稳定值的 86.5%；当 $t=3T$、$4T$、$5T$ 时，曲线分别上升到稳定值的 95%、98.2% 和 99.3%。故一般取 $t_s = (3 \sim 4)T$。

2. 最大超调量 $\sigma\%$

一阶系统的单位阶跃响应为非周期响应，故系统无振荡、无超调，$\sigma\% = 0$。

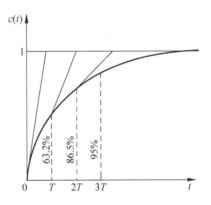

图 3-3-3　一阶系统的单位阶跃响应曲线

3. 稳态误差 e_{ss}

由于 $c(t)$ 的终值为 1，因而系统在阶跃输入时稳态误差为零。

3.3.3　一阶系统的单位脉冲响应

单位脉冲信号的拉普拉斯变换为 1，所以系统的输出为

$$C(s) = G(s)R(s) = \frac{1}{Ts+1} \quad (3\text{-}3\text{-}5)$$

其拉普拉斯反变换为

$$c(t) = \frac{1}{T}e^{-\frac{t}{T}} \qquad t \geqslant 0 \qquad (3\text{-}3\text{-}6)$$

一阶系统的单位脉冲响应为一单调下降的指数曲线,如图 3-3-4 所示。单位脉冲响应在 $t=0$ 时等于 $1/T$,它与单位阶跃响应在 $t=0$ 时的变化率相等。这说明单位脉冲响应是单位阶跃响应的导数,而单位阶跃响应就是单位脉冲响应的积分。这个结果进一步说明了单位阶跃信号作为典型输入信号的重要性和代表性。

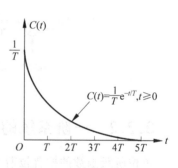

图 3-3-4 一阶系统的脉冲响应曲线

3.3.4 一阶系统的单位斜坡响应

单位斜坡信号 $r(t)=t$ 的拉普拉斯变换为 $R(s)=1/s^2$,所以系统的输出为

$$C(s) = G(s) \cdot R(s) = \frac{1}{Ts+1} \cdot \frac{1}{s^2} = \frac{1}{s^2} - \frac{T}{s} + \frac{T^2}{Ts+1} \qquad (3\text{-}3\text{-}7)$$

其拉普拉斯反变换为

$$c(t) = C_{ss} + C_{tt} = (t-T) + Te^{-\frac{t}{T}} \qquad t \geqslant 0 \qquad (3\text{-}3\text{-}8)$$

式中,稳态分量 $C_{ss}=t-T$,表示一个与单位斜坡信号斜率相同、时间滞后 T 的斜坡函数。暂态分量 $C_{tt}=Te^{-\frac{t}{T}}$ 按指数规律衰减至零。

因为

$$C_{tt} = Te^{-\frac{t}{T}} \qquad (3\text{-}3\text{-}9)$$

所以一阶系统的斜坡响应稳态误差为

$$e_{ss} = \lim_{t \to \infty}[r(t) - c(t)]$$
$$= \lim_{t \to \infty} T(1 - e^{-\frac{t}{T}}) = T \qquad (3\text{-}3\text{-}10)$$

从提高斜坡响应的精度来看,要求一阶系统的时间常数 T 要小,即时间常数 T 越小,响应速度越快,跟踪误差越小,输出信号滞后于输入信号的时间也越短。同样地,单位斜坡响应就是单位阶跃响应的积分。图 3-3-5 为一阶系统的单位斜坡响应。

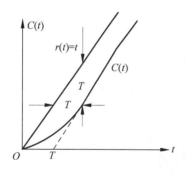

图 3-3-5 一阶系统的斜坡响应曲线

3.3.5 一阶系统的单位加速度响应

单位加速度信号 $r(t)=\frac{1}{2}t^2$ 的拉普拉斯变换为 $R(s)=1/s^3$,所以系统的输出为

$$C(s) = G(s)R(s) = \frac{1}{Ts+1} \cdot \frac{1}{s^3} = \frac{1}{s^3} - \frac{T}{s^2} + \frac{T^2}{s} - \frac{T^3}{Ts+1} \qquad (3\text{-}3\text{-}11)$$

其拉普拉斯反变换为

$$c(t) = C_{ss} + C_{tt} = \frac{1}{2}t^2 - Tt + T^2(1 - e^{-\frac{t}{T}}) \qquad t \geqslant 0 \qquad (3\text{-}3\text{-}12)$$

因为

$$e(t) = r(t) - c(t) = Tt - T^2(1 - e^{-\frac{t}{T}}) \qquad (3\text{-}3\text{-}13)$$

所以一阶系统的加速度响应稳态误差为

$$e_{ss} = \lim_{t\to\infty}[r(t) - c(t)] = \lim_{t\to\infty}[Tt - T^2(1 - e^{-\frac{t}{T}})] = \infty \tag{3-3-14}$$

这就意味着，对于一阶系统来说，不能实现对加速度输入信号的跟踪。

比较一阶系统对上述信号的输出响应可以发现，脉冲响应、阶跃响应、斜坡响应之间也存在同样的对应关系。这表明，系统对某种输入信号导数的响应，等于对该输入信号响应的导数；反之，系统对某种输入信号积分的响应，等于系统对该输入信号响应的积分。这是线性定常系统的一个重要特征，它不仅适用于一阶线性定常系统，也适用于高阶线性定常系统。因此，在后面的分析中，将主要研究系统的单位阶跃响应。

【例 3-3-1】 已知系统结构图如图 3-3-6 所示，其中，$G(s) = \dfrac{10}{0.2s+1}$，加上 K_0, K_H 环节，使 t_s 减小为原来的 0.1 倍，且总放大倍数不变，求 K_0, K_H。

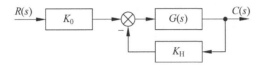

图 3-3-6　系统结构图

解：依题意，要使闭环系统 $t_s^* = 0.1 \times 0.2 = 0.02$，且闭环增益为 10。

$$\Phi(s) = K_0 \frac{G(s)}{1 + K_H G(s)} = K_0 \frac{\dfrac{10}{0.2s+1}}{1 + \dfrac{10K_H}{0.2s+1}}$$

$$= \frac{10K_0}{0.2s + 1 + 10K_H} = \frac{10K_0/(1+10K_H)}{\dfrac{0.2}{1+10K_H}s + 1}$$

令 $\begin{cases} T = \dfrac{0.2}{1+10K_H} = 0.02 \\ K = \dfrac{10K_0}{1+10K_H} = 10 \end{cases}$ 联立解出 $\begin{cases} K_H = 0.9 \\ K_0 = 10 \end{cases}$

3.4　二阶系统的时域分析

如果控制系统的运动方程为二阶微分方程，或者自动控制系统的特征方程 s 的最高阶次为二次，则该系统为二阶系统。例如 RLC 网络、机械平移系统和位置随动系统。

Terms and Concepts

The second-order system is described by second-order differential equations. Second-order systems are important because their behavior is very different from first-order systems and may exhibit feature such as oscillatory response, or overshoot. Moreover, their analysis generally is helpful to form basis for the understanding of analysis and design techniques.

已知 RLC 网络如第 2 章中所示,其运动方程为

$$LC\frac{d^2c(t)}{dt^2} + RC\frac{dc(t)}{dt} + c(t) = r(t) \tag{3-4-1}$$

传递函数为

$$G(s) = \frac{C(s)}{R(s)} = \frac{1}{LCs^2 + RCs + 1} = \frac{1/LC}{s^2 + (R/L)s + 1/LC} \tag{3-4-2}$$

令 $\omega_n^2 = 1/LC$,则 $\omega_n = 1/\sqrt{LC}$ 称为无阻尼振荡频率或自然频率,属于闭环系统的固有频率;令 $2\xi\omega_n = R/L$,则 $\xi = \frac{R}{2}\sqrt{\frac{C}{L}}$ 称为二阶系统的阻尼比(Damping Ratio)或相对阻尼系数,量纲为 1。传递函数变形为

$$G(s) = \frac{C(s)}{R(s)} = \frac{\omega_n^2}{s^2 + 2\xi\omega_n s + \omega_n^2} \tag{3-4-3}$$

令式(3-4-3)的分母多项式为零,得到二阶系统的特征方程

$$s^2 + 2\xi\omega_n s + \omega_n^2 = 0 \tag{3-4-4}$$

其特征根为

$$s_{1,2} = -\xi\omega_n \pm \omega_n\sqrt{\xi^2 - 1} \tag{3-4-5}$$

由上可知,二阶系统的动态特性取决于 ξ 和 ω_n 这两个参数。其中,根据阻尼比 ξ 的取值,这两个极点可以是实数或者一对共轭复数,即其特征根和瞬态响应也有很大的差异。

Terms and Concepts

Critical damping: The case where damping is on the boundary between underdamping and overdamping.

Damp ratio: A measure of damping. A dimensionless number for the second-order characteristic equation.

3.4.1 二阶系统的单位阶跃响应

1. 无阻尼($\xi = 0$, Underdamped)

当 $R = 0$,即 $\xi = 0$ 时,系统具有一对共轭虚根 $s_{1,2} = \pm j\omega_n$,对应的单位阶跃响应为

$$c(t) = 1 - \cos\omega_n t \tag{3-4-6}$$

式(3-4-6)表明系统在无阻尼时,响应是一个纯正弦信号,电路中将发生不衰减的电磁振荡现象,振荡频率为 ω_n。

2. 欠阻尼($0 < \xi < 1$, Underdamping)

当 $0 < R < R_c$($R_c = 2\sqrt{L/C}$),即 $0 < \xi < 1$ 时,系统具有一对共轭复根,即

$$s_{1,2} = -\xi\omega_n \pm j\omega_n\sqrt{1-\xi^2} = -\xi\omega_n \pm j\omega_d \tag{3-4-7}$$

式中,$\omega_d = \omega_n\sqrt{1-\xi^2}$ 称为有阻尼振荡频率(或有阻尼自然频率)。

对于单位阶跃输入信号,$R(s) = 1/s$,则系统输出的拉普拉斯变换为

$$C(s) = \frac{\omega_n^2}{s(s+\xi\omega_n-j\omega_d)(s+\xi\omega_n+j\omega_d)} \qquad (3\text{-}4\text{-}8)$$

对式(3-4-8)取拉普拉斯反变换,得到对应的单位阶跃响应为

$$c(t) = \mathcal{L}^{-1}[C(s)] = 1 - \frac{1}{\sqrt{1-\xi^2}} e^{-\xi\omega_n t} \sin\left(\omega_d t + \arctan\frac{\sqrt{1-\xi^2}}{\xi}\right) \quad t \geqslant 0 \qquad (3\text{-}4\text{-}9)$$

由式(3-4-9)可知,欠阻尼二阶系统的单位阶跃响应有稳态响应分量和瞬态响应分量组成,等式右边第一项为稳态响应分量1,第二项为瞬态响应分量,是一个幅值按指数规律衰减的阻尼正弦振荡,振荡频率为 ω_d,而 ω_d 正好由极点的虚部给出,衰减速度则由常数决定,正好由极点的实部给出。从电路中分析,当 $0 < R < R_c (R_c = 2\sqrt{L/C})$,即 $0 < \xi < 1$ 时,电路中将发生衰减振荡的电磁现象,R 越大,电磁振荡衰减越快。

3. 临界阻尼($\xi = 1$, Critical Damping)

当 $R \neq 0$,即 $\xi = 1$ 时,系统具有一对重负实根,即

$$s_{1,2} = -\xi\omega_n \pm \omega_n\sqrt{\xi^2-1} = -\omega_n \qquad (3\text{-}4\text{-}10)$$

对于单位阶跃输入信号,$R(s)=1/s$,则系统输出的拉普拉斯变换为

$$C(s) = \frac{\omega_n^2}{s(s+\omega_n)^2} \qquad (3\text{-}4\text{-}11)$$

对式(3-4-11)取拉普拉斯反变换得对应的单位阶跃响应为

$$c(t) = \mathcal{L}^{-1}[C(s)] = 1 - e^{-\omega_n t}(1+\omega_n t) \quad t \geqslant 0 \qquad (3\text{-}4\text{-}12)$$

由式(3-4-12)可知,临界阻尼二阶系统的单位阶跃响应由稳态响应分量和瞬态响应分量组成,等式右边第一项为稳态响应分量1,第二项为瞬态响应分量,是一个衰减的过程。从电路中分析,当 $R \neq 0$ 时,将对电磁振荡起阻尼作用。R 越大,阻尼越大,故 R 称为 RLC 电路的阻尼系数,$R=R_c$ 称为临界阻尼系数。而 $\xi = \frac{R}{2}\sqrt{\frac{C}{L}} = \frac{R}{R_c} = \frac{\text{实际阻尼系数}}{\text{临界阻尼系数}}$,故 ξ 称为相对阻尼系数或阻尼比。

4. 过阻尼($\xi > 1$, Overdamp/Overdamping)

当 $R > R_c$,即 $\xi > 1$ 时,系统具有一对不相等的负实根,即

$$s_{1,2} = -\xi\omega_n \pm \omega_n\sqrt{\xi^2-1} \qquad (3\text{-}4\text{-}13)$$

对于单位阶跃输入信号,$R(s)=1/s$,则系统输出的拉普拉斯变换为

$$C(s) = \frac{\omega_n^2}{(s+\xi\omega_n+\omega_n\sqrt{\xi^2-1})(s+\xi\omega_n-\omega_n\sqrt{\xi^2-1})} \cdot \frac{1}{s} \qquad (3\text{-}4\text{-}14)$$

对式(3-4-14)取拉普拉斯反变换,得到对应的单位阶跃响应为

$$c(t) = \mathcal{L}^{-1}[C(s)] = 1 + \frac{\omega_n}{2\sqrt{\xi^2-1}}\left(\frac{e^{-s_1 t}}{s_1} - \frac{e^{-s_2 t}}{s_2}\right) \quad t \geqslant 0 \qquad (3\text{-}4\text{-}15)$$

式中,$s_1 = (\xi+\sqrt{\xi^2-1})\omega_n, s_2 = (\xi-\sqrt{\xi^2-1})\omega_n$。

由式(3-4-15)可知,过阻尼二阶系统的单位阶跃响应由稳态响应分量和瞬态响应分量组成,等式右边第一项为稳态响应分量1,第二项为瞬态响应分量,是两个指数衰减过程的叠加,因而瞬态响应是单调的衰减过程。从电路中分析,当 $R > R_c(R_c = 2\sqrt{L/C})$,即 $\xi > 1$

时,电路中将发生单调变化的电磁现象。

表 3-4-1 表示当 ξ 为不同值时,其相应系统极点的数值与阶跃响应曲线。

表 3-4-1 ξ 不同取值时的二阶系统的单位阶跃响应

阻 尼 比	特 征 根	单位阶跃响应
无阻尼 ($\xi=0$)	一对共轭纯虚根 $s_{1,2}=\pm j\omega_n$	等幅振荡过程
欠阻尼 ($0<\xi<1$)	一对实部为负的共轭负根 $s_{1,2}=-\xi\omega_n\pm j\omega_n\sqrt{1-\xi^2}$	衰减振荡过程
临界阻尼 ($\xi=1$)	两个相等的负实根 $s_{1,2}=-\omega_n$	非周期过程
过阻尼 ($\xi>1$)	两个不相等的负实根 $s_{1,2}=-\xi\omega_n\pm\omega_n\sqrt{\xi^2-1}$	非周期过程

3.4.2 欠阻尼二阶系统的动态性能分析

通常,控制系统的性能指标是通过其单位阶跃响应的特征量来描述的。为了定量地评价二阶系统的控制质量,需要进一步分析 ξ 和 ω_n 对系统单位阶跃响应的影响,并定义二阶系统单位阶跃响应的一些特征量作为评价系统的性能指标。

在实际工程控制中,除了一些不允许产生振荡的系统之外,通常希望系统的响应过程在具有适当的振荡特性情况下,能有较短的调整时间和较高的响应速度。在设计二阶系统时,一般取 $\xi=0.4\sim0.8$,即系统工作在欠阻尼状态下。小的 $\xi(\xi<0.4)$ 会造成系统瞬态响应的严重超调,而大的 $\xi(\xi>0.8)$ 会使系统的响应变得缓慢。因此,针对二阶系统欠阻尼的工作响应,来确定二阶系统的性能指标和定量关系的推导,如图 3-4-1 所示。控制系统的单位阶跃响应一般来说是与初始条件有关的,为了便于比较各种系统的控制质量,通常假设系统的初始条件为零。

假设系统为欠阻尼系统 ($0<\xi<1$),此时系统的单位阶跃响应为

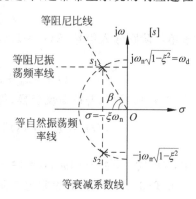

图 3-4-1 欠阻尼二阶系统各特征参量间的关系

$$c(t) = 1 - e^{-\xi\omega_n t}\left(\cos\omega_d t + \frac{\xi}{\sqrt{1-\xi^2}}\sin\omega_d t\right)$$

$$= 1 - \frac{1}{\sqrt{1-\xi^2}}e^{-\xi\omega_n t}\sin(\omega_d t + \beta) \quad t \geqslant 0 \tag{3-4-16}$$

其中,β 为共轭复数对负实轴的张角,称为阻尼角或滞后角,$\beta = \arctan\dfrac{\sqrt{1-\xi^2}}{\xi} = \arccos\xi = \arcsin\sqrt{1-\xi^2}$。由式(3-4-16)可得,响应特性完全由 ξ 和 ω_n 这两个特征参量决定。

1. 上升时间 t_r

根据上升时间的定义,令 $c(t_r) = 1$,可得到

$$c(t_r) = 1 - \frac{1}{\sqrt{1-\xi^2}}e^{-\xi\omega_n t_r}\sin(\omega_d t_r + \beta) = 1 \tag{3-4-17}$$

因为 $e^{-\xi\omega_n t_r} \neq 0$,所以

$$t_r = \frac{\pi - \beta}{\omega_d} = \frac{\pi - \arccos\xi}{\omega_n\sqrt{1-\xi^2}} \tag{3-4-18}$$

由式(3-4-18)可知,当阻尼比 ξ 一定时,阻尼角 β 不变,系统的响应速度与 ω_n 成反比。在 ω_d 一定时,阻尼比越小,上升时间就越短。

2. 峰值时间 t_p

根据定义,对系统的单位阶跃响应求导,并令该导数为零,即可求得峰值时间。

$$\left.\frac{dc(t)}{dt}\right|_{t=t_p} = 0 \tag{3-4-19}$$

因为峰值时间对应第一个峰值的时间,所以

$$t_p = \frac{\pi}{\omega_d} = \frac{\pi}{\omega_n\sqrt{1-\xi^2}} \tag{3-4-20}$$

峰值时间 t_p 等于阻尼振荡周期的一半。同时,峰值时间 t_p 与阻尼振荡频率 ω_d 成反比。当阻尼比一定时,ω_d 越大,t_p 越短,响应速度越快。当 ω_n 一定时,阻尼比越小,t_p 越短,响应速度也越快。

3. 最大超调量 $\sigma\%$

当 $t = t_p$ 时,$c(t)$ 有最大值 $c_{\max}(t) = c(t_p)$,也就是说超调量发生在峰值时间。并且单位阶跃响应的稳态值为 $c(\infty) = 1$,这样得到响应的最大值为

$$c(t_p) = 1 - \frac{1}{\sqrt{1-\xi^2}}e^{-\frac{\pi\xi}{\sqrt{1-\xi^2}}}\sin(\pi + \beta) = 1 + e^{-\frac{\pi\xi}{\sqrt{1-\xi^2}}} \tag{3-4-21}$$

所以

$$\sigma\% = e^{-\frac{\pi\xi}{\sqrt{1-\xi^2}}} \times 100\% \tag{3-4-22}$$

由式(3-4-22)可知,超调量 $\sigma\%$ 仅由阻尼比 ξ 来

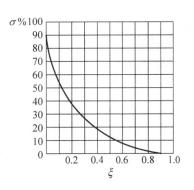

图 3-4-2 超调量 $\sigma\%$ 和 ξ 的关系曲线

决定,而与自然频率 ω_n 无关。由图 3-4-2 和表 3-4-2 可知,阻尼比越大,超调量越小,平稳性越好。

表 3-4-2 典型阻尼比对应的超调量

ξ	0.3	0.4	0.5	0.6	0.68	0.707	0.8	0.9
$\sigma\%$	37.2%	25%	16.3%	9%	5%	4.3%	1.5%	0.2%

在自动控制系统设计中,通常取 $\xi=0.4\sim0.8$,相应的 $\sigma\%=25.4\%\sim1.5\%$。在这样的条件下,系统的总体性能良好。

4. 调节时间 t_s

从调节时间的定义来看,由于 t_s 既出现在指数上,又出现在正弦函数内,故调节时间的表达式很难确定。根据

$$|c(t)-c(\infty)| \leqslant \Delta \cdot c(\infty) \quad t \geqslant t_s \qquad (3\text{-}4\text{-}23)$$

(Δ 等于 1/2 误差带宽度)求得调节时间的计算公式为

$$t_s \approx \frac{1}{\xi\omega_n}\ln\frac{1}{\Delta\sqrt{1-\xi^2}} \qquad (3\text{-}4\text{-}24)$$

在 ξ 较小时,若取 $\Delta=5\%$ 时,得到 $t_s \approx \frac{3}{\xi\omega_n}=3T$。若取 $\Delta=2\%$ 时,得到 $t_s \approx \frac{4}{\xi\omega_n}=4T$。

由式(3-4-24)可知,调节时间与系统的阻尼比和自然频率的乘积成反比。通常阻尼比 ξ 是根据最大允许超调量的要求来决定的,也就是说,在不改变最大允许超调量的前提下,通常通过调整自然频率就可以改变瞬时响应的持续时间。

5. 振荡次数 N

振荡次数 N 表示在调节时间内,系统响应的振动次数,用数学公式表示为

$$N=\frac{t_s}{2\pi/\omega_d}=\frac{\omega_d t_s}{2\pi} \geqslant \frac{1}{\xi\omega_n}\ln\frac{1}{\Delta\sqrt{1-\xi^2}} \qquad (3\text{-}4\text{-}25)$$

当考虑 $\Delta=5\%$ 时,有

$$N=\frac{3\sqrt{1-\xi^2}}{2\pi\xi} \qquad (3\text{-}4\text{-}26)$$

当考虑 $\Delta=2\%$ 时,有

$$N=\frac{2\sqrt{1-\xi^2}}{\pi\xi} \qquad (3\text{-}4\text{-}27)$$

通常 N 取整数。

6. 稳态误差

欠阻尼系统在阶跃信号作用下的稳态误差恒为零。

现在讨论一下上升时间和调节时间之间的关系。对于较小的 ξ 将可以得到较短的上升时间,而较快的调节时间则需要较大的 ξ。因此在设计问题中为了得到满意的结果,ξ 的取值需要做出折中的处理。如果需要同时考虑最大超调量的要求时,一般取阻尼比 $\xi=0.4\sim0.8$。工程上常取 $\xi=0.707$ 作为设计依据,称为二阶工程最佳。此时,超调量为 4.3%,调节时间最短(5% 的误差标准)。

【例 3-4-1】 二阶系统如图 3-4-3 所示,其中 $\xi=0.6$,$\omega_n=5\text{rad/s}$,$r(t)=1(t)$,求 t_r,t_p,t_s,$\sigma\%$ 和 N。

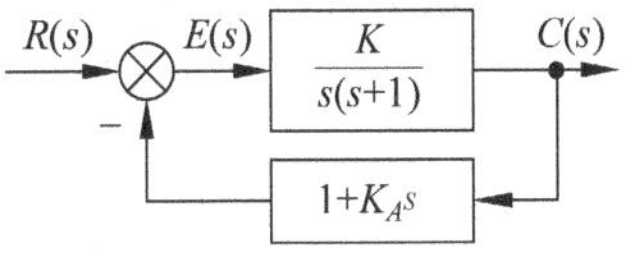

解: $\sqrt{1-\xi^2}=\sqrt{1-0.6^2}=0.8$

$\omega_d=\omega_n\sqrt{1-\xi^2}=5\times0.8=4$

$\xi\omega_n=0.6\times5=3$

$\phi=\arctan\dfrac{\sqrt{1-\xi^2}}{\xi}=\arctan\dfrac{0.6}{0.8}=0.93\text{rad}$

$t_r=\dfrac{\pi-\beta}{\omega_d}=\dfrac{\pi-0.93}{4}=0.55\text{s}$

$t_p=\dfrac{\pi}{\omega_d}=\dfrac{3.14}{4}=0.785\text{s}$

$\sigma\%=e^{-\frac{\pi\xi}{\sqrt{1-\xi^2}}}\times100\%=e^{-\frac{3.14\times0.6}{0.8}}\times100\%=9.5\%$

$t_s\approx\dfrac{3}{\xi\omega_n}=1\text{s}$ ($\Delta=5\%$)

$t_s\approx\dfrac{4}{\xi\omega_n}=1.33\text{s}$ ($\Delta=2\%$)

$N=\dfrac{\omega_d t_s}{2\pi}=\dfrac{1.33}{2\times0.785}=0.8$ ($\Delta=2\%$)

$N=\dfrac{\omega_d t_s}{2\pi}=\dfrac{1}{2\times0.785}=0.6$ ($\Delta=5\%$)

图 3-4-3 二阶系统结构图

【例 3-4-2】 根据图 3-4-4 的过渡过程曲线确定质量 M、黏性摩擦系数 f 和弹簧刚度 K 的值。

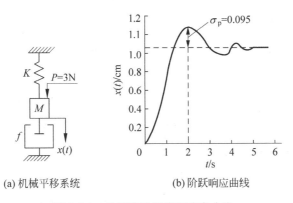

(a) 机械平移系统   (b) 阶跃响应曲线

图 3-4-4 机械平移系统和响应曲线

解: 由图可知,$\sigma_p=0.095$,$t_p=2\text{s}$,$x(\infty)=\lim\limits_{t\to\infty}x(t)=0.01\text{m}$

根据题意,得

$$M\dfrac{d^2x}{dt^2}+f\dfrac{dx}{dt}+Kx=P$$

则有

$$\dfrac{X(s)}{P(s)}=\dfrac{1}{Ms^2+fs+K}=\dfrac{1}{K}\cdot\dfrac{K/M}{s^2+(f/M)s+K/M}$$

根据公式,得

$$\omega_n^2 = K/M, 2\xi\omega_n = f/M$$

当 $p(t) = 3 \cdot 1(t)$ 时,有

$$X(s) = \frac{1}{Ms^2 + fs + K} \cdot \frac{3}{s}$$

$$x(\infty) = \lim_{t\to\infty} x(t) = \lim_{s\to 0} s \cdot X(s) = \lim_{s\to 0} s \cdot \frac{1}{Ms^2 + fs + K} \cdot \frac{3}{s} = \frac{3}{K} = 0.01$$

则

$K = 300 \text{N/m}$

$\sigma_p = 0.095 \Rightarrow \xi = 0.6$

$t_p = 2 \Rightarrow \omega_n = 1.96 \text{rad/s}$

$\omega_n^2 = K/M \Rightarrow 300/M = 1.96^2 \Rightarrow M = 78 \text{kg}$

$2\xi\omega_n = f/M \Rightarrow 2 \times 0.6 \times 1.96 = f/78 \Rightarrow f = 180 \text{N} \cdot \text{s/m}$

3.5 高阶系统时域分析法概述

按照前述章节中关于系统阶次的定义,通常将三阶及三阶以上的系统称为高阶系统。从严格意义上讲,大多数实际的控制系统都属于高阶系统,对于这些系统,从理论上进行定量时域分析比较复杂,计算机仿真一般是有效的解决办法之一。这里仅对高阶系统时间响应进行简单的定性分析。一般认为,高阶系统具有如下的传递函数

$$G_B(s) = \frac{C(s)}{R(s)} = \frac{K(s^m + b_{m-1}s^{m-1} + \cdots + b_1 s + b_0)}{s^n + a_{n-1}s^{n-1} + \cdots + a_1 s + a_0}$$

$$= \frac{K(s^m + b_{m-1}s^{m-1} + \cdots + b_1 s + b_0)}{\prod_{j=1}^{q}(s + p_j) \prod_{k=1}^{r}(s^2 + 2\xi_k\omega_k s + \omega_k^2)} \quad m \leqslant n, q + 2r = n$$

设输入信号为单位阶跃函数,则

$$C(s) = G_B(s) R(s) = \frac{K(s^m + b_{m-1}s^{m-1} + \cdots + b_1 s + b_0)}{s \prod_{j=1}^{q}(s + p_j) \prod_{k=1}^{r}(s^2 + 2\xi_k\omega_k s + \omega_k^2)} \quad (3\text{-}5\text{-}1)$$

如果其各极点互不相同,则式(3-5-1)可展开成

$$C(s) = \frac{a}{s} + \sum_{j=1}^{q} \frac{a_j}{s + p_j} + \sum_{k=1}^{r} \frac{\beta_k(s + \xi_k\omega_k) + \gamma_k(\omega_k\sqrt{1-\xi_k^2})}{(s + \xi_k\omega_k)^2 + (\omega_k\sqrt{1-\xi_k^2})^2}$$

经拉普拉斯逆变换,得

$$c(t) = a + \sum_{j=1}^{q} a_j e^{-p_j t} + \sum_{k=1}^{r} \beta_k e^{-\xi_k\omega_k t} \cdot \cos(\omega_k\sqrt{1-\xi_k^2} t)$$

$$+ \sum_{k=1}^{r} \gamma_k e^{-\xi_k\omega_k t} \cdot \sin(\omega_k\sqrt{1-\xi_k^2} t) \quad (3\text{-}5\text{-}2)$$

由式(3-5-2)可见,一般的高阶系统的动态响应是由一些一阶惯性环节和二阶振荡环节的响应函数叠加组成的。当所有极点均具有负实数时,除了常数 a,其他各项随着时间 $t \to \infty$ 而衰减为零,即系统是稳定的。

某些高阶系统通过合理的简化,可以用低阶系统近似,以下两种情况可以作为简化的依据。

(1) 由式(3-5-1)和式(3-5-2)可见,系统极点的负实部越是远离虚轴,该极点对应的项在动态响应中衰减得越快;反之,距虚轴最近的闭环极点对应着动态响应中衰减最慢的项,该极点对(或极点)对动态响应起主导作用,称为"主导极点"。一般工程上当极点 A 距离虚轴的距离大于 5 倍的极点 B 距虚轴的距离时,且附近无其他的零点或极点,分析系统时可忽略极点 A。

(2) 闭环传递函数中,如果负实部的零点和极点在数值上相近,则可将该零点和极点一起消掉,称为"偶极子相消"。

综上所述,高阶系统时间响应可分为稳态分量和瞬态分量,有以下结论。

(1) 瞬态分量的各个运动模态衰减的快慢取决于对应的极点和虚轴的距离,离虚轴越远的极点对应的运动模态衰减得越快。

(2) 各模态所对应的系数和初相角取决于零点和极点的分布。

① 若某一极点越靠近零点,且远离其他极点和原点,则相应的系数越小。

② 若一对零极点相距很近,该极点对应的系数就非常小。

③ 若某一极点远离零点,它越靠近原点或其他极点,则相应的系数越大。

(3) 系统的零点和极点共同决定了系统响应曲线的形状。

(4) 对系统响应起主要作用的极点称为主导极点。

(5) 非零初始条件时的响应由零初始条件时的响应和零输入响应组成。

Reading Material

Under certain circumstances, a higher order transfer function maybe approximated to a lower order transfer function. This approximation is based on the concept of dominant poles. We will use numerical examples to illustrate this concept.

Firstly, by mapping the poles and zeros of the system transfer function in the complex s-plane, we can draw useful conclusions about the transient behavior of a dynamic system. Each negative real pole of the transfer function will contribute an exponentially decaying transient term. If any positive real pole is present, it gives rise to an exponentially increasing term thereby rendering the system unstable. Complex poles, if present, always occur in conjugate pairs. Each complex pole pair gives rise to an exponentially decaying sinusoidal waveform if the real part of the poles is negative. An exponentially increasing sinusoidal waveform would result if the real part of the complex pole pair is positive (unstable response). Whether a particular high order transfer function can be approximated or not by a lower order transfer function depends on the pole-zero distribution in the s-plane.

If a single negative real pole is distinctly closer to the imaginary axis, and all other poles and zeros are at least six times further away into the negative half of the s-plane, then the system can be approximated by a first order transfer function having the above nearest pole as its pole. If instead of a real pole, a pair of complex poles are distinctly near the imaginary axis, then the system can be approximated by a second order transfer function(complex time lag) having the above two complex poles as its poles. If a pole of a transfer function is very near a zero of the same transfer function, then they cancel one another. They can be disregarded in analysing the transient response.

3.6 控制系统的稳定性分析

在控制系统的分析研究中,最重要的问题是系统的稳定性问题,它是保证系统能够正常运行的首要条件。控制系统在实际运行中,总会受到外界或内部的一些因素扰动,例如负载和能源的波动、系统参数的变化、环境条件的改变等,都会使被控制量偏离原来的平衡工作状态,并随时间的推移而发散。因此,不稳定的系统是无法正常工作的。如何分析系统的稳定性并提出保证系统稳定的措施,就成了自动控制理论的基本任务之一。

Reading Material

When the analysis and design of control systems are considered, stability is of utmost importance. From a practical point of view, an unstable system is of little value. In practical operation, almost all control systems are subject to extraneous or inherent disturbances, such as fluctuation of load or power source, variation of system parameters or circumstance.

Many physical systems are inherently open loop unstable. Introducing feedback is useful to stabilize the unstable plant and then adjust the transient performance with an appropriate controller. For open-loop stable plants, feedback is still used to improve the system performance.

(1) When considering the design and analysis of feedback control system, stability is of utmost importance.

(2) A stable system is defined as a system with a bounded(limited) system response.

Or: a stable system is a dynamic system with a bounded response and to a bounded input(Bounded Input Bounded Output, BIBO).

(3) The concept of stability can be illustrated by considering a right circular cone placed on a plane horizontal surface, see Fig. 3-6-1.

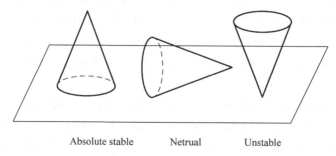

Fig. 3-6-1 The concept of stability

(4) The location in the s-plane of the poles of a system indicates the resulting transient response.

3.6.1 稳定的基本概念和稳定的充分必要条件

1. 稳定的基本概念

在自动控制理论中,有多种稳定性的定义,本书只讨论其中最常见的一种,即渐近稳定性系统。图 3-6-2(a)为一个单摆系统(Pendulum System)示意图,设在外界作用下,单摆由原平衡点 A 偏移到新的位置 A',偏移角 ϕ。当外界扰动力取消后,单摆在重力作用下围绕原平衡点 A 反复振荡,经过一段时间单摆因受介质阻碍作用使其回到原平衡点 A,故称 A 为稳定平衡点。图 3-6-2(c)中,小球超出了 C、D 范围后系统就不再是线性的,故可以认为该系统在线性范围内是稳定的。

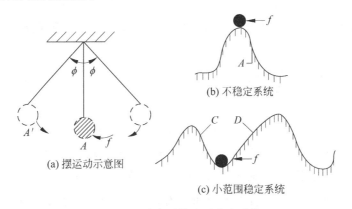

图 3-6-2 稳定系统和不稳定系统

根据李雅普诺夫稳定性理论(Liapunov Stability Theory),线性控制系统的稳定性可叙述如下:若线性控制系统在初始扰动的影响下,其动态过程随时间的推移逐渐衰减并趋于零(原平衡工作点),则称系统渐近稳定,简称稳定;反之,若在初始扰动的影响下,系统的动态过程随时间的推移而发散,则称系统不稳定。

2. 线性系统稳定的充要条件

线性系统的稳定性是系统在外界扰动消失后,自身具有的一种恢复能力,也就是说系统的稳定性取决于系统自身的固有特性,即系统的结构和参数,而与外界条件(输入信号)无关。

设线性定常系统在初始条件为零时,输入一个理想单位脉冲 $\delta(t)$,这相当于系统在零平衡状态,受到一个外界扰动信号的作用下,如果 $t \to \infty$ 时,系统的输出响应 $c(t)$ 收敛到原来的零平衡状态,即

$$\lim_{t \to \infty} c(t) = 0 \tag{3-6-1}$$

则称该系统是稳定的。

Terms and Concepts

In a mathematical manner, the definition of stability may be stated: if the zero-input of the linear time-invariant system, subject to nonzero initial condition, approaches zeros as time approaches infinity, i.e.,

$$\lim_{t \to \infty} c(t) = 0$$

then the system is said to be stable; otherwise the system is unstable.

设系统的闭环传递函数为

$$\Phi(s) = \frac{b_m s^m + b_{m-1} s^{m-1} + \cdots + b_1 s + b_0}{a_n s^n + a_{n-1} s^{n-1} + \cdots + a_1 s + a_0} \quad n \geqslant m \tag{3-6-2}$$

其特征方程为 $a_n s^n + a_{n-1} s^{n-1} + \cdots + a_1 s + a_0 = 0$。如果特征方程的所有根互不相同，且有 q 个实数根 P_i 和 r 对共轭复数根，则在单位脉冲函数 $\delta(t)$ 的作用下，系统输出量的拉普拉斯变换可表示为

$$C(s) = \frac{K_r \prod_{j=1}^{m}(s - Z_j)}{\prod_{i=1}^{q}(s - P_i) \prod_{k=1}^{r}(s^2 + 2\xi_k \omega_{nk} s + \omega_{nk}^2)} \cdot 1 \tag{3-6-3}$$

式中，k 为常数；P_i 为特征方程的实根，$q + 2r = n$。将式(3-6-3)用部分分式法展开并进行拉普拉斯反变换得

$$c(t) = \sum_{i=1}^{q} A_i e^{P_i t} + \sum_{k=1}^{r} e^{-\xi_k \omega_{nk} t}(B_k \cos \omega_{dk} t + C_k \sin \omega_{dk} t) \tag{3-6-4}$$

式中，$\omega_{dk} = \omega_{nk}\sqrt{1 - \xi^2}$。

式(3-6-4)表明：①当系统特征方程的根都具有负实部时，各瞬态分量都是衰减的，且有 $\lim_{t \to \infty} c(t) = 0$，此时系统是稳定的。②如果特征根中有一个或一个以上具有正实部，则该根对应的瞬态分量是发散的，此时有 $\lim_{t \to \infty} c(t) \to \infty$，系统是不稳定的。③如果特征根中具有一个或一个以上的零实部根，而其余的特征根具有负实部，则 $c(t)$ 趋于常数或作等幅振荡，即 $\lim_{t \to \infty} c(t) = k$ 时系统处于稳定和不稳定的临界状态，常称为临界稳定状态（不属于渐近稳定）。对于大多数实际系统，当它处于临界状态时，也是不能正常工作的，所以临界稳定的系统在工程上属于不稳定系统。在经典控制理论中，只有渐近稳定的系统才称为稳定系统，否则称为不稳定系统。

由此可见，线性定常系统稳定的充分必要条件是：闭环系统特征方程的所有根都具有负实部，或者说闭环传递函数的所有极点均位于 s 平面的左半部分（不包括虚轴）。

Terms and Concepts

A necessary and sufficient condition for a feedback system to be stale is that all the poles of the system transfer function have negative real parts.

The closed-loop poles close to imaginary axis dominate those father away, and in many cases the response may be approximated by the response of the closed-loop poles closest to the imaginary axis. Fig. 3-6-3 is the good region of the close-loop poles in s-plane.

3.6.2　代数稳定判据

由以上讨论可知，控制系统稳定的充要条件是其特征方程的根均具有负实部。因此，为了判别系统的稳定性，就要求出系统特征方程的根，并检验它们是否都具有负实部。但是，

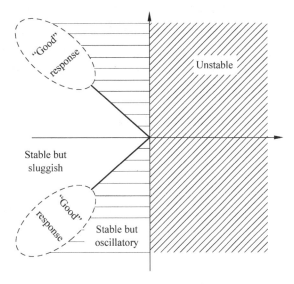

Fig. 3-6-3　The good region of the close-loop poles in s-plane

这种求解系统特征方程的方法,对低阶系统尚可以进行,而对高阶系统,求根就很烦琐。因此,人们希望寻求一种不需要求解特征方程而能判别系统稳定性的间接代数方法,这种方法称为代数稳定判据(Algebraic Criterion of Stability),主要有劳斯稳定判据和赫尔维兹判据。

Terms and Concepts

Routh-Hurwitz Stability Criterion will enable us to find out whether a feedback control system is stable or not from its characteristic function, $1+G(s)H(s)$, without having to actually factories the characteristic function. Routh-Hurwitz method is based on ordering the coefficients of the characteristic equation into an array as follows.

When we construct Routh's Array, four distinct cases arise.

Case 1: No elements of the first column is zero.

Case 2: Zero in the 1st column while some other elements of the row containing the zero are nonzero.

Case 3: Zero in the 1st column and other elements of the row containing the zero are also zero.

Case 4: Characteristic equation has repeated roots on the $j\omega$ axis.

1. 劳斯稳定判据

劳斯稳定判据(Routh's Stability Criterion)利用特征方程的各项系数进行代数运算,得出全部特征根具有负实部的条件,以此作为判别系统是否稳定的依据。

设线性系统的特征方程为

$$a_n s^n + a_{n-1} s^{n-1} + \cdots + a_1 s + a_0 = 0 \qquad (3\text{-}6\text{-}5)$$

系统稳定的必要条件是其特征方程的各项系数均为正,即 $a_i > 0 (i=0,1,2,\cdots,n)$,及是否有缺项。根据必要条件,在判据系统的稳定性时,可事先检查系统特征方程的系数是否都大于零，若有任何系数是负数或等于零，则系统是不稳定的。但是，当特征方程满足稳定的必要条件时，并不意味着系统一定是稳定的，要判断系统是否稳定还需检验其是否满足系统稳定的充分条件。稳定的充分条件是：劳斯阵列的第一列元素均大于零，则系统稳定。

应用劳斯判据分析系统的稳定性时，可按照下面方法进行。将系统特征方程各项系数组成劳斯阵列表，如表 3-6-1 所示。

表 3-6-1 劳斯阵列表

对应项	第 1 列	第 2 列	第 3 列	…
s^n	a_n	a_{n-2}	a_{n-4}	
s^{n-1}	a_{n-1}	a_{n-3}	a_{n-5}	
s^{n-2}	$c_{13}=\dfrac{a_{n-1}a_{n-2}-a_n a_{n-3}}{a_{n-1}}$	$c_{23}=\dfrac{a_{n-1}a_{n-4}-a_n a_{n-5}}{a_{n-1}}$	$c_{33}=\dfrac{a_{n-1}a_{n-6}-a_n a_{n-7}}{a_{n-1}}$	
s^{n-3}	$c_{14}=\dfrac{c_{13}a_{n-3}-a_{n-1}c_{23}}{c_{13}}$	$c_{24}=\dfrac{c_{13}a_{n-5}-a_{n-1}c_{33}}{c_{13}}$	$c_{34}=\dfrac{c_{13}a_{n-3}-a_{n-1}c_{43}}{c_{13}}$	
s^{n-4}	c_{15}	c_{25}	c_{35}	
⋮				
s^2	$c_{1,n-1}$	$c_{2,n-1}$		
s^1	$c_{1,n}$			
s^0	$c_{1,n+1}=a_0$			

由劳斯阵列表结构可知，其线性系统稳定的劳斯判据可归结为以下几点。

1) 劳斯阵列表第一列所有系数均不为零的情况

如果劳斯阵列表中第一列的系数都具有相同的符号，则系统是稳定的，否则系统是不稳定的。劳斯阵列表中第一列系数符号改变的次数等于该特征方程式的根在 s 右半平面的个数，即不稳定根的个数等于劳斯阵列表中第一列系数符号改变的次数。

【例 3-6-1】 已知系统的特征方程为
$$s^5+6s^4+14s^3+17s^2+10s+2=0$$
试用劳斯判据分析系统的稳定性。

解：列劳斯阵列表

s^5	1	14	10
s^4	6	17	2
s^3	$\dfrac{6\times14-1\times17}{6}=\dfrac{67}{6}$	$\dfrac{6\times10-1\times2}{6}=\dfrac{58}{6}$	
s^2	$\dfrac{\dfrac{67}{6}\times17-6\times\dfrac{58}{6}}{\dfrac{67}{6}}=\dfrac{791}{67}$	$\dfrac{\dfrac{67}{6}\times2-6\times0}{\dfrac{67}{6}}=2$	
s^1	$\dfrac{\dfrac{791}{67}\times\dfrac{58}{6}-\dfrac{67}{6}\times2}{\dfrac{791}{67}}=\dfrac{6150}{791}$		
s^0	2		

劳斯阵列表第一列的系数符号相同，故系统是稳定的。由于判别系统是否稳定只与劳斯阵列表中第一列系数的符号有关，而把劳斯阵列表中某一行系数同乘以一个正数不会改变第一列系数的符号，所以为简化运算，常把劳斯阵列表的某一行同乘以一个正数后，再继续运算。

本例中，劳斯阵列表可按照如下方法计算。

s^5	1	14	10	
s^4	6	17	2	
s^3	67	58		同乘以 6
s^2	791	134		同乘以 67
s^1	36900			同乘以 791
s^0	134			

由于第一列系数的符号相同，故本系统稳定，结论与前面的一致。

【例 3-6-2】 已知系统的特征方程为
$$s^4+2s^3+s^2+s+1=0$$
试用劳斯判据分析系统的稳定性。

解：列劳斯阵列表

s^4	1	1	1
s^3	2	1	0
s^2	$\dfrac{2\times1-1\times1}{2}=\dfrac{1}{2}$	$\dfrac{2\times1-1\times0}{2}=1$	
s^1	$\dfrac{1\times1-2\times2}{1}=-3$		
s^0	$\dfrac{-3\times1-1\times0}{-3}=1$		

由于劳斯阵列表第一列的系数变号两次，一次由 1/2 变为 -3，另一次由 -3 变为 1，故特征方程有两个根在 s 平面右半部分，故系统是不稳定的。

2) 劳斯阵列表某行的第一列系数等于零，而其余各项不全为零的情况

解决的办法是用一个很小的正数 ε 代替第一列的零项，然后按照通常方法计算劳斯阵列表中的其余项。而劳斯阵列表中第一列系数的变号次数不依赖于 ε。

【例 3-6-3】 已知系统的特征方程为
$$s^4+s^3+2s^2+2s+5=0$$
试用劳斯判据分析系统的稳定性。

解：列劳斯阵列表

s^4	1	2	5
s^3	1	2	0
s^2	$0\approx\varepsilon$	5	
s^1	$(2\varepsilon-5)/\varepsilon$		
s^0	5		

当 ε 的取值足够小时，$(2ε-5)/ε$ 将取负值，故劳斯阵列表第一列系数变号两次，由劳斯判据可知，特征方程有两个根具有正实部，系统是不稳定的。

3) 劳斯阵列表某行所有系数均为零的情况

如果劳斯阵列表中某一行(如第 K 行)各项为零，这说明在 s 平面内存在以原点为对称的特征根。解决方法是利用第 $K-1$ 行的系数构成辅助方程，使辅助方程 $A(s)=0$。然后求辅助方程对 s 的导数，将其系数代替原全部为零的第 K 行，继续计算劳斯阵列。辅助方程给出了特征方程中成对根对称分布在 s 平面内的位置及数目，显然根的数目是偶数，辅助方程的次数总是偶数。

【例 3-6-4】 已知系统的特征方程为

$$s^5+s^4+4s^3+24s^2+3s+63=0$$

试用劳斯判据分析系统的稳定性。

解：列劳斯阵列表

s^5	1	4	3	
s^4	1	24	63	
s^3	-20	-60	0	
s^2	21	63	0	构造辅助方程 $A(s)=21s^2+63=0$
s^1	0	0	0	辅助方程求导 $A(s)=42s=0$
s^0				

由上表看出，s^1 行的各项全为零，为了求出 s^1 和 s^0 各行，由 s^2 行的各项系数构成辅助方程 $A(s)$。此辅助方程对 s 求导，用 42 代替零行的第一个系数，则劳斯阵列表变为

s^5	1	4	3
s^4	1	24	63
s^3	-20	-60	0
s^2	21	63	0
s^1	42	0	
s^0	63		

由上表可见，第一列有两次符号变化，因此有两个实部为正的根在 s 平面右半平面，故系统是不稳定的。另外，由辅助方程

$$A(s)=21s^2+63=0$$

求得 $s_{1,2}=\pm j\sqrt{3}$。这是一对共轭虚根，从这一点来看，系统也是不稳定的。

由上可知，运用劳斯判据，不仅可以判定系统是否稳定，还可以用来分析系统参数的变化对稳定性产生的影响，从而给出使系统稳定的参数范围。

【例 3-6-5】 已知系统的结构图如图 3-6-4 所示。当 $\xi=0.2$ 时，$\omega_n=86.6$。试确定 K 为何值时，系统稳定。

解：根据图 3-6-4 写出开环传递函数为

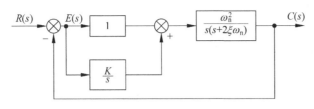

图 3-6-4　系统结构图

$$G(s)=\frac{\omega_n^2(s+K)}{s^2(s+2\xi\omega_n)}$$

其闭环传递函数为

$$\phi(s)=\frac{\omega_n^2(s+K)}{s^3+2\xi\omega_n s^2+\omega_n^2 s+K\omega_n^2}$$

特征方程为

$$s^3+2\xi\omega_n s^2+\omega_n^2 s+K\omega_n^2=0$$

将 $\xi=0.2, \omega_n=86.6$ 代入特征方程,得

$$s^3+34.6s^2+7500s+7500K=0$$

由特征方程列劳斯阵列表

s^3	1	7500	0
s^2	34.6	7500K	
s^1	$(34.6\times 7500-7500K)/34.6$		
s^0	7500K		

要使系统稳定,必须满足

$$\begin{cases}\dfrac{34.6\times 7500-7500K}{34.6}>0\\ 7500K>0\end{cases}$$

解得

$$0<K<34.6$$

在系统的分析中,劳斯判据可以根据系统特征方程的系数来确定系统的稳定性,同时还能确定反馈系统稳定时开环增益或某些参数的取值范围。但是,它并不能指出系统是否具有满意的动态过程,同时也不能提供改善系统稳定性的方法和途径。

Terms and Concepts

Routh-Hurwitz criterion: A criterion for determining the stability of a system by examining the characteristic equation of the transfer function. The criterion states that the number of roots of the characteristic equation with positive real parts is equal to the number of change of sign of the coefficients in the first column of the Routh array.

2. 赫尔维兹判据

赫尔维兹判据(Hurwitz Criterion)也是根据特征方程的系数来判别系统的稳定性。设

线性系统的特征方程为
$$a_n s^n + a_{n-1} s^{n-1} + \cdots + a_1 s + a_0 = 0$$
以特征方程式的各项系数组成如下行列式

$$\Delta_n = \begin{vmatrix} a_{n-1} & a_n & 0 & 0 & 0 & 0 & \cdots \\ a_{n-3} & a_{n-2} & a_{n-1} & a_n & 0 & 0 & \cdots \\ a_{n-5} & a_{n-4} & a_{n-3} & a_{n-2} & a_{n-1} & a_n & \cdots \\ a_{n-7} & a_{n-6} & a_{n-5} & a_{n-4} & a_{n-3} & a_{n-2} & \cdots \\ \vdots & \vdots & \vdots & \vdots & \vdots & \vdots & a_n \end{vmatrix} \quad (3\text{-}6\text{-}6)$$

赫尔维兹判据指出,系统稳定的充要条件是,上述行列式及其各阶顺序主子式 $\Delta_i (i=1, 2, \cdots, n-1)$ 均大于零,即

$$\begin{aligned} \Delta_1 &= a_{n-1} > 0 \\ \Delta_2 &= \begin{vmatrix} a_{n-1} & a_n \\ a_{n-3} & a_{n-2} \end{vmatrix} > 0 \\ &\vdots \\ \Delta_n &> 0 \end{aligned} \quad (3\text{-}6\text{-}7)$$

【例 3-6-6】 已知系统的特征方程为
$$3s^4 + 10s^3 + 5s^2 + s + 2 = 0$$
试用赫尔维兹判据判断系统的稳定性。

解:系统的行列式
$$\Delta_4 = \begin{vmatrix} 10 & 3 & 0 & 0 \\ 1 & 5 & 10 & 3 \\ 0 & 2 & 1 & 5 \\ 0 & 0 & 0 & 2 \end{vmatrix}$$

$$\Delta_1 = 10 > 0$$

$$\Delta_2 = \begin{vmatrix} 10 & 3 \\ 1 & 5 \end{vmatrix} > 0$$

$$\Delta_3 = \begin{vmatrix} 10 & 3 & 0 \\ 1 & 5 & 10 \\ 0 & 2 & 1 \end{vmatrix} = -153 < 0$$

由赫尔维兹判据可知,该系统不稳定。

Reading Material

Once the block diagrams of control systems are developed, the next step is to carry out analysis. There are two types of analysis: quantitative and qualitative. In quantitative analysis, we are interested in the exact response of control systems due to a specific excitation. In qualitative analysis, we are interested in general properties of control systems. We discuss first the former and then the latter.

Control systems are inherently time-domain systems, so the introduced specifications are natural and have simple physical interpretation. For example, in pointing a telescope at a star, the steady-state performance (accuracy) is the main concern; the specifications on the rise time, overshoot, and settling time are not critical. However, in aiming missiles at an aircraft, both accuracy and speed of response are important. In the design of an aircraft, the specification is often given as shown in Fig. 3-6-5. It is required that the step response of the system be confined to the region shown. This region is obtained by a compromise between the comfort or physical limitations of the pilot and the maneuverability of the aircraft. In the design of an elevator, any appreciable overshoot is undesirable. Different applications have different specifications.

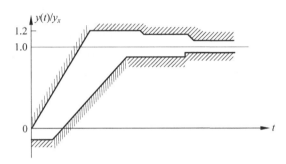

Fig. 3-6-5   Allowable step response

A system is said to be sluggish if its rise time and settling time are large. If a system is designed for a fast response, or to have a small rise time and a small settling time, then the system may exhibit a large overshoot, as can been see from Fig. 3-6-5. Thus, the requirements on the rise time and overshoot are often conflicting and must be reached by compromise.

The steady-state response of $G_o(s)$ depends only on a number of coefficients of $G_o(s)$, thus the steady-state performance can easily be incorporated into the design. The transient response of $G_o(s)$ depends on both its poles and zeros. Except for some special cases, no simple relationship exists between the specifications and pole-zero locations. Therefore, designing a control system to meet transient specifications is not as simple as designing one to meet steady-state specifications.

3.7 控制系统稳态误差的分析及计算

控制系统时间响应的特征可以用瞬态响应和稳态响应表示。系统的稳态分量反映系统跟踪控制信号的准确度或抑制扰动信号的能力,用稳态误差来描述。稳态误差适用于衡量特定类型的输入施加与控制系统时系统稳态精度的一种判别标准,具有不可避免性。稳态误差的来源有多种,在系统的分析和设计中,稳态误差是一项重要的性能指标,它与元器件的不灵敏、零点漂移、老化及各种传动机械的间隙、摩擦等因素有关,习惯上称这些由非线性因素造成的稳态误差为静差,静差不属于本书讨论范畴之内。本章主要讨论由系统本身的结构、输入信号的形式、参数及外作用等因素所引起的稳态误差。

Terms and Concepts

Disturbance signal: An unwanted input signal that affects the system's output signal.

Error signal: The difference between the desired output and the actual output.

Steady-state error: The error when the time period is large and the transient response has decayed, leaving the continuous response.

System sensitivity: The ratio of the change in the system transfer function to the change of a process transfer function(or parameter)for a small incremental change.

As a function of the time, the error signal consists of two parts: The transient component and the steady-state component. Since the steady-state error is meaningful only for the stable systems, the transient component must approach zero as time approaches infinite. Hence, the steady-state error of a feedback control system is defined as the error when time approaches infinity.

3.7.1 稳态误差的定义

系统的稳态误差(Steady-State Error)是指在稳定条件下(即对于稳定系统)输入加入后经过足够长的时间,其瞬态响应已经衰减到足够小时,稳态响应的期望值与实际值之间的误差,以 e_{ss} 来表示稳态误差,即

$$e_{ss} = \lim_{t \to \infty} e(t) \tag{3-7-1}$$

系统的输出误差为

$$e(t) = y_d(t) - y(t) \tag{3-7-2}$$

其中,$y(t)$ 表示输出实际值,$y_d(t)$ 表示输出期望值。

如果 $H(s)=1$,输出量的期望值为输入量 $x(t)$,则输出误差即为

$$e(t) = x(t) - y(t) \tag{3-7-3}$$

当 $H(s) \neq 1$ 时,$x(t)$ 和 $y(t)$ 可能具有完全不同的量纲,因此不能应用式(3-7-3)。可定义输出误差为

$$e(t) = x(t) - b(t) \tag{3-7-4}$$

或表示成

$$E(s) = X(s) - B(s) = X(s) - H(s)Y(s) = X(s) - H(s)G(s)E(s) \tag{3-7-5}$$

即

$$E(s) = \frac{X(s)}{1 + H(s)G(s)} \tag{3-7-6}$$

根据拉普拉斯变换的终值定理(Final-Value Theorem),可以得到稳态误差的表达式

$$e_{ss} = \lim_{t \to \infty} e(t) = \lim_{s \to 0} sE(s) = \lim_{s \to 0} \frac{sX(s)}{1 + H(s)G(s)} \tag{3-7-7}$$

式(3-7-7)表明,系统稳态误差取决于输入信号 $X(s)$ 的形式以及开环传递函数 $H(s)G(s)$ 的结构。对于一个给定的稳态系统,当输入信号 $X(s)$ 形式确定时,系统是否存在稳态误差就取决于开环传递函数 $H(s)G(s)$ 所描述的系统结构。因此,按照控制系统跟踪不同输入信号的能力来进行系统分类是很有必要的。

【例 3-7-1】 单位负反馈系统的开环传递函数为

$$G(s) = \frac{K(0.5s+1)}{s(s+1)(2s+1)}$$

确定对于单位斜坡输入的稳态误差。

解：闭环系统的特征方程为

$$\Delta(s) = 2s^3 + 3s^2 + (1+0.5K)s + K = 0$$

由于二阶赫尔维兹行列式为

$$D_2 = \begin{vmatrix} 3 & K \\ 2 & 1+0.5K \end{vmatrix}$$

系统稳定的充要条件为 $D_2 > 0$，即 $0 < K < 6$。

由于 $R(s) = 1/s^2$，误差函数为

$$E(s) = \phi_e(s)R(s) = \frac{s(s+1)(2s+1)}{s(s+1)(2s+1) + (1+0.5s)K} \cdot \frac{1}{s^2}$$

因此，在 $0 < K < 6$ 的情况下，稳态误差为

$$e_{ss} = \lim_{s \to 0} sE(s) = \frac{1}{K}$$

3.7.2 系统类型与稳态误差

令系统开环传递函数(Open-Loop Transfer Function)为

$$H(s)G(s) = \frac{K \prod_{j=1}^{m}(\tau_j s + 1)}{s^v \prod_{i=1}^{n-v}(T_i s + 1)} \quad n \geqslant m \tag{3-7-8}$$

其中，K 为系统的开环增益(Open-Loop Gain)；τ_j 和 T_i 为时间常数(Time Constant)；v 为开环传递函数在 s 平面坐标原点上的极点个数，即开环传递函数中串联的积分环节个数。为了分析稳态误差与系统结构的关系，可以根据开环传递函数积分环节来规定控制系统的类型，系统分类如下。

(1) $v=0$ 称为 0 型系统，或者称为有差系统。

(2) $v=1$ 称为 Ⅰ 型系统，或者称为一阶无差系统。

(3) $v=2$ 称为 Ⅱ 型系统，或者称为二阶无差系统。

(4) $v>2$，除复合控制外，Ⅱ 型以上的系统，实际上很难使之稳定，所以这种类型的系统在控制工程中一般不会碰到，在此不作讨论。

令

$$H_0(s)G_0(s) = \frac{\prod_{j=1}^{m}(\tau_j s + 1)}{\prod_{i=1}^{n-v}(T_i s + 1)} \tag{3-7-9}$$

当 $s \to 0$ 时，有 $H_0(s)G_0(s) \to 1$，则式(3-7-8)近似为

$$H(s)G(s) = \frac{K}{s^v} H_0(s)G_0(s) = \frac{K}{s^v} \tag{3-7-10}$$

这样系统稳态误差计算通式可表示为

$$e_{ss} = \lim_{s \to 0} sE(s) = \lim_{s \to 0} \frac{sX(s)}{1 + K/s^v} = \frac{\lim_{s \to 0}[s^{v+1}X(s)]}{K + \lim_{s \to 0} s^v} \quad (3\text{-}7\text{-}11)$$

式(3-7-11)表明，稳态误差 e_{ss} 与系统的类型、系统的增益以及输入类型有关。下面分别讨论阶跃信号、斜坡信号和加速度信号的稳态误差情况。

3.7.3 给定输入信号下的稳态误差计算

1. 阶跃输入

令 $x(t) = A \cdot u(t)$，A 为常量，即 $X(s) = A/s$，则系统的稳态误差为

$$e_{ss} = \lim_{s \to 0} \frac{sX(s)}{1 + H(s)G(s)} = \frac{A}{1 + \lim_{s \to 0} H(s)G(s)} = \frac{A}{1 + K_p} \quad (3\text{-}7\text{-}12)$$

式中，$K_p = \lim_{s \to 0} H(s)G(s) = \lim_{s \to 0} \frac{K}{s^v}$ 为静态位置误差系数(Static Position Error Constant)，K_p 的大小反映系统在阶跃输入下的稳态精度。K_p 越大，e_{ss} 就越小，所以说 K_p 反映了系统跟踪阶跃输入的能力。各种类型系统的静态位置误差系数为

$$K_p = \begin{cases} K, & v = 0 \\ \infty, & v \geq 1 \end{cases} \quad (3\text{-}7\text{-}13)$$

从而得到各种类型系统的稳态误差为

$$e_{ss} = \begin{cases} \dfrac{A}{1 + K} = 常数, & v = 0 \\ 0, & v \geq 1 \end{cases} \quad (3\text{-}7\text{-}14)$$

式(3-7-14)表明，由于0型系统中没有积分环节，它对阶跃输入的稳态误差为一定值，误差的大小与系统的开环放大系数 K 成反比，K 越大，稳态误差越小。对于实际系统来说，通常是允许存在稳态误差的，但不允许超过规定的指标。为了降低稳态误差，可在稳定条件允许的前提下，增大系统的开环放大系数，若要求系统对阶跃输入作用下不存在稳态误差，则必须选用Ⅰ型或高于Ⅰ型的系统。

2. 斜坡输入

令 $x(t) = A \cdot t$，A 为常量，即 $X(s) = A/s^2$，则系统的稳态误差为

$$\begin{aligned} e_{ss} &= \lim_{s \to 0} \frac{sX(s)}{1 + H(s)G(s)} = \lim_{s \to 0} \frac{A}{s + sH(s)G(s)} \\ &= \frac{A}{\lim_{s \to 0} sH(s)G(s)} = \frac{A}{K_v} \end{aligned} \quad (3\text{-}7\text{-}15)$$

式中，$K_v = \lim_{s \to 0} sH(s)G(s) = \frac{K}{s^{v-1}}$ 为静态速度误差系数(Static Velocity Error Constant)，K_v 的大小反映系统在斜坡输入下的稳态精度。K_v 越大，e_{ss} 就越小，所以说 K_v 反映了系统跟踪斜坡输入的能力。各种类型系统的静态速度误差系数为

$$K_v = \begin{cases} 0, & v=0 \\ K, & v=1 \\ \infty, & v \geqslant 2 \end{cases} \tag{3-7-16}$$

从而得到各种类型系统的稳态误差为

$$e_{ss} = \begin{cases} \infty, & v=0 \\ \dfrac{A}{K} = 常数, & v=1 \\ 0, & v \geqslant 2 \end{cases} \tag{3-7-17}$$

式(3-7-17)表明,在单位斜坡输入作用下,0 型系统的稳态误差为∞,即不能跟踪。Ⅰ型系统稳态时能跟踪斜坡输入,但存在一定值的稳态误差,且误差与开环放大系数 K 成反比。为了使稳态误差不超过规定值,可以增大系统的 K 值。在稳态时,系统的输出量与输入信号虽以同一速度变化,但前者在位置上要落后后者一个常量。Ⅱ型或高于Ⅱ型系统的稳态误差总为零,即稳态时总能准确跟踪斜坡信号。因此,对于单位斜坡输入,要使系统的稳态误差为一定值或为零,系统必须有足够的积分环节。

3. 加速度输入

令 $x(t) = \dfrac{1}{2}At^2$,A 为常量,即 $X(s) = A/s^3$,则系统的稳态误差为

$$\begin{aligned} e_{ss} &= \lim_{s \to 0} \frac{sX(s)}{1+H(s)G(s)} = \lim_{s \to 0} \frac{A}{s^2 + s^2 H(s)G(s)} \\ &= \frac{A}{\lim\limits_{s \to 0} s^2 H(s)G(s)} = \frac{A}{K_a} \end{aligned} \tag{3-7-18}$$

式中,$K_a = \lim\limits_{s \to 0} s^2 H(s)G(s) = \dfrac{K}{s^{v-2}}$ 为静态加速度误差系数(Static Acceleration Error Constant),K_a 的大小反映系统在加速度输入下的稳态精度。K_a 越大,e_{ss} 就越小,所以说 K_a 反映了系统跟踪加速度输入的能力。各种类型系统的静态加速度误差系数为

$$K_a = \begin{cases} 0, & v=0,1 \\ K, & v=2 \\ \infty, & v \geqslant 3 \end{cases} \tag{3-7-19}$$

从而得到各种类型系统的稳态误差为

$$e_{ss} = \begin{cases} \infty, & v=0,1 \\ \dfrac{A}{K} = 常数, & v=2 \\ 0, & v \geqslant 3 \end{cases} \tag{3-7-20}$$

式(3-7-20)表明,在加速度输入作用下,0 型和Ⅰ型系统的稳态误差为∞,即说明 0 型和Ⅰ型系统不能跟踪加速度输入信号。只有Ⅱ型系统能跟踪,它的稳态误差为一定值,且误差与开环放大系数成反比。对高于Ⅱ型的系统,其稳态误差为零。但是此时要使系统稳定则比较困难。

各种输入信号作用下的稳态误差如表 3-7-1 所示。

表 3-7-1 各种输入信号作用下的稳态误差

系统类型	静态误差系数			输入信号		
	K_p	K_v	K_a	$A \cdot u(t)$	$A \cdot t$	$\frac{1}{2}At^2$
0 型	K	0	0	$\frac{A}{1+K}$	∞	∞
Ⅰ 型	∞	K	0	0	$\frac{A}{K}$	∞
Ⅱ 型	∞	∞	K	0	0	$\frac{A}{K}$
Ⅲ 型	∞	∞	∞	0	0	0

由表 3-7-1 可以看出,为了减少给定误差,可以增加前向通道上的积分环节个数或增大系统的开环放大系数。若给定的输入信号是上述典型信号的线性组合,则系统相应的稳态误差就由叠加原理求出。例如,若输入信号为

$$x(t) = A + Bt + \frac{1}{2}Ct^2$$

则系统的总稳态误差为

$$e_{ss} = \frac{A}{1+K_p} + \frac{B}{K_v} + \frac{C}{K_a} \qquad (3\text{-}7\text{-}21)$$

综上所述,稳态误差系数 K_p、K_v 和 K_a 描述了系统对减小和消除稳态误差的能力,因此它们是系统稳态特性的一种表示方法。提高开环放大系数 K 或增加开环传递函数中的积分环节数,都可以达到减小或消除系统稳态误差的目的。但是,这两种方法都受到系统稳定性的限制。因此,对于系统的准确性和稳定性必须统筹兼顾、全面衡量。

3.7.4 扰动作用下的稳态误差计算

稳态误差可以分为两种。一种是当系统仅仅受到输入量的作用而没有任何扰动时的稳态误差,称为给定稳态误差;另一种是输入信号为零,而由扰动量作用于系统上时所引起的稳态误差,称为扰动稳态误差,反映了系统抗干扰能力的强弱。当线性系统既受到输入信号作用,同时又受到扰动作用时,系统的稳态误差是这两项误差之和,如图 3-7-1 所示。

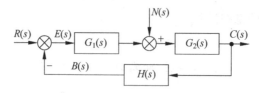

图 3-7-1 扰动输入作用下的系统结构图

扰动输入可以作用在系统的不同位置,因此,即使系统对于某种形式给定输入的稳态误差为零,但对同一形式扰动输入的稳态误差则不一定为零。根据线性系统的叠加原理,以图 3-7-1 所示系统来讨论由扰动输入所产生的稳态误差。按照前面给出的误差信号的定

义可得扰动输入引起的误差为

$$E(s) = R(s) - B(s) = -H(s)C(s) \qquad (3\text{-}7\text{-}22)$$

而此时系统的输出为

$$C(s) = \frac{G_2(s)}{1 + G_1(s)G_2(s)H(s)}N(s) \qquad (3\text{-}7\text{-}23)$$

所以

$$E(s) = -\frac{G_2(s)H(s)}{1 + G_1(s)G_2(s)H(s)}N(s) = \Phi_{en}(s) \cdot N(s) \qquad (3\text{-}7\text{-}24)$$

式中,$\Phi_{en}(s) = -\frac{G_2(s)H(s)}{1 + G_1(s)G_2(s)H(s)}$,称为扰动输入作用下系统的误差传递函数。此时,系统的稳态误差为

$$e_{ssn} = \lim_{t \to \infty} e(t) = \lim_{s \to 0} -\frac{sG_2(s)H(s)}{1 + G_1(s)G_2(s)H(s)} \cdot N(s) \qquad (3\text{-}7\text{-}25)$$

当 $\lim_{s \to 0} G_1(s)G_2(s)H(s) \gg 1$ 时,式(3-7-25)可近似表示为

$$e_{ssn} = \lim_{s \to 0} -\frac{sN(s)}{G_1(s)} \qquad (3\text{-}7\text{-}26)$$

由式(3-7-26)可知,扰动信号作用下产生的稳态误差 e_{ssn} 除了与扰动信号的形式有关外,还与扰动作用点之前的传递函数(即参数)有关,但与扰动作用点之后的传递函数无关。

3.7.5 减少稳态误差的方法

由前面的讨论可知,采取以下措施可以改善系统的稳态精度。

(1) 增大系统开环增益或扰动作用点之前系统的前向通道增益,但增加太大可能引起系统稳定性下降。

(2) 在系统的前向通道或主反馈通道设置串联积分环节,可以消除系统在特定输入信号和特定扰动作用下的稳态误差。

(3) 采用串联控制抑制内回路扰动。

(4) 采用复合控制方法。

Reading Material

What is meant by stability of a dynamic system? A stable system is a dynamic system with a bounded response to a bounded input. How do we find out whether a system is stable or not? The stability of a linear dynamic system is directly related to the location of the poles of system transfer function in the complex s-plane. For stability, all poles of system transfer function must lie in the left half of the complex s-plane.

As we have seen in chapter 1, introduction of the negative feedback control loop to dynamic systems brings about several benefits. Provision of feedback control on the other hand might result in an unstable system under some circumstances. An unstable feedback control system is generally of no practical value. The issue of ensuring the stability of a feedback control system is therefore central to control system design.

The stability of a dynamic system is directly related to the location of the poles of the system transfer function in the s-plane. By careful choice of the closed-loop parameters, the designer is required to ensure that the resulting transfer function of the closed loop system has the poles in the desired locations of the s-plane. Provision of feedback control to a dynamic system may sometimes (depending on the characteristics of the plant being controlled) result in a system transfer function having one or more of its poles with positive real part. Such a system will be useless and unsafe. Therefore our top priority when designing control systems is to make sure that we don't end up with an unstable system.

3.8 MATLAB 在本章中的应用

时域分析法是控制系统重要的分析方法之一。在 MATLAB 环境下,提供了多种函数和工具箱用于时域分析。下面主要采用英文阅读材料的方式,首先介绍在 MATLAB 下如何建立控制系统的传递函数模型,例如第 1 章介绍过的单输入单输出模型(SISO Model)等。

Reading Material

1. SISO Transfer Function Models

A continuous-time SISO(Single-Input/Single-Output) transfer function

$$G(s) = \frac{n(s)}{d(s)}$$

is characterized by its numerator and denominator, both polynomials of the Laplace variable s.

There are two ways to specify SISO transfer functions.

1) Using the tf command

To specify a SISO transfer function model $G(s) = n(s)/d(s)$ using the tf command, type

```
h = tf(num,den)
```

where num and den are row vectors listing the coefficients of the polynomials $n(s)$ and $d(s)$, respectively, when these polynomials are ordered in descending powers of s. The resulting variable h is a TF object containing the numerator and denominator data.

For example, you can create the transfer function $h(s) = s/(s^2 + 2s + 10)$ by typing

```
h = tf([1 0],[1 2 10])
```

Note the customized display used for TF objects.

2) As rational expressions in the Laplace variable s

You can also specify transfer functions as rational expressions in the Laplace variable s by

(1) Defining the variable s as a special TF model

```
s = tf('s');
```

(2) Entering your transfer function as a rational expression in s

For example, once s is defined with tf as in (1)

```
H = s/(s^2 + 2*s +10);
```

Produces the same transfer function as

```
h = tf([1 0],[1 2 10]);
```

Note: You need only define the variable s as a TF model once. All of the subsequent models you create using rational expressions of s are specified as TF objects, unless you convert the variable s to zpk.

2. MIMO Transfer Function Models

MIMO (Multiple-Input/Multiple-Output) transfer functions are two-dimensional arrays of elementary SISO transfer functions. There are several ways to specify MIMO transfer function models, including.

☐ Concatenation of SISO transfer function models
☐ Using tf with cell array arguments

Consider the rational transfer matrix

$$\boldsymbol{H}(s) = \begin{bmatrix} \dfrac{s-1}{s+1} \\ \dfrac{s+2}{s^2+4s+5} \end{bmatrix}$$

You can specify $H(s)$ by concatenation of its SISO entries. For instance,

```
h11 = tf([1 -1],[1 1]);
h21 = tf([1 2],[1 4 5]);
```

or, equivalently,

```
s = tf('s')
h11 = (s-1)/(s+1);
h21 = (s+2)/(s^2+4*s+5);
```

then, it can be concatenated to form $\boldsymbol{H}(s)$.

```
H = [h11; h21]
```

This syntax mimics standard matrix concatenation and tends to be easier and more readable for MIMO systems with many inputs and/or outputs. Alternatively, to define MIMO transfer functions using tf, you need two cell arrays (say, N and D) to represent the sets of numerator and denominator polynomials, respectively.

For example, for the rational transfer matrix $\boldsymbol{H}(s)$, the two cell arrays N and D should contain the row-vector representations of the polynomial entries of

$$N(s) = \begin{bmatrix} s-1 \\ s+2 \end{bmatrix} \quad D(s) = \begin{bmatrix} s+1 \\ s^2+4s+5 \end{bmatrix}$$

You can specify this MIMO transfer matrix $H(s)$ by typing

```
N = {[1 1];[1 2]};        % cell array for N(s)
D = {[1 1];[1 4 5]};      % cell array for D(s)
H = tf(N,D)
```

Notice that both N and D have the same dimensions as H. For a general MIMO transfer matrix $H(s)$, the cell array entries $N\{i,j\}$ and $D\{i,j\}$ should be row-vector representations of the numerator and denominator of $H_{ij}(s)$, the ijth entry of the transfer matrix $H(s)$.

3. Pure Gains

You can use tf with only one argument to specify simple gains or gain matrices as TF objects. For example,

$$G = \text{tf}([1\ 0;2\ 1])$$

produces the gain matrix

$$G = \begin{bmatrix} 1 & 0 \\ 2 & 1 \end{bmatrix}$$

while

$$E = \text{tf}$$

creates an empty transfer function.

4. Zero-Pole-Gain Models

This section explains how to specify continuous-time SISO and MIMO zero-pole-gain models. The specification for discrete-time zero-pole-gain models is a simple extension of the continuous-time case.

1) SISO Zero-Pole-Gain Models

Continuous-time SISO zero-pole-gain models are of the form

$$G(s) = \frac{C(s)}{R(s)} = k\frac{(s-z_1)(s-z_2)\cdots(s-z_m)}{(s-p_1)(s-p_2)\cdots(s-p_n)} \quad n \geqslant m$$

where k is a real-or complex-valued scalar (the gain), and $z_1, z_2, \cdots, z_m$ and $p_1, p_2, \cdots, p_n$ are the real or complex conjugate pairs of zeros and poles of the transfer function $H(s)$. This model is closely related to the transfer function representation: the zeros are simply the numerator roots, and the poles, the denominator roots.

There are two ways to specify SISO zero-pole-gain models:

☐ Using the zpk command

☐ As rational expressions in the Laplace variable s

The syntax to specify ZPK models directly using zpk is

```
h = zpk(z,p,k)
```

where z and p are the vectors of zeros and poles, and k is the gain. This produces a ZPK object h that encapsulates the z, p, and k data. For example, typing

```
h = zpk(0, [1-i 1+i 2], 2)
```

produces

```
Zero/pole/gain:
        2s
    --------------
(s-2)(s^2+2s+2)
```

You can also specify zero-pole-gain models as rational expressions in the laplace variable s by

(1) Defining the variable s as a ZPK model.

```
s = zpk('s')
```

(2) Entering the transfer function as a rational expression in s.

For example, once s is defined with ZPK,

```
H = 2*s/((s-2)*(s^2+2*s+2))
```

returns the same ZPK model as

```
h = zpk([0], [2 1-i 1+i], 2);
```

Note: You need only define the ZPK variable s once. All subsequent rational expressions of s will be ZPK models, unless you convert the variable s to TF.

2) MIMO Zero-Pole-Gain Models

Just as with TF models, you can also specify a MIMO ZPK model by concatenation of its SISO entries.

You can also use the command zpk to specify MIMO ZPK models. The syntax to create a p-by-m MIMO zero-pole-gain model using zpk is

```
H = zpk(Z,P,K)
```

where

- Z is the p-by-m cell array of zeros($Z\{i,j\}$=zeros of $H_{ij}(s)$)
- P is the p-by-m cell array of poles($P\{i,j\}$=poles of $H_{ij}(s)$)
- K is the p-by-m matrix of gains($K\{i,j\}$=gain of $H_{ij}(s)$)

For example, typing

```
Z = { [ ], 5;[1-i 1+i] [ ] };
P = { 0, [1 1];[1 2 3], [ ] };
K = [1 3;2 0];
H = zpk(Z,P,K)
```

creates the two-input/two-output zero-pole-gain model

$$H(s) = \begin{bmatrix} \dfrac{-1}{s} & \dfrac{3(s+5)}{(s+1)^2} \\ \dfrac{2(s^2-2s+2)}{(s-1)(s-2)(s-3)} & 0 \end{bmatrix}$$

Notice that you use [ ] as a place-holder in Z(or P) when the corresponding entry of $H(s)$ has no zeros(or poles).

由上可见,在 MATLAB 环境下,可以采取多种方式对自动控制系统建模。下面以直流电机为例,以英文阅读材料的方式说明整个建模过程。

Reading Material

SISO Example: the DC Motor

A simple model of a DC motor (Fig. 3-8-1) driving an inertial load shows the angular rate of the load, $\omega(t)$, as the output and applied voltage, $v_{app}(t)$, as the input. The ultimate goal of this example is to control the angular rate by varying the applied voltage. This picture shows a simple model of the DC motor.

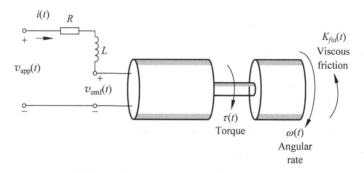

Fig. 3-8-1　A simple model of the DC motor

1. A Simple Model of a DC Motor Driving an Inertial Load

In this model, the dynamics of the motor itself are idealized, for instance, the magnetic field is assumed to be constant. The resistance of the circuit is denoted by R and the self-inductance of the armature by L. If you are unfamiliar with the basics of DC motor modeling, consult any basic text on physical modeling. The important thing here is that with this simple model and basic laws of physics, it is possible to develop differential equations that describe the behavior of this electromechanical system. In this example, the relationships between electric potential and mechanical force are Faraday's law of induction and Ampere's law for the force on a conductor moving through a magnetic field.

2. Mathematical Derivation

The torque τ seen at the shaft of the motor is proportional to the current i induced by the applied voltage

$$\tau(t) = K_m i(t)$$

where K_m, the armature constant, is related to physical properties of the motor, such as magnetic field strength, the number of turns of wire around the conductor coil, and so on. The back (induced) electromotive force, v_{emf}, is a voltage proportional to the angular rate ω seen at the shaft

$$v_{emf}(t) = K_b \omega(t)$$

where K_b, the emf constant, also depends on certain physical properties of the motor.

The mechanical part of the motor equations is derived using Newton's law, which states that the inertial load J times the derivative of angular rate equals the sum of all the torques about the motor shaft. The result is this equation

$$J \frac{d\omega}{dt} = \sum \tau_i = -K_f \omega(t) + K_m i(t)$$

where $K_f \omega$ is a linear approximation for viscous friction.

Finally, the electrical part of the motor equations can be described by

$$v_{app}(t) - v_{emf}(t) = L \frac{di}{dt} + Ri(t)$$

or, solving for the applied voltage and substituting for the back emf

$$v_{app}(t) = L \frac{di}{dt} + Ri(t) + K_b \omega(t)$$

This sequence of equations leads to a set of two differential equations that describe the behavior of the motor, the first for the induced current

$$\frac{di}{dt} = -\frac{R}{L} i(t) - \frac{K_b}{L} \omega(t) + \frac{1}{L} v_{app}(t)$$

and the second for the resulting angular rate

$$\frac{d\omega}{dt} = -\frac{1}{J} K_f \omega(t) + \frac{1}{J} K_m i(t)$$

3. State-Space Equations for the DC Motor

Given the two differential equations derived in the last section, you can now develop a state-space representation of the DC motor as a dynamic system. The current i and the angular rate ω are the two states of the system. The applied voltage, $v_{app}(t)$, is the input to the system, and the angular velocity is the output.

$$\frac{d}{dt} \begin{bmatrix} i \\ \omega \end{bmatrix} = \begin{bmatrix} \dfrac{R}{L} & -\dfrac{R_b}{L} \\ \dfrac{K_m}{J} & -\dfrac{K_f}{J} \end{bmatrix} \cdot \begin{bmatrix} i \\ \omega \end{bmatrix} + \begin{bmatrix} \dfrac{1}{L} \\ 0 \end{bmatrix} \cdot v_{app}(t)$$

$$y(t) = \begin{bmatrix} 0 & 1 \end{bmatrix} \cdot \begin{bmatrix} i \\ \omega \end{bmatrix} + \begin{bmatrix} 0 \end{bmatrix} \cdot v_{app}(t)$$

4. Constructing SISO Models

Once you have a set of differential equations that describe your plant, you can construct SISO models using simple commands in the Control System Toolbox.

The following sections discuss

- ☐ Constructing a state-space model of the DC motor.
- ☐ Converting between model representations.
- ☐ Creating transfer function and zero/pole/gain models.

5. Constructing a State-Space Model of the DC Motor

Listed below are nominal values for the various parameters of a DC motor.

```
R = 2.0            % Ohms
L = 0.5            % Henrys
Km = 0.015         % torque constant
Kb = 0.015         % emf constant
Kf = 0.2           % Nms
J = 0.02           % kg.m^2/s^2
```

Given these values, you can construct the numerical state-space representation using the *ss* function.

```
A = [-R/L -Kb/L; Km/J -Kf/J];
B = [1/L; 0];
C = [0 1];
D = [0];
sys_dc = ss(A,B,C,D)
```

This is the output of the last command.

```
a =
        x1      x2
  x1    -4     -0.03
  x2    0.75   -10
b =
        u1
  x1    2
  x2    0
c =
        x1   x2
  y1    0    1
d =
        u1
  y1    0
```

6. Converting Between Model Representations

Now that you have a state-space representation of the DC motor, you can convert to other model representations, including transfer function (TF) and zero/pole/gain (ZPK) models.

(1) Transfer Function Representation. You can use tf to convert from the state-space representation to the transfer function. For example, use this code to convert to the transfer function representation of the DC motor.

```
sys_tf = tf(sys_dc)
```

Transfer function：

```
       1.5
------------------
s^2 + 14 s + 40.02
```

(2) Zero/Pole/Gain Representation. Similarly, the zpk function converts from state-space or transfer function representations to the zero/pole/gain format. Use this code to convert from the state-space representation to the zero/pole/gain form for the DC motor.

```
sys_zpk = zpk(sys_dc)
```

Zero/pole/gain：

```
         1.5
------------------
(s + 4.004)(s + 9.996)
```

Note：The state-space representation is best suited for numerical computations. For highest accuracy, convert to state space prior to combining models and avoid the transfer function and zero/pole/gain representations, except for model specification and inspection.

以上学习了在 MATLAB 工作环境下，自动控制系统的建模过程。这里通过两个例子，实现时域分析控制，讨论其控制效果。

【例 3-8-1】 已知连续系统的传递函数为

$$G(s) = \frac{C(s)}{R(s)} = \frac{3s^4 + 2s^3 + 5s^2 + 4s + 6}{s^5 + 3s^4 + 4s^3 + 2s^2 + 7s + 2}$$

要求：
(1) 求出该系统的零、极点及增益；
(2) 绘出其零、极点图，判断系统稳定性。

解：

```
% This program create a transfer function and then finds/displays its poles, zeros and gain
NUM = [3,2,5,4,6];
DEN = [1,3,4,2,7,2];
[z,p,k] = tf2zp(NUM,DEN)
pzmap(NUM,DEN);
title('Poles and zeros map');
You should get the following response.
z =   0.4019 + 1.1965i
      0.4019 - 1.1965i
     -0.7352 + 0.8455i
```

$$
\begin{aligned}
&\quad\quad -0.7352-0.8455\mathrm{i}\\
&p = -1.7680+1.2673\mathrm{i}\\
&\quad\quad -1.7680-1.2673\mathrm{i}\\
&\quad\quad 0.4176+1.1130\mathrm{i}\\
&\quad\quad 0.4176-1.1130\mathrm{i}\\
&\quad\quad -0.2991\\
&k = 3
\end{aligned}
$$

【例 3-8-2】 应用 MATLAB 分析系统的动态特性。在 MATLAB 中提供了求取连续系统的单位阶跃响应 step，以及任意输入下的仿真函数 lsim。

已知典型二阶系统的传递函数为

$$G(s)=\frac{C(s)}{R(s)}=\frac{\omega_n^2}{2s^2+2\xi\omega_n s+\omega_n^2}$$

其中 $\omega_n=6$，绘制系统在 $\xi=0.1,0.2,\cdots,1.0,2.0$ 时的单位阶跃响应。

Consider the transfer function of a typical second-order system as follows.

解：可执行如下程序：

```
% This program plots a curve of step response
wn = 6;
kosi = [0.1,0.2,1.0,2.0];
figure(1)
hold on
for kos = kosi
    num = wn.^2;
    den = [2,2*kos*wn,wn.^2];
    step(num,den);
end;
title('Step Response');
hold off
```

仿真运行结果如图 3-8-2 所示。

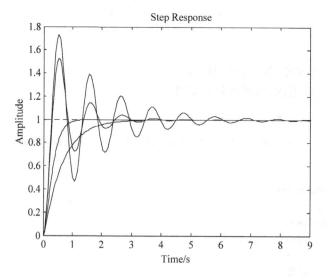

图 3-8-2　例 3-8-2 运行结果

本章小结

(1) 自动控制系统一般不但要具有足够的稳定性,还应该有较高的稳态控制精度和较快的响应过程。为了评价这三方面的性能,必须定义几个反映稳、准、快三方面性能的指标。

(2) 一阶系统和二阶系统是时域分析法重点分析的两类系统。在本章中,详细介绍了一阶系统和二阶系统的单位阶跃响应、单位斜坡响应和单位脉冲响应。以二阶系统为主,介绍了各种时域性能指标的求取和对系统性能的影响。对于高阶系统,如果其特性近似于一阶或二阶系统,则可在一定条件下,将其降为一阶或二阶系统,然后按一阶或二阶系统作近似分析。而对于一般的高阶系统,可用劳斯判据来判断系统稳定性,用终值定理来计算稳态误差。

(3) 系统平稳性、快速性和稳态精度对系统参数的要求是矛盾的,在系统参数的选择无法同时满足几方面时,可采用在前向通道中加比例微分环节及增加微分负反馈等措施来改善系统动态性能,使系统能同时满足几方面的要求。

(4) 系统能正常工作的首要条件是,系统是稳定的。可采用劳斯判据来判断系统的稳定性。若系统结构不稳定,则可通过用比例反馈来包围有积分作用的环节,从而改变环节的积分性质,以及在前向通道中增加比例微分环节等方法,使系统变为结构稳定系统。

(5) 稳态误差是衡量系统控制精度的性能指标。稳态误差可分为有给定的信号引起的误差以及有扰动信号引起的误差两种。稳态误差也可用误差系数来表达。系统的稳态误差主要是由积分环节的个数和开环增益来确定的。为了提高精度等级,可增加积分环节的数目,为了减小有限误差,可增加开环增益。但这两种方法都会使系统的稳定性变差,甚至导致系统不稳定。而采用补偿的方法,则可在保证稳定性的前提下减小稳态误差。

Summary and Outcome Checklist

In this topic we have been concerned with the definition and usefulness of quantitative measures of performance of feedback control systems. We learn that there is strong correlation between the system transient response and the locations in the s-plane of the poles of the closed-loop transfer function. For systems having second-order transfer function, valuable relationships have been developed between the performance specifications and the two characteristic parameters, damping ratio, ξ, and natural frequency, ω_n. Relying on the notion of dominant poles, these relationships are considered applicable to those higher order systems that have a dominant pair of complex poles.

While the introduction of the negative feedback control loop to dynamic systems brings about several benefits, it might result in an unstable system under some circumstances. Such an unstable system will be useless and unsafe. The issue of ensuring the stability of the feedback control system is therefore central to control system design.

The stability of a dynamic system is directly related to the location of the poles of the system transfer function in the s-plane. By careful choice of the closed-loop parameters, the designer is required to ensure that the resulting transfer function of the closed loop system has the poles in the desired locations of the s-plane.

The fundamental reasons for using feedback control, despite its additional cost and complexity are:

(1) decrease in the sensitivity of the system to variations in the parameters of the plant/process being controlled.

(2) ease of control and adjustment of the transient response of the system.

(3) reduction of the impact of disturbance and noise signals within the system.

(4) reduction of the steady-state error of the system.

Tick the box for each statement with which you agree.

☐ I am able to determine the poles and zeros of a given transfer function and map them in the complex s-plane.

☐ I am able to determine, by the partial fraction method, the time response of a given transfer function to standard test input signals such as the step, impulse, ramp, and sinusoid.

☐ I am able to distinguish between the transient, and steady state, responses of a dynamic system.

☐ I can explain what performance measures (or specifications) are, why they are important to the design process, and what makes a particular measure a good candidate to be a performance measure.

☐ I can list the performance measures commonly used by control practitioners.

☐ I can describe the correlation that exist between the transient response of a system and the location of the system poles and zeros in the s-plane.

☐ I can explain, with the help of an illustrative example, the notion of dominant pole/poles of a transfer function.

☐ I can express the transfer function of 1st, and 2nd order systems in their respective standard form and identify the characteristic parameters.

☐ I can write down the time response of 1st, and 2nd order systems (in terns of their respective characteristic parameters) to standard test input signals such as the step, impulse, ramp, and sinusoid.

☐ By examining the transfer function of higher order systems (order $>$ 3), I can determine whether the transfer function can/cannot be approximated by a lower order transfer function.

☐ I can, if feasible, approximate a higher order transfer function by a lower order transfer function.

☐ I can predict the time response of a control system if its transfer function is known.

- ☐ I can describe the quantitative relationships that exist between the various performance measures and the characteristic parameters of systems having the standard 2nd order transfer function.
- ☐ I can explain why, when designing control systems, designers often attempt to achieve a second order type transfer function for the control system.
- ☐ I can explain the meaning of the term sensitivity.
- ☐ I am able to determine the sensitivity of a feedback control systems to variations in the parameters of the system.
- ☐ Given the transfer function of a system, I am able to determine its poles and zeros and map them in the complex s-plane.
- ☐ I am able to determine the transient response and the steady state error of a feedback control system for standard input test signals.
- ☐ I am able to estimate the effect of disturbances on the performance of a feedback control system.
- ☐ I can list the four main benefits of feedback control.
- ☐ I can explain the concept of stability in the context of linear dynamic systems.
- ☐ I can state the relationship between stability and the s-plane location of the poles of the system transfer function.
- ☐ I am able to determine whether a system is stable or not by applying the Routh-Hurwitz stability criterion.
- ☐ I can explain the meaning of the term relative stability, and analyse the relative stability of a dynamic system by applying Routh-Hurwitz stability criterion.
- ☐ I am able to determine how the locations of poles of system transfer function in the s-plane change as a parameter of the system is varied.

习题

3-1 设系统的初始条件为零,其微分方程式如下：
(1) $0.2\dot{c}(t)=2r(t)$
(2) $0.04\ddot{c}(t)+0.24\dot{c}(t)+c(t)=r(t)$

试求：两个系统的单位冲激响应及单位阶跃响应,并求系统(2)的过渡过程及最大超调量 σ_p、峰值时间 t_p 过渡过程时间 t_s。

3-2 已知某单位反馈系统的闭环传递函数为
$$\Phi(s)=\frac{C(s)}{R(s)}=\frac{15.36(s+6.25)}{(s^2+2s+2)(s+6)(s+8)}$$

试近似计算系统的单位阶跃响应性能指标：最大超调量 $\delta\%$、调节时间 t_s、稳态误差 e_{ss}。

3-3 典型二阶系统的单位阶跃响应为
$$c(t)=1-1.25e^{-1.2t}\sin(1.6t+53.1°)$$

试求系统的最大超调量 σ_p、峰值时间 t_p、过渡过程时间 t_s。

3-4 系统零初始条件下的单位阶跃响应为
$$c(t)=1+0.2e^{-60t}-1.2e^{-10t}$$
(1) 试求该系统的闭环传递函数。
(2) 试确定阻尼比 ξ 与阻尼自振角频率 ω_n。

3-5 设系统的闭环传递函数为
$$\frac{C(s)}{R(s)}=\frac{\omega_n^2}{s^2+2\xi\omega_n s+\omega_n^2}$$
为使系统阶跃响应有5%最大超调量和2s的过渡过程时间,试求 ξ 和 ω_n。

3-6 对由如下闭环传递函数表示的三阶系统
$$\frac{C(s)}{R(s)}=\frac{816}{(s+2.74)(s+0.2+j0.3)(s+0.2-j0.3)}$$
说明该系统是否有主导极点。如有,求出该极点。

3-7 已知二阶系统的闭环传递函数为 $\phi(s)=\dfrac{C(s)}{R(s)}=\dfrac{\omega_n^2}{s^2+2\xi\omega_n s+\omega_n^2}$,试在同一 $[s]$ 平面上画出对应题3-7图中3条单位阶跃响应曲线的闭环极点相对位置,并简要说明。图中 t_{s1}、t_{s2} 分别是曲线①、曲线②的过渡过程时间, t_{p1}、t_{p2}、t_{p3} 分别是曲线①、②、③的峰值时间。

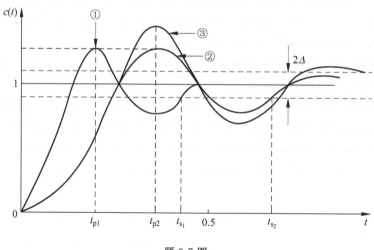

题 3-7 图

3-8 设单位反馈系统其开环传递函数为
$$G(s)=\frac{K}{s(1+s/3)(1+s/6)}$$
若要求闭环特征方程的根的实部均小于 -1,问 K 值应取在什么范围内?如果要求实部均小于 -2,情况又如何?

3-9 系统方程如下,试用劳斯判据确定系统正实部根的个数。
(1) $\Delta(s)=s^4+5s^3+10s^2+20s+24=0$
(2) $\Delta(s)=s^3-3s+2=0$
(3) $\Delta(s)=s^6+s^5+6s^4+5s^3+9s^2+4s+4=0$
(4) $\Delta(s)=s^5+2s^4-s-2=0$

(5) $\Delta(s) = s^5 + s^4 + 2s^3 + 2s^2 + s + 1 = 0$

3-10 已知单位反馈系统的开环传递函数为 $G(s) = \dfrac{10K}{s(1+T_1 s)(1+T_2 s)}$，式中 $T_1 = 0.1(s)$，$T_2 = 0.5(s)$，输入信号为 $r(t) = 2 + 0.5t$。

(1) 求 $K=1$ 时的系统稳态误差。

(2) 是否可以选择某一合适的 K，使得系统稳态误差为 0.025？

3-11 系统如题 3-11 图所示，试求下列各系统的静态误差函数 k_p，k_v 和 k_a，以及输入 $r(t) = 6t^2 \times 1(t)$ 时的静态误差。

3-12 设随动系统的微分方程为

$$T_M T_a \frac{d^3 c(t)}{dt^3} + T_M \frac{d^2 c(t)}{dt^2} + \frac{dc(t)}{dt} + Kc(t) = Kr(t)$$

其中，$c(t)$ 为系统输出量；$r(t)$ 为系统输入量；T_M 为电动机机电时间常数；T_a 为电动机电磁时间常数；K 为系统开环增益；初始条件全部为零。试讨论：

(1) T_a、T_M 与 K 之间关系对系统稳定性的影响。

(2) 当 $T_a = 0.01$，$T_M = 0.1$，$K = 500$ 时，可否忽略 T_a 的影响？在什么情况下 T_a 的影响可忽略？

3-13 系统结构如题 3-13 图所示。若系统以 $\omega = 2\text{rad/s}$ 频率持续振荡，试确定相应的 K 和 a 的值。

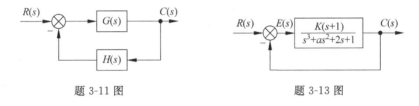

题 3-11 图　　　　　　　　　　题 3-13 图

3-14 设控制系统如题 3-14 图所示。其中扰动信号 $n(t) = 1(t)$。试问：是否可以选择某一合适的 K_1 值，使系统在扰动作用下的稳态误差值为 $e_{\text{ssn}} = -0.099$？

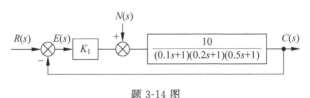

题 3-14 图

3-15 控制系统的误差还有一种定义方法，这就是：无论是对于单位反馈系统还是非单位反馈系统，误差均定义为系统的输入量与输出量之差，即

$$e(t) \triangleq r(t) - c(t)$$

现在设闭环系统的传递函数为

$$G_c(s) = \frac{b_m s^m + b_{m-1} s^{m-1} + \cdots + b_1 s + b_0}{s^n + a_{n-1} s^{n-1} + \cdots + a_1 s + a_0} \quad (n \geq m)$$

试证：系统在单位斜坡函数输入作用下，不存在稳态误差的条件是 $a_0 = b_0$ 和 $a_1 = b_1$。

3-16 设系统结构图如题 3-16 图所示。如果要求系统的超调量等于 15%，峰值时间等于 0.8s,试求增益 K_1 和 K_t,同时确定在此条件下系统的延时时间、上升时间和调节时间。

3-17 系统结构图如题 3-17 图所示。

(1) 已知 $G_1(s)$ 的单位阶跃响应为 $1-e^{-2t}$,试求 $G_1(s)$;

(2) 当 $G_1(s)=\dfrac{1}{s+2}$,且 $r(t)=10\times 1(t)$ 时,试求:①系统的稳态输出;②系统的峰值时间 t_p,超调量 $\delta\%$,调节时间 t_s 和稳态误差 e_{ss};③概括绘出系统的输出响应 $c(t)$ 曲线。

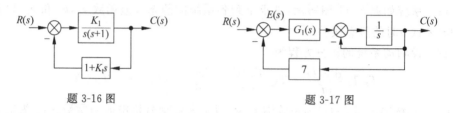

题 3-16 图　　　　　　　　　题 3-17 图

3-18 系统结构图如题 3-18 图所示,试问:

(1) 为确保系统稳定,如何取 K 值?

(2) 为使系统特征根全部位于 s 平面 $s=-1$ 的左侧,K 应取何值?

(3) 当 $r(t)=2t+2$ 时,要求系统稳定误差 $e_{ss}\leqslant 0.25$,K 应取何值?

3-19 控制系统结构图如图 3-19 所示:

(1) 试确定参数 K_1,K_2 使系统极点配置在 $\lambda_{1,2}=-5\pm j5$;

(2) 设计 $G_1(s)$,使 $r(t)$ 作用下的稳态误差恒为零。

(3) 设计 $G_2(s)$,使 $n(t)$ 作用下的稳态误差恒为零。

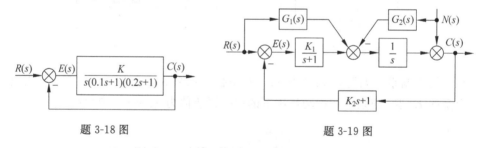

题 3-18 图　　　　　　　　　题 3-19 图

3-20 (西北工业大学硕士研究生入学考试试题)某单位反馈的三阶系统(开环无零点);开环增益 $k\in(0,5)$ 时系统稳定,此时在 $r(t)=1(t)$ 作用下系统无稳态误差;当 $k=5$ 时,系统单位阶跃响应呈现频率 $\omega=\sqrt{6}$ 的等幅振荡。试求:

(1) 由上述条件确定系统的传递函数。

(2) 确定系统主导极点位于 $\beta=60°$ 线($\xi=0.5$)时,全部 3 个极点的位置,并由主导极点估算系统的动态性能指标 $\sigma\%$ 和 t_s。

3-21 (中国科技大学硕士研究生入学考试试题)单位负反馈控制系统如题 3-21 图所示,求:

(1) 试确定闭环稳定的反馈系数 K_b 的取值范围。

(2) 若已确定系统的一个闭环极点为 -5,试求 K_b 的取值和其余的闭环极点。

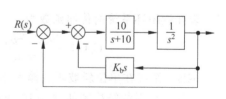

题 3-21 图

(3) 根据题(2)得到的系统配置,采用时域方法分析系统的动态性能和稳定性能。

3-22 (南京航空航天大学硕士研究生入学考试试题)已知某系统结构如题 3-22(a)图所示,试求:

(1) 当反馈通道传递函数 $H(s)=1$ 时,其开环系统单位阶跃响应曲线如题 3-22(b)图所示,试确定系统的增益 K 阻尼比 ξ 和自然频率 ω_n。

(2) 若要求系统的阻尼比提高到 ξ',而保持系统增益 K 自然频率 ω_n 不变,试设计反馈通道的传递函数 $H(s)$。

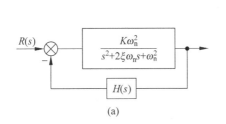 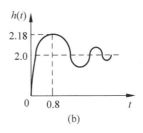

题 3-22 图

3-23 (浙江大学硕士研究生入学考试试题)设单位负反馈系统的开环传递函数为 $G(s)=\dfrac{K}{s(Ts+1)}$,要求所有的特征根位于 s 平面上,$s=-1\pm j\omega$ 的左侧,且阻尼比 $\xi \geqslant 0.5$。

(1) 在平面上用阴影线表示特征根的分布情况。

(2) 求出 K,T 的取值范围并在 $K=T$ 直角平面表示。

第 4 章

根轨迹法

4.1 引言

在第 3 章中，主要介绍自动控制系统的时域分析。控制系统的稳定性和时间响应的基本特征由闭环系统特征方程的根，即闭环极点所决定。时间响应的大小由系统的闭环零点和闭环极点共同决定。因此，研究闭环特征根在复平面上的分布对于分析系统的性能有着重要的意义。进而，可以利用改变系统的闭环零点和极点的分布来间接地设计控制系统，以期达到理想的性能要求。

然而，对于高阶系统来说，手工求解闭环特征方程的根的过程是非常复杂的。尤其是当系统的参数（如开环增益、开环零点和开环极点等）发生变化时，系统特征方程的系数会发生变化，这时闭环特征根就需要重复计算，而且还不能直观地看出系统参数变化对闭环特征根分布的影响趋势。控制系统的设计者通常希望借助某种较为简单的分析方法，当已知的开环系统某个参数发生变化时，可以很明确地看出闭环特征根的变化趋势，从而能够判断系统稳定性，预测闭环控制系统的性能，并且能够对于控制系统进行校正，以获得希望的性能指标。

1948 年，伊万斯（W. R. Evans）提出了一种求解系统特征根的简单方法——根轨迹法，在控制系统的分析和设计中得到了广泛的应用。当开环增益或其他参数发生变化时，其全部数值对应的闭环极点均可在根轨迹图上简便地确定。因为系统的稳定性由系统闭环极点唯一确定，而系统的稳态性能和动态性能又与闭环零、极点在 s 平面上的位置密切相关，所以，根轨迹图不仅可以直接给出闭环系统时间响应的全部信息，而且可以根据系统性能指标的要求调整参数以及系统开环零、极点的位置，便于系统的分析与综合。除此之外，用根轨迹法还可方便简单地求解高阶代数方程的根。

Reading Material

As we have seen in chapter 3, the relative stability and the transient performance of a closed loop control system are directly related to the location of the roots of the characteristic equation in the s-plane. It is frequently necessary to adjust one or more system's parameters in order to obtain suitable root locations. Therefore it is worthwhile to determine how the roots of the characteristic equation of a given system migrate in the s-plane as the parameters are varied.

The locus of roots in the *s*-plane can be determined by a graphical method. A graph of the locus of roots as one system parameter varies is known as a root locus plot. In this section, we will learn practical techniques for obtaining a sketch of a root locus plot by hand. We also consider computer generated root locus plots and illustrate their effectiveness in the design process.

本章的主要内容：
- 根轨迹的基本概念
- 根轨迹绘制的基本规则
- 广义根轨迹
- 系统性能的根轨迹法分析
- MATLAB 在本章中的应用

4.2 根轨迹的基本概念

1. 根轨迹概念

根轨迹简称根迹，它是指当开环系统某一参数从零到无穷变化时，闭环极点在 s 平面上变化的轨迹。

Terms and Concepts

Root locus is the plot of all the poles of the closed-loop transfer function in the complex *s*-plane as one of the parameters appearing in the characteristic equation(say, K) is given different positive values in the range zero to infinity.

(1) Here we assume that numerical values of all the remaining parameters of the characteristic equation are either known in advance or assigned a fixed set of values.

(2) From the root locus plot it is very easy to see how changes in the value of the parameter affect the system transient response.

(3) By inspecting the root locus plot, we may be able to select an optimum value for the parameter K.

为了说明根轨迹的基本概念，以图 4-2-1 所示系统为例，该系统开环函数为

$$G(s) = \frac{K}{s(0.5s+1)} = \frac{K^*}{s(s+2)} \tag{4-2-1}$$

定义根轨迹增益

$$K^* = 2K \tag{4-2-2}$$

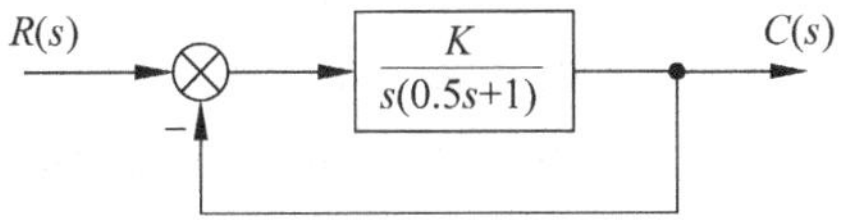

图 4-2-1 控制系统结构图

闭环传递函数为

$$\Phi(s) = \frac{C(s)}{R(s)} = \frac{K^*}{s^2 + 2s + K^*} \tag{4-2-3}$$

闭环特征方程为

$$D(s) = s^2 + 2s + K^* \tag{4-2-4}$$

则特征根为

$$s_1 = -1 + \sqrt{1 - K^*} \tag{4-2-5}$$

$$s_2 = -1 - \sqrt{1 - K^*} \tag{4-2-6}$$

当系统参数 K^*（或 K）从零变化到无穷大时，可以用解析的方法求出闭环极点的全部数值，闭环极点的变化情况如表 4-2-1 所示。

表 4-2-1　K^*、K 变化时系统的特征根

K^*	K	s_1	s_2
0	0	0	-2
0.5	0.25	-0.3	-1.7
1	0.5	-1	-1
2	1	$-1+j$	$-1-j$
5	2.5	$-1+j2$	$-1-j2$
$\vdots$	$\vdots$	$\vdots$	$\vdots$
∞	∞	$-1+j\infty$	$-1-j\infty$

将这些数值标注在复平面上，并连成光滑的粗实线，如图 4-2-2 所示，粗实线就成为系统的根轨迹，根轨迹上的箭头表示随着 K 值的增加根轨迹的变化趋势，而标注的数值则代表与闭环极点位置相应的开环增益 K 的数值。

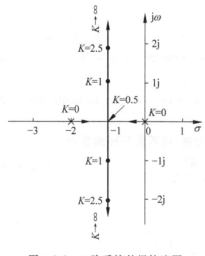

图 4-2-2　二阶系统的根轨迹图

当增益 K 从零到无穷变化时，特征方程的根在 s 平面上移动的轨迹如图 4-2-2 所示。当 $K=0$ 时，$s_1=0$，$s_2=-2$，此时，系统的闭环极点与系统的开环极点相同。将这两个根用符号"×"在 s 平面上标注出来，如图 4-2-2 所示。以后我们用"×"表示当 $K=0$ 时特征方程的根，即开环极点。用"○"表示系统的开环零点。①当 $0<K<1$ 时，两个极点 s_1 和 s_2 都是负实数极点，随 K 值的增大，s_1 减小，s_2 增大，s_1 从原点开始沿负实轴向左移动，s_2 从 -2 开始沿负实轴向右移动。因此，从原点到 $(-2,j0)$ 点这段负实轴是根轨迹的一部分。这时，系统处于过阻尼状态，其阶跃响应是非周期的。②当 $K=1$ 时，$s_1=s_2=-1$，特征方程有两个重实根。这时系统处于临界阻尼状态，其阶跃响应仍然是非周期的。③当 $K>1$ 时，$s_{1,2} = -1 \pm \sqrt{K-1}$，特征方程有两个共轭复数根，其实部为 -1，不随 K 值变化，虚部的数值则随 K 值的增大而变化。s_1 从 $(-1,j0)$ 开始沿直线向上移动，s_2 从 $(-1,j0)$ 开始沿直线向下移

动。当 K 从零变化到无穷大时，闭环特征方程的根在复平面上移动的轨迹如图 4-2-2 所示。

图 4-2-2 所示的根轨迹是由解析的方法得到的，对于阶次较高的控制系统，用解析的方法绘制显然是不适用的。我们希望能有简便的图解方法，可以根据已知的开环传递函数迅速绘出闭环系统的根轨迹。为此，需要研究闭环零、极点与开环零、极点之间的关系。

2. 根轨迹方程

控制系统如图 4-2-3 所示。它的闭环传递函数为

$$G_c(s) = \frac{G(s)}{1+G(s)H(s)} \quad (4\text{-}2\text{-}7)$$

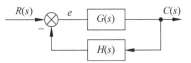

图 4-2-3 控制系统结构图

闭环特征方程为

$$1+G(s)H(s) = 0 \quad (4\text{-}2\text{-}8)$$

设系统的开环传递函数为

$$G(s)H(s) = K\frac{b_m s^m + b_{m-1}s^{m-1}+\cdots+b_1 s+1}{a_n s^n + a_{n-1}s^{n-1}+\cdots+a_1 s+1} \quad (4\text{-}2\text{-}9)$$

或写成

$$G(s)H(s) = K^* \frac{\prod_{j=1}^{m}(s-z_j)}{\prod_{i=1}^{n}(s-p_i)} \quad (4\text{-}2\text{-}10)$$

式(4-2-9)中的 K 是系统的开环增益。式(4-2-10)中的 z_j 和 p_i 分别是开环传递函数的零点和极点，K^* 是将分子和分母分别写成因子相乘的形式时提取的系数，称作根轨迹增益，它与系统开环增益的关系为

$$K = K^* \frac{\prod_{j=1}^{m}(-z_j)}{\prod_{i=1}^{n}(-p_i)} \quad (4\text{-}2\text{-}11)$$

将特征方程写成如下形式

$$G(s)H(s) = K^* \frac{\prod_{j=1}^{m}(s-z_j)}{\prod_{i=1}^{n}(s-p_i)} = -1 \quad (4\text{-}2\text{-}12)$$

式(4-2-12)为根轨迹方程，它是一复数方程，由于复数方程两边的幅值和相角分别相等，因此可将式(4-2-12)用以下两个方程描述，即

幅值条件：$$K^* = -\frac{\prod_{i=1}^{n}(s-p_i)}{\prod_{j=1}^{m}(s-z_j)} \quad (4\text{-}2\text{-}13)$$

相角条件：$$\sum_{j=1}^{m}\angle(s-z_j) - \sum_{i=1}^{n}\angle(s-p_i) = (2k+1)\pi \quad k=0,\pm1,\pm2,\cdots \quad (4\text{-}2\text{-}14)$$

分别将式(4-2-13)和式(4-2-14)称作幅值条件和相角条件，满足幅值条件和相角条件

的 s 值就是特征方程的根，即系统的闭环极点。当 K^* 从零到无穷变化时，特征方程的根在复平面上变化的轨迹就是根轨迹。实际上，只要满足相角条件的点都是根轨迹上的点，当 K^* 值确定以后，可根据幅值条件在根轨迹上确定相应的闭环极点，除了开环增益 K（或根轨迹增益）外，系统其他参数变化时对闭环特征方程根的影响也可通过根轨迹表示出来，只要将特征方程整理后，将可变参数标在相应的位置上，就可利用相角条件绘制出根轨迹来。

Reading Material

In plotting the root locus, we do not place any restriction on the value of K. K can be any position number in the range zero to infinity $(0 \rightarrow \infty)$. Hence we can always satisfy the magnitude condition by choosing the numerical value of K appropriately. We can draw the root locus by identifying all values of $s(=s_p$, say) in the complex s-plane that would satisfy the angle condition. The value of K that would place the closed loop pole at a particular point on the root locus is then determined by applying the magnitude condition.

4.3 根轨迹绘制的基本规则

由 4.2 节可知，当 K^* 从零到无穷变化时，依据相角条件，可以在复平面上找到满足 K^* 变化时的所有闭环极点，绘出系统的根轨迹，但是这种方法比较烦琐。

Reading Material

Root Locus Method is a graphical technique for sketching the locus of the closed loop poles in the s-plane as one of the system parameters is varied. And it shows the influence of the parameter on stability and transient response of the system. So, it is a powerful tool for the analysis and design of feedback control systems. Root locus plot, as an approximate hand sketch, can provide qualitative information concerning the stability and performance of a feedback control system. MATLAB can be used to generate accurate root locus plots.

本节通过研究根轨迹和开环零极点的关系，利用根轨迹的性质可以减少绘图工作量，较迅速地画出根轨迹的大致形状和变化趋势。首先讨论以根轨迹增益作为参变量时的 180°根轨迹的绘制法则。

1. 根轨迹的起点和终点

【**法则 1**】 根轨迹的起点和终点。根轨迹起始于开环极点，终止于开环零点。如果开环零点数目 m 小于开环极点数目 n，则有 $(n-m)$ 条根轨迹终止于 s 平面无穷远处。

根轨迹的起点是指 $K^*=0$ 时特征根在 s 平面上的位置，根轨迹的终点是指 $K^* \rightarrow \infty$ 时特征根在 s 平面上的位置。

【**证明**】 式(4-2-12)中 K^* 可以从零变到无穷。当 $K^*=0$ 时，有

$$s = p_i \quad i = 1, 2, \cdots, n$$

说明 $K^*=0$ 时,闭环特征方程式的根就是开环传递函数 $G(s)H(s)$ 的极点,所有根轨迹必起于开环极点。将式(4-2-12)改写为

$$\frac{\prod_{j=1}^{m}(s-z_j)}{\prod_{i=1}^{n}(s-p_i)}=-\frac{1}{K^*} \tag{4-3-1}$$

当 $K^*\to\infty$ 时,由式(4-3-1)可得

$$s=z_j \quad j=1,2,\cdots,m$$

所以根轨迹必然终止于开环零点。

在实际系统中,开环传递函数分子多项式次数 m 一般小于分母多项式次数 n,即满足不等式 $m\leqslant n$,这时有 $(n-m)$ 条根轨迹的终点将在无穷远处。若系统有 $(n-m)$ 个无穷大的开环零点,则系统的开环零点数和开环极点数就相同了。

2. 根轨迹的分支数、对称性和连续性

【法则 2】 根轨迹的分支数、对称性和连续性。根轨迹是对称于实轴的连续曲线,其分支数等于系统开环极点数 n。

1) 根轨迹的连续性

用代数定理可以证明,式(4-2-12)中参数 K^* 连续变化,特征方程的根便连续变化,即根轨迹曲线是连续曲线。

2) 根轨迹的对称性

特征方程的系数由实际物理系统参数所决定,所以一定是实数,特征方程若有复数根,必是共轭复根。由于根轨迹是闭环特征根的集合,所以根轨迹是对称于实轴的曲线。

3) 根轨迹的分支数

根轨迹在 s 平面上的分支数必与控制系统特征方程式的阶次相一致,即根轨迹的分支数等于开环极点数,也等于闭环极点数。

Terms and Concepts

Locus is a path or trajectory that is traced out as a parameter is changed.

Number of separate loci is equal to the number of poles of the transfer function, assuming that the number of the poles is greater than or equal to the number of zeros of the transfer function.

3. 根轨迹的渐近线

【法则 3】 根轨迹的渐近线就是确定当开环零点数目 m 小于开环极点数目 n 时,$(n-m)$ 条根轨迹沿什么方向趋于 s 平面无穷远处。渐近线可认为是 $K^*\to\infty$、$s\to\infty$ 时的渐近线。

当开环极点有限极点数 n 大于有限零点数 m 时,有 $(n-m)$ 条根轨迹分支沿着与实轴交角为 φ_a、交点为 σ_a 的一组渐近线趋向无穷远处,且有

$$\varphi_a=\frac{(2l+1)\pi}{n-m} \quad l=0,1,2,\cdots,n-m-1 \tag{4-3-2}$$

$$\sigma_a = \frac{\sum_{i=1}^{n} p_i - \sum_{j=1}^{m} z_j}{n-m} \tag{4-3-3}$$

若开环传递函数无零点,取 $\sum z_j = 0$。

【证明】 渐近线就是 s 值很大时的根轨迹,因此渐近线也一定对称于实轴。将开环传递函数写成多项式的形式,得

$$G(s)H(s) = K^* \frac{\prod_{j=1}^{m}(s-z_j)}{\prod_{i=1}^{n}(s-p_i)} = K^* \frac{s^m + b_1 s^{m-1} + \cdots + b_{m-1}s + b_m}{s^n + a_1 s^{n-1} + \cdots + a_{n-1}s + a_n}$$

令

$$b_1 = -\sum_{j=1}^{m} z_j \qquad a_1 = -\sum_{i=1}^{n} p_i$$

当 s 值很大时,上式可近似为

$$G(s)H(s) = \frac{K^*}{s^{n-m} + (a_1 - b_1)s^{n-m-1}} \tag{4-3-4}$$

由式(4-2-8)得渐近线方程

$$s^{n-m}\left(1 + \frac{a_1 - b_1}{s}\right) = -K^* \tag{4-3-5}$$

又有

$$s\left(1 + \frac{a_1 - b_1}{s}\right)^{\frac{1}{n-m}} = (-K^*)^{\frac{1}{n-m}} \tag{4-3-6}$$

根据二项式定理

$$\left(1 + \frac{a_1 - b_1}{s}\right)^{\frac{1}{n-m}} = 1 + \frac{a_1 - b_1}{(n-m)s} + \frac{1}{2!} \times \frac{1}{n-m}\left(\frac{1}{n-m} - 1\right)\left(\frac{a_1 - b_1}{s}\right)^2 + \cdots$$

在 s 值很大时,有

$$\left(1 + \frac{a_1 - b_1}{s}\right)^{\frac{1}{n-m}} = 1 + \frac{a_1 - b_1}{(n-m)s} \tag{4-3-7}$$

将式(4-3-7)代入式(4-3-6),渐近线方程整理为

$$s\left[1 + \frac{a_1 - b_1}{(n-m)s}\right] = (-K^*)^{\frac{1}{n-m}}$$

现在以 $s = \sigma + j\omega$ 代入式,得

$$\left(\sigma + \frac{a_1 - b_1}{n-m}\right) + j\omega = \sqrt[n-m]{K^*}\left[\cos\frac{(2k+1)\pi}{n-m} + j\sin\frac{(2k+1)\pi}{n-m}\right] \tag{4-3-8}$$

其中,$k = 0, 1, \cdots, n-m-1$。

令实部和虚部分别相等,则有

$$\sigma + \frac{a_1 - b_1}{n-m} = \sqrt[n-m]{K^*}\cos\frac{(2k+1)\pi}{n-m}$$

$$\omega = \sqrt[n-m]{K^*}\sin\frac{(2k+1)\pi}{n-m}$$

从最后两个方程中解出

$$\sqrt[n-m]{K^*} = \frac{\omega}{\sin\varphi_a} = \frac{\sigma - \sigma_a}{\cos\varphi_a} \quad (4\text{-}3\text{-}9)$$

$$\omega = (\sigma - \sigma_a)\tan\varphi_a \quad (4\text{-}3\text{-}10)$$

式中 φ_a、σ_a 得证。

在 s 平面上，式(4-3-10)代表直线方程，它与实轴的交角为 φ_a，交点为 σ_a。当 k 取不同值时，可得 $(n-m)$ 个 φ_a 角，而 σ_a 不变，因此根轨迹渐近线是 $(n-m)$ 条与实轴交点为 σ_a，交角为 φ_a 的一组射线。

Reading Material

Actually, in the real application, the angles of the asymptotes will only depend on the value of $n-m$. So, we have the following rules for your reference.

(1) $n-m=1$：1 asymptote：$-\pi$

(2) $n-m=0$：no asymptote

(3) $n-m=2$：2 asymptote：$\pm\dfrac{\pi}{2}$

(4) $n-m=3$：3 asymptote：$\pm\dfrac{\pi}{3}$

【**例 4-3-1**】 已知系统的开环传递函数为

$$G(s)H(s) = \frac{K^*(s+2)}{s^2(s+1)(s+4)}$$

试画出该系统根轨迹的渐近线。

解：对于该系统有 $n=4$，$m=1$，$n-m=3$；3 条渐近线与实轴交点位置为 $\sigma_a=-1$，它们与实轴正方向的交角分别是

$$\varphi_a = \frac{(2k+1)\pi}{n-m} = \frac{\pi}{3} \quad k=0$$

$$\varphi_a = \frac{(2k+1)\pi}{n-m} = \pi \quad k=1$$

$$\varphi_a = \frac{(2k+1)\pi}{n-m} = -\frac{\pi}{3} \quad k=2$$

渐近线如图 4-3-1 所示。

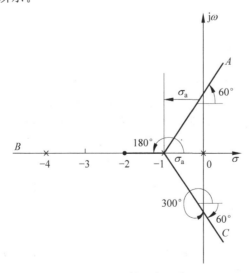

图 4-3-1 根轨迹渐近线

Terms and Concepts

Asymptote is the path the root locus follows as the parameter becomes very large and approaches infinity. The number of asymptotes is equal to the number of poles minus the number of zeros.

Asymptote centroid is the center of the linear asymptotes.

4. 根轨迹在实轴上的分布

【法则 4】 根轨迹在实轴上的分布。实轴上的某一区域,若右边开环实数零、极点个数之和为奇数,则该区域必是根轨迹。共轭复数开环极点、零点对确定实轴上的根轨迹无影响。

设 $G(s)H(s) = \dfrac{k(s-z_1)}{(s-p_1)(s-p_2)(s-p_3)}$,已知 p_1, p_2 是共轭复数极点,开环零点、极点在 s 平面上的位置如图 4-3-2 所示。在 s 平面实轴上取试验点,用相角条件检查该试验点是否在根轨迹上。首先在 z_1 与 p_3 之间选取试验点,则有

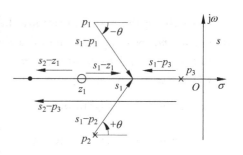

图 4-3-2 确定实轴上的根轨迹

$$\angle G(s)H(s) = \angle(s_1 - z_1) - \angle(s_1 - p_1) - \angle(s_1 - p_2) - \angle(s_1 - p_3)$$
$$= 0° - (-\theta) - \theta - 180° = -180°$$
$$\angle G(s)H(s) = \angle(s_2 - z_1) - \angle(s_2 - p_1) - \angle(s_2 - p_2) - \angle(s_2 - p_3)$$
$$= \angle(s_2 - z_1) - \angle(s_2 - p_3) = 180° - 180° = 0°$$

可见 s_2 不是根轨迹上的点。

Terms and Concepts

Root locus segments on the real axis are the root loci lying in a section of the real axis to the left of an odd number of poles and zeros.

5. 根轨迹的会合点和分离点

【法则 5】 根轨迹的分离点与分离角的坐标应满足方程(4-3-14)或方程(4-3-16)。

若干支根轨迹在复平面上某一点相遇后又分开,称该点为分离(会合)点。通常,当根轨迹分支在实轴上相交后走向复平面时,习惯上称该相交点为根轨迹的分离点;反之,当根轨迹分支由复平面走向实轴时,它们在实轴上的交点称为会合点,如图 4-3-3 所示。

证明 设系统开环传递函数为

$$G(s)H(s) = K^* \dfrac{N(s)}{D(s)} \quad (4\text{-}3\text{-}11)$$

其中,$N(s) = \prod\limits_{j=1}^{m}(s-z_j), D(s) = \prod\limits_{i=1}^{n}(s-p_i)$。

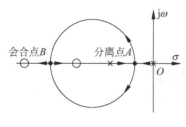

图 4-3-3 根轨迹分离点与会合点

闭环特征方程
$$f(s) = D(s) + K^* N(s) = 0 \tag{4-3-12}$$
得
$$K^* = -\frac{D(s)}{N(s)} \tag{4-3-13}$$

设特征方程有二重根 s_1，则有
$$f(s) = D(s) + K^* N(s) = (s-s_1)^2 p(s)$$
式中，$p(s)$ 是 s 的 $(n-2)$ 次多项式。

重根时满足
$$\frac{\mathrm{d}f(s)}{\mathrm{d}s} = \frac{\mathrm{d}}{\mathrm{d}s}[D(s) + K^* N(s)] = 2(s-s_1)p(s) + (s-s_1)^2 \frac{\mathrm{d}p(s)}{\mathrm{d}s}$$

所以重根、分离点和会合点满足下列方程
$$\frac{\mathrm{d}f(s)}{\mathrm{d}s} = 0 \tag{4-3-14}$$
及
$$\frac{\mathrm{d}D(s)}{\mathrm{d}s} + K^* \frac{\mathrm{d}N(s)}{\mathrm{d}s} = 0 \tag{4-3-15}$$

将式(4-3-13)代入式(4-3-15)得
$$N(s)\frac{\mathrm{d}D(s)}{\mathrm{d}s} - D(s)\frac{\mathrm{d}N(s)}{\mathrm{d}s} = 0$$
即
$$\frac{\mathrm{d}}{\mathrm{d}s}\left(\frac{D(s)}{N(s)}\right) = 0 \tag{4-3-16}$$

式(4-3-14)和式(4-3-16)是分离点和会合点应满足的方程。在它们的根中，经检验确实处于实轴的根轨迹上，并使 K^* 为正数的根，才是实际的分离点或会合点。

【例 4-3-2】 单位反馈系统的开环传递函数为
$$G(s)H(s) = \frac{k}{s(s+1)(s+2)}$$
试确定实轴上的根轨迹区间，并计算根轨迹的分离（会合）点和分离角。

解： 令 $s(s+1)(s+2)=0$，解得 3 个开环极点 $p_1=0$，$p_2=-1$，$p_3=-2$。则根轨迹分支数等于 3，3 条轨迹起点分别是 $(0,\mathrm{j}0)$、$(-1,\mathrm{j}0)$、$(-2,\mathrm{j}0)$，终点均为无穷远处。实轴上的根轨迹为 $(-\infty,-2]$ 段及 $[-1,0]$ 段。

由已知条件得
$$N(s)=1, \quad D(s)=s(s+1)(s+2)=s^3+3s^2+2s$$
则
$$f(s)=D(s)+K^* N(s)=s^3+3s^2+2s+k=0$$

由 $\dfrac{\mathrm{d}f(s)}{\mathrm{d}s}=3s^2+6s+2=0$ 解得
$$s_1=-0.422, \quad s_2=-1.578$$

根据根轨迹在实轴上的分布，可知 s_1 是实轴上的分离点，s_2 不是根轨迹上的点，故应舍去。

Terms and Concepts

Breakaway point is the point on the real axis where the locus departs from the real axis of the s-plane.

6. 根轨迹的出射角和入射角

【法则 6】 根轨迹的出射角和入射角。当开环零、极点处于复平面上时,根轨迹离开开环极点处的切线方向与正实轴的夹角称为出射角,以 θ_{p_i} 标志;根轨迹进入开环零点处切线方向与正实轴的夹角称为入射角,以 φ_{z_i} 表示,如图 4-3-4 所示。始于开环复数极点处的根轨迹出射角的一般表达式为

$$\theta_{p_1} = \pm 180° + \sum_{j=1}^{m} \angle(p_1 - z_j) - \sum_{i=2}^{n} \angle(p_1 - p_i) \quad (4\text{-}3\text{-}17)$$

止于开环复数零点处的根轨迹入射角的一般表达式为

$$\varphi_{z_1} = \pm 180° + \sum_{i=1}^{n} \angle(z_1 - p_i) - \sum_{j=2}^{m} \angle(z_1 - z_j) \quad (4\text{-}3\text{-}18)$$

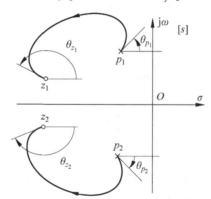

图 4-3-4 根轨迹的出射角与入射角

7. 根轨迹与虚轴的交点

【法则 7】 根轨迹与虚轴的交点。若根轨迹与虚轴相交,说明控制系统有位于虚轴上的闭环极点,即交点上的 K^* 值和 ω 值可用劳斯判据确定,也可令闭环特征方程中的 $s = \mathrm{j}\omega$,然后分别令其实部和虚部为零求得。

令 $s = \mathrm{j}\omega$,代入特征方程,即

$$1 + G(\mathrm{j}\omega)H(\mathrm{j}\omega) = 0$$

由特征方程两端的实部和虚部分别相等,有

$$\mathrm{Re}[1 + G(\mathrm{j}\omega)H(\mathrm{j}\omega)] = 0 \quad (4\text{-}3\text{-}19)$$

$$\mathrm{Im}[1 + G(\mathrm{j}\omega)H(\mathrm{j}\omega)] = 0 \quad (4\text{-}3\text{-}20)$$

联立式(4-3-19)和式(4-3-20)即可求出与虚轴交点处 K^* 值和 ω 值。

8. 根之和

【法则 8】 根之和。系统的闭环特征方程在 $n > m$ 的情况下,可有不同形式的表示

$$\prod_{i=1}^{n}(s-p_i) + K^* \prod_{j=1}^{m}(s-z_j) = s^n + a_1 s^{n-1} + \cdots + a_{n-1}s + a_n$$

$$= \prod_{i=1}^{n}(s-s_i) = s^n + \left(-\sum_{i=1}^{n}s_i\right)s^{n-1} + \cdots + \prod_{i=1}^{n}(-s_i)$$

$$= 0$$

式中，s_i 为闭环特征根。

当 $n-m \geq 2$ 时，特征方程第二项系数与 K^* 无关，无论 K^* 取何值，开环 n 个极点之和总是等于闭环特征方程 n 个根之和，即

$$\sum_{i=1}^{n}s_i = \sum_{i=1}^{n}p_i \tag{4-3-21}$$

在开环极点确定的情况下，这是一个不变的常数。所以，当开环增益 K 增大时，若闭环某些根在 s 平面上向左移动，则另一部分根必向右移动。

此法则对判断根轨迹的走向是很有用的。

Terms and Concepts

Parameter design is a method of selecting one or two parameter using the root locus method. Root locus is the locus or path of the roots traced out on the s-plane as a parameter is changed. Root locus method is the method for determining the locus of roots of the characteristic equation $1+KP(s)=0$ as K varies from 0 to infinity.

Root sensitivity is the sensitivity of the roots as a parameter changes from its normal value. The root sensitivity is the incremental change in the divided by the proportional change of the parameter.

9. 绘制根轨迹的基本规则

以上 8 条法则是绘制根轨迹图所必须遵循的基本规则，可以根据这 8 条法则，在 s 平面上绘制出大致的根轨迹图。此外，还需注意以下几点规范画法。

(1) 根轨迹的起点(开环极点 p_i)用符号"×"标示；根轨迹的终点(开环零点 z_i)用符号"○"标示。

(2) 根轨迹由起点到终点是随系统开环根轨迹增益值的增加而运动的，要用箭头"→"标示根轨迹运动的方向。

(3) 要标出一些特殊点的 K^* 值，如起点($K^*=0$)、终点($K^* \to \infty$)；根轨迹在实轴上的分离点 $d(K^*=K_d^*)$；与虚轴的交点 $\omega_c(K^*=K_c^*)$。还有一些要求标出的闭环极点 s_1 及其对应的开环根轨迹增益 K_1，也应在根轨迹图上标出，以便于进行系统的分析与综合。

Reading Material

The following features of the root locus enable us to hand sketch the root locus.

(1) The root loci are symmetrical with respect to the horizontal(real) axis.

(2) Number of separate loci is equal to the number of open loop poles(n).

(3) If number of open loop poles $n >$ number of open loop zeros m, then m sections

(or branches) of root loci start ($K=0$) at m poles and terminate (as $K\to\infty$) at the m zeros.

(4) The remaining $n-m$ branches of the root loci start ($K=0$) at the remaining $n-m$ poles, and proceed to infinity (as $K\to\infty$) along asymptotes centered at σ_A and with angles φ_A.

The centroid of the asymptotes is given by

$$\sigma_A = \frac{-p_1 - p_2 - \cdots - p_n - (-z_1 - z_2 - \cdots - z_m)}{n-m} \qquad (4\text{-}3\text{-}22)$$

and the angle of the asymptotes is

$$\varphi_A = \frac{-180° \pm k \times 360°}{n-m} \qquad (4\text{-}3\text{-}23)$$

(5) Segments of the real axis that have, on their RHS, an odd number of real open loop poles and real open loop zeros form part of the root loci.

(6) We can determine the points at which the root loci crosses the imaginary axis (if it does so) either by applying the Routh-Hurwitz criterion, or by substituting $s=j\omega$ (a purely imaginary number) and then solve for ω and K by equating the real part and imaginary parts of the resulting equation separately to zero.

(7) We can determine the points (if any) at which the root loci break away from the real axis and go into the complex plane (or break into the real axis from the complex plane).

Rearrange the characteristic equation,

$$1 + K\frac{(s+z_1)(s+z_2)\cdots(s+z_m)}{(s+p_1)(s+p_2)\cdots(s+p_n)} = 0 \qquad (4\text{-}3\text{-}24)$$

in the alternate form

$$K = -\frac{(s+p_1)(s+p_2)\cdots(s+p_n)}{(s+z_1)(s+z_2)\cdots(s+z_m)} \qquad (4\text{-}3\text{-}25)$$

Differentiate the above equation with respect to s, and equate $\dfrac{dK}{ds}$ to zero and solve the resulting equation to obtain the break points.

(8) We can determine the angle of departure of the locus from a complex pole (if such poles exist). Angle of departure from a complex pole $= k \times 360° + 180° +$ sum of angles of vectors (lines) from all zeros to the complex pole $-$ sum of angles of all vectors (lines) from all remaining poles to the complex pole.

(9) Similarly, we can determine the angle of arrival of the locus at a complex zero (if such zeros exist).

Angle of arrival at a complex zero $= k \times 360° +$ sum of angles of vectors (lines) from all poles to the complex zero $-$ sum of angles of all vectors (lines) from all remaining zeros to the complex zero.

Now, hand sketching of the root locus can now be done. If needed, you may check whether a particular point in the s-plane is a point on the root loci or not by applying the angle condition. Having drawn the root locus, you may choose any point on the gain, K, that would place a closed loop pole on the chosen point of the root loci by applying the magnitude condition.

【例 4-3-3】 负反馈系统的开环传递函数为 $G(s) = \dfrac{k(s+1)}{s^2(0.1s+1)}$,绘制根轨迹。

解:三条根轨迹,起始于 $0,0,-10$;一条终止于 -1,另两条趋于无穷远。

(1) 渐近线与实轴的交点

$$\sigma_a = \dfrac{-10+1}{2} = -4.5$$

渐近线与实轴正方向的夹角为 $\dfrac{\pi}{2}, \dfrac{3\pi}{2}$;实轴上 $[-10,-1]$ 是根轨迹。

(2) 求分离点与会合点

$$\dfrac{d}{ds}\left[\dfrac{s+1}{s^2(0.1s+1)}\right] = 0$$

$$\Rightarrow \dfrac{d(s+1)}{ds}(0.1s^3+s^2) - (s+1)\dfrac{d(0.1s^3+s^2)}{ds} = 0$$

$$\Rightarrow s(2s^2+13s+20) = 0$$

$$\Rightarrow s_1 = -2.5, s_2 = -4, s_3 = 0$$

根轨迹如图 4-3-5 所示。

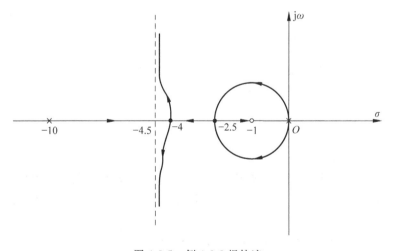

图 4-3-5 例 4-3-3 根轨迹

【例 4-3-4】 单位负反馈系统的开环传递函数为

$$G(s) = \dfrac{k(s+1)}{s(s-3)}$$

(1) 绘制系统的根轨迹。
(2) 证明实数轴以外部分的根轨迹是圆。
(3) 求出使闭环系统稳定的 k 值范围,并求出阶跃响应有衰减的振荡分量和无振荡分量时 k 值的范围。

(4) 求出 $k=10$ 时系统的闭环极点和单位阶跃响应,并说明此时系统的单位阶跃响应是否有超调。

解:(1) 根轨迹有两个分支,分别起始于 $0,3$,终止于 -1 和无穷远。实轴上 $[0,3]$,$(-\infty,-1]$ 是根轨迹。分离点和会合点求解如下:

$$\frac{\mathrm{d}(s+1)}{\mathrm{d}s}(s^2-3s)-(s+1)\frac{\mathrm{d}(s^2-3s)}{\mathrm{d}s}=0$$

$$\Rightarrow s^2+2s-3=0$$

$$\Rightarrow s_1=1, s_2=-3$$

都在根轨迹上。

特征方程为

$$s^2+(k-3)s+k=0$$

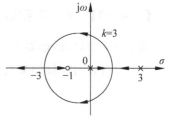

图 4-3-6 例 4-3-4 根轨迹

当 $k=3$ 时,特征根为纯虚数 $\pm\mathrm{j}\sqrt{3}$;根轨迹如图 4-3-6 所示。

(2) 设

$$s=\sigma+\mathrm{j}\omega$$

$$\angle G=\arctan\frac{\omega}{\sigma+1}-\arctan\frac{\omega}{\sigma}-\arctan\frac{\omega}{\sigma-3}=-180°$$

$$\Rightarrow 180°+\arctan\frac{\omega}{\sigma+1}=\arctan\frac{\omega}{\sigma}+\arctan\frac{\omega}{\sigma-3}$$

$$\Rightarrow \frac{\omega}{\sigma+1}=\frac{\frac{\omega}{\sigma}+\frac{\omega}{\sigma-3}}{1-\frac{\omega}{\sigma}\cdot\frac{\omega}{\sigma-3}}$$

$$\Rightarrow \sigma^2+2\sigma+\omega^2=3 \Rightarrow (\sigma+1)^2+\omega^2=2^2$$

这是以 $(-1,\mathrm{j}0)$ 为圆心,以 2 为半径的圆。

设

$$\mathrm{Im}G(s)=0$$

同样可以证明上述结论。

若开环传递函数有两个实数极点和一个零点(或两个实数零点和一个极点),只要零点(或极点)不在这两个极点(或零点)之间,则实数以外的根轨迹是以零点(或极点)为圆心的圆。

(3) 当 $k>3$ 时,特征根全都具有负实部,闭环稳定。

将 $s=-3$ 代入特征方程可求出 $k=9$。当 $k\geqslant 9$ 时,特征根全是负实数,瞬态响应中无振荡分量。当 $3<k<9$ 时阶跃响应有衰减的振荡分量。

(4) $k=10$,特征方程为

$$s^2-3s+10s+10=(s+2)(s+5)=0$$

闭环极点为 $-2,-5$。单位阶跃响应 $C(s)$ 为

$$C(s)=\frac{G(s)}{1+G(s)}R(s)=\frac{10(1+s)}{s(s+2)(s+5)}=\frac{1}{s}+\frac{1.67}{s+2}-\frac{2.67}{s+5}$$

$$c(t)=1+1.67\mathrm{e}^{-2t}-2.67\mathrm{e}^{-5t}$$

由计算或数字仿真可知,单位阶跃响应在 $t=0.5\mathrm{s}$ 时有最大超调 40%。系统无振荡分量但有超调的原因是系统有 $s=-1$ 的零点。

4.4 广义根轨迹

前面介绍的普通根轨迹或一般根轨迹的绘制规则是以开环根轨迹增益 K^* 为可变参数绘制的,大多数系统都属于这种情况。但有时,为了分析系统方便起见,或着重研究某个系统参数(如时间常数、反馈系数等)对系统性能的影响,也常常以这些参数作为可变参数绘制根轨迹,称为广义根轨迹。

1. 参数根轨迹

我们把以非开环根轨迹增益作为可变参数绘制的根轨迹称为参数根轨迹。

绘制参数根轨迹的法则与绘制常规根轨迹的法则完全相同。只要在绘制参数根轨迹之前,引入等效单位反馈系统和等效传递函数概念,则常规根轨迹的所有绘制法则,均适用于参数根轨迹的绘制。为此,需要对闭环特征方程
$$1+G(s)H(s)=0$$
进行等效变换,将其写为
$$A\frac{P(s)}{Q(s)}=-1 \tag{4-4-1}$$
其中,A 是系统中任意的变化参数,而 $P(s)$ 和 $Q(s)$ 为两个与 A 无关的首一多项式。显然
$$Q(s)+AP(s)=1+G(s)H(s)=0 \tag{4-4-2}$$
根据式(4-4-2),可得等效单位反馈系统,其等效开环传递函数为
$$G_1(s)H_1(s)=A\frac{P(s)}{Q(s)} \tag{4-4-3}$$

利用式(4-4-3)画出的根轨迹,就是参数 A 变化时的参数根轨迹。需要强调指出,等效开环传递函数是根据式(4-4-2)得来的,因此"等效"的含义仅在闭环极点相同的这一点上成立,而闭环极点一般是不同的。由于闭环零点对系统动态性能有影响,所以由闭环零、极点分布来分析和估算系统性能时,可以采用参数根轨迹上的闭环极点,但必须采用原来闭环系统的零点。这一处理方法和结论,对于绘制开环零极点变化时的根轨迹同样适用。

【**例 4-4-1**】 已知系统的开环传递函数为
$$G(s)=\frac{2}{s(s+1)(Ts+1)}$$
试绘制以时间常数 T 为可变参数的根轨迹。

解:(1) 系统的特征方程为
$$s(s+1)(Ts+1)+2=0$$
上式可改写为
$$s^2+s+2+Ts^2(s+1)=0$$
用 s^2+s+2 除等式两边得

$$1+\frac{Ts^2(s+1)}{s^2+s+2}=0$$

令

$$G_1(s)=\frac{Ts^2(s+1)}{s^2+s+2}=0$$

则有

$$1+G_1(s)=0$$

称 $G_1(s)$ 为系统的等效开环传递函数。在等效开环传递函数中,除时间常数 T 取代了普通根轨迹中开环根轨迹增益 K 的位置外,其形式与绘制普通根轨迹的开环传递函数完全一致,这样便可根据绘制普通根轨迹的 8 条基本规则来绘制参数根轨迹。

(2) 系统特征方程的最高阶次是 3,由法则 1 和法则 2 知,该系统有三条连续且对称于实轴的根轨迹,根轨迹的终点($T=\infty$)是等效开环传递函数的三个零点,即 $z_1=z_2=0$。本例中,系统的等效开环传递函数的零点数 $m=3$,极点数 $n=2$,即 $m>n$。在前面已经指出,这种情况在实际物理系统中一般不会出现,然而在绘制参数根轨迹时,其等效开环传递函数却常常出现这种情况。

与 $n>m$ 情况类似,这时可认为有 $(m-n)$ 条根轨迹起始于 s 平面的无穷远处(无限极点)。因此,本例的三条根轨迹的起点($T=0$)分别为 $p_1=-0.5+j0.866$,$p_2=-0.5-j0.866$ 和无穷远处(无限极点)。

由法则 3 知,实轴上的根轨迹是实轴上 -1 至 $-\infty$ 线段。

由法则 6 可求出两个起始角分别为

$$\theta_{p_1}=-180°+60°+120°+120°-90°=30°$$

$$\theta_{p_2}=-\theta_{p_1}=-30°$$

由法则 7 可求出根轨迹与虚轴的两个交点,将 $s=j\omega$ 代入特征方程得

$$-jT\omega^3-(T+1)\omega^2+j\omega+2=0$$

由此得到虚部方程和实部方程分别为

$$\begin{cases}\omega-T\omega^3=0\\2-(T+1)\omega^2=0\end{cases}$$

解虚部方程得 ω 的合理值为 $\omega_c=\pm\sqrt{\dfrac{1}{T}}$,代入实部方程求得 $T_c=1\text{s}$,所以 $\omega_c=\pm 1$ 为根轨迹与虚轴的两个交点。

由根轨迹图可知,当时间常数 $T=T_c=1\text{s}$ 时,系统处于临界稳定状态;当 $T>1\text{s}$ 时,根轨迹在 s 平面右半部,系统不稳定。由此可知,参数根轨迹在研究非开环根轨迹增益对系统性能的影响是很方便的。

由上面的例子,可将绘制参数根轨迹的方法归纳为以下两个步骤。

(1) 先根据系统的特征方程 $1+G(s)H(s)=0$ 求出系统的等效开环传递函数 $G_1(s)H_1(s)$,使 $G_1(s)H_1(s)$ 与绘制普通根轨迹的开环传递函数有相同的形式,即

$$G_1(s)H_1(s)=K^{*\prime}\frac{\prod\limits_{j=1}^{m}|s-z_j|}{\prod\limits_{i}^{n}|s-p_i|} \tag{4-4-4}$$

其中，$K^{*'}$为除开环根轨迹增益K^*以外的任何参数，它是绘制参数根轨迹的可变参数。

(2) 根据绘制普通根轨迹的7条基本规则和等效开环传递函数$G_1(s)H_1(s)$绘制出系统的参数根轨迹。

2. 正反馈系统的根轨迹

正反馈系统的根轨迹方程为
$$G(s)H(s) = 1 \qquad (4\text{-}4\text{-}5)$$
由此可得到绘制正反馈系统根轨迹的幅值条件和相角条件分别为
$$|G(s)H(s)| = 1 \qquad (4\text{-}4\text{-}6)$$
$$\angle G(s)H(s) = 2k\pi, \quad k = 0, \pm 1, \pm 2, \cdots \qquad (4\text{-}4\text{-}7)$$

与负反馈系统根轨迹的幅值条件和相角条件相比较可知，正反馈系统和负反馈系统的幅值条件相同；负反馈系统的根轨迹遵循$180°$相角条件，而正反馈系统的根轨迹遵循$0°$相角条件，故正反馈系统根轨迹又称为零度根轨迹。由于相角条件不同，在绘制正反馈系统根轨迹时，需对前面介绍的绘制负反馈系统普通根轨迹的8条基本规则中与相角条件有关的三条规则作相应修改，这些规则如下。

(1) 正反馈系统根轨迹的渐近线与实轴正方向的夹角应为
$$\varphi_a = \frac{(2k+1)\pi}{n-m} \quad k = 0, 1, 2, \cdots, n-m-1 \qquad (4\text{-}4\text{-}8)$$

(2) 正反馈系统在实轴上的根轨迹是那些在其右侧的开环实零点和开环实极点的总数为偶数或零的线段。

(3) 正反馈系统的起始角为φ_a，终止角为
$$\theta_{pl} = -\sum_{\substack{j=1 \\ j \neq l}}^{m} \angle(s - z_j) + \sum_{i=1}^{n} \angle(s - p_j) \qquad (4\text{-}4\text{-}9)$$

除以上三条外，其余规则与$180°$根轨迹相同。

【例 4-4-2】 正反馈系统的开环传递函数为
$$G(s)H(s) = \frac{K^*}{(s^2+2s+2)(s^2+2s+5)}$$
绘制系统的根轨迹，并求出使系统稳定的K的取值范围。

解：按零度根轨迹的画法规则，系统的开环极点为
$$p_{1,2} = -1 \pm \mathrm{j}$$
$$p_{3,4} = -1 \pm 2\mathrm{j}$$
实轴上的根轨迹为$(-\infty, \infty)$，根轨迹有4条渐近线。

渐近线与实轴的交点、夹角分别为
$$\sigma_a = -1$$
$$\varphi_a = 0°, 90°, 180°, 270°$$
起始角为

$$\theta_{P_1}=90°, \theta_{P_2}=-90°, \theta_{P_3}=270°, \theta_{P_4}=270°$$

由分离点方程

$$\frac{1}{d+1+j2}+\frac{1}{d+1-j2}+\frac{1}{d+1+j}+\frac{1}{d+1-j}=0$$

得根轨迹的分离点

$$d=-1$$

系统的特征方程为

$$D(s)=s^4+4s^3+11s^2+14s+10-K^*=0$$

由劳斯判据可知,当 $K^*=10$ 时,闭环系统临界稳定,根轨迹与虚轴的交点为 $s=0$,根轨迹如图 4-4-1 所示。

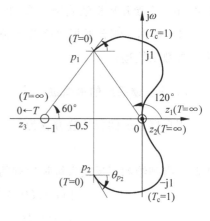

图 4-4-1 系统根轨迹图

4.5 系统性能的根轨迹法分析

自动控制系统的稳定性,由它的闭环极点唯一确定,其动态性能与系统的闭环极点和零点在 s 平面上的分布有关。因此,确定控制系统闭环极点和零点在 s 平面上的分布,特别是从已知的开环零、极点的分布确定闭环零、极点的分布,是对控制系统进行分析必须首先要解决的问题。第 3 章介绍的解析法是解决的方法之一,即求出系统特征方程的根。解析法虽然比较精确,但对四阶以上的高阶系统是很困难的。

4.5.1 根轨迹分析法概述

根轨迹法是解决上述问题的另一途径,它是在已知系统的开环传递函数零、极点分布的基础上,研究某一个和某些参数的变化对系统闭环极点分布的影响的一种图解方法。由于根轨迹图直观、完整地反映系统特征方程的根在 s 平面上分布的大致情况,所以通过一些简单的作图和计算,就可以看到系统参数的变化对系统闭环极点的影响趋势,这对分析研究控制系统的性能和提出改善系统性能的合理途径都具有重要意义。

【例 4-5-1】 已知单位反馈系统的开环传递函数为

$$G(s)H(s)=\frac{K^*}{s(s+1)(s+2)(s+3)}$$

试根据系统的根轨迹方法分析系统的稳定性和计算闭环主导极点具有阻尼比 $\zeta=0.5$ 时系统的动态性能指标。

解:(1) 先根据系统的开环传递函数和绘制根轨迹的基本规则绘制出系统的根轨迹图,如图 4-5-1 所示。

系统的特征方程是

$$s(s+1)(s+2)(s+3)+K^*=s^4+6s^3+11s^2+6s+K^*=0$$

由法则 1 和法则 2 知该系统有 4 条连续且对称于实轴的根轨迹,起点分别是系统的 4 个开环极点,即 $p_1=0, p_2=-1, p_3=-2, p_4=-3$,且 4 条根轨迹都趋向无穷远处。

由法则 3 知实轴上的根轨迹是 $0\sim-1$ 线段和 $-2\sim$

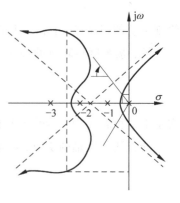

图 4-5-1 系统根轨迹图

−3 线段。

由法则 4 可求出 4 条渐近线与实轴的交点为 −1.5,它们与实轴正方向的夹角分别是 $\pm 45°,\pm 135°$。

由法则 6 可求出根轨迹与实轴的两个交点(分离点)分别为 $d_1=-0.38, d_2=-2.62$。

由劳斯判据求根轨迹与虚轴的交点,先根据特征方程列出劳斯表:

s^4	1	11	K^*
s^3	6	6	0
s^2	10	K^*	0
s^1	$\dfrac{60-6K^*}{10}$	0	
s^0	K^*		

(2) 系统稳定性分析

由根轨迹图知,有两条从 s 平面左半部穿过虚轴进入 s 平面右半部,它们与虚轴的交点 $\omega_c=\pm 1$,且交点处对应的临界开环根轨迹增益 $K_c^*=10$。由开环根轨迹增益 K_c^* 与系统开环放大系数 K 之间的关系可求出系统稳定的临界开环放大系数 $K_c=10/6=1.67$。

(3) 系统动态性能

在根轨迹图上,首先求出满足阻尼比 $\zeta=0.5$ 时系统的主导极点 s_1、s_2 的位置(假定 s_1、s_2 满足主导极点的条件)。方法是作等阻尼比线 OA,使 OA 与实轴负方向的夹角为 $\beta = \arccos\zeta = \arccos 0.5 = 60°$,等阻尼比线 OA 与根轨迹的交点 s_1,即为满足阻尼比 $\zeta=0.5$ 系统的一个闭环极点(即系统特征方程的一个根)。测得 s_1 在 s 平面上的坐标位置为

$$s_1 = -0.3 + j0.52$$

由根轨迹的对称性得到另一共轭复数极点为

$$s_2 = -0.3 - j0.52$$

由幅值条件可求出闭环极点所对应的系统开环根轨迹增益 K_{r1}^* 为

$$K_{r1}^* = |s_1||s_1+1||s_1+2||s_1+3| = 6.35$$

将 s_1、s_2 和 K_{r1}^* 代入特征方程,由根和系数之间的关系很容易得到另外两个闭环极点 s_3、s_4,它们也是一对共轭复数极点

$$s_{3,4} = -2.7 \pm j3.37$$

由此可计算出

$$\frac{\mathrm{Re}[s_{3,4}]}{\mathrm{Re}[s_{1,2}]} = \frac{-2.7}{-0.3} = 9$$

共轭复数极点 $s_{3,4}$ 与虚轴的距离是共轭复数极点 $s_{1,2}$ 与虚轴的距离的 9 倍,且闭环极点附近无闭环零点,这说明 s_1、s_2 满足主导极点的条件,该系统可近似成由闭环主导极点构成的一个二阶系统,其闭环传递函数为

$$\phi(s) = \frac{\omega_n^2}{s^2+2\zeta\omega_n s+\omega_n^2} = \frac{s_1 s_2}{(s-s_1)(s-s_2)} = \frac{0.36}{s^2+0.6s+0.36}$$

此时,对应的系统开环增益为

$$K_v = \frac{K_{r1}}{6} = 1.06$$

系统的动态性能可根据二阶系统的性能指标公式计算。

调整时间

$$t_s = \frac{3+\ln\frac{1}{\sqrt{1-\zeta^2}}}{\omega_n \zeta} = \frac{3+\ln\frac{1}{\sqrt{1-0.5^2}}}{0.6 \times 0.5} = 10.5$$

超调量

$$\sigma_p\% = e^{-\frac{\zeta\pi}{\sqrt{1-\zeta^2}}} = e^{-\frac{0.5\pi}{\sqrt{1-0.5^2}}} = 16.3\%$$

峰值时间

$$t_p = \frac{\pi}{\omega_n\sqrt{1-\zeta^2}} = \frac{\pi}{0.6\sqrt{1-0.5^2}} = 6.04$$

系统的暂态特性取决于闭环零、极点的分布,因而和根轨迹的形状密切相关。而根轨迹的形状又取决于开环零、极点的分布。那么开环零、极点对根轨迹形状的影响如何,这是单变量系统根轨迹法的一个基本问题。知道了闭环极点以及闭环零点(通常闭环零点是容易确定的),就可以对系统的动态性能进行定性分析和定量计算。

4.5.2 增加开环极点对控制系统的影响

大量实例表明:增加位于 s 左半平面的开环极点,将使根轨迹向右半平面移动,系统的稳定性能降低。例如,设系统的开环传递函数为

$$G_k(s) = \frac{K^*}{s(s+a_1)} \quad a_1 > 0 \tag{4-5-1}$$

则可绘制系统的根轨迹,如图 4-5-2(a)所示。若增加一个开环极点 $p_3 = a_2$,根据这时的开环传递函数

$$G_{k1}(s) = \frac{K_1}{s(s+a_1)(s+a_2)} \quad a_2 > 0 \tag{4-5-2}$$

可绘制系统的根轨迹,如图 4-5-2(b)所示。

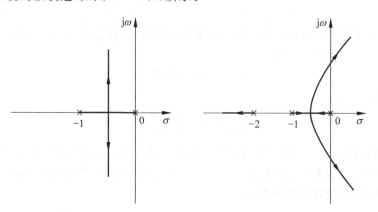

(a) 系统的根轨迹   (b) 增加开环极点后系统的根轨迹

图 4-5-2 增加开环极点对根轨迹的影响

由图 4-5-2 可见:增加开环极点,使根轨迹的复数部分向右半平面弯曲。若取 $a_1=1$、$a_2=2$,则渐近线的倾角由原来的 $\pm 90°$ 变为 $\pm 60°$;分离点由原来的 -0.5 向右移至 -0.422;与分离点相对应的开环增益,由原来的 0.25(即 $K^*=0.5\times0.5=0.25$)减少到

0.19(即 $K_1^* = \frac{1}{2} \times 0.422 \times 0.578 \times 1.578 = 0.19$)。这意味着，对于具有同样振荡倾向的控制系统，增加开环极点后会使开环增益值下降。一般来说，增加的开环极点越靠近虚轴，其影响越大，使根轨迹向右半平面的弯曲就越严重，因而系统稳定性能的降低便越明显。

4.5.3 增加开环零点对控制系统的影响

一般来说，开环传递函数 $G(s)H(s)$ 增加零点，相当于引入微分作用，使根轨迹向 s 平面的左半部分移动，将提高系统的稳定性。例如，图 4-5-2(b)表示式(4-5-1)增加一个零点 $z = -2$ 的根轨迹(并设 $a_1 = 1$)，轨迹向 s 平面左半部分移动，且成为一个圆，结果使控制系统的稳定性提高，如图 4-5-3(a)所示。图 4-5-3(b)是式(4-5-1)增加一对共轭复数零点的根轨迹。

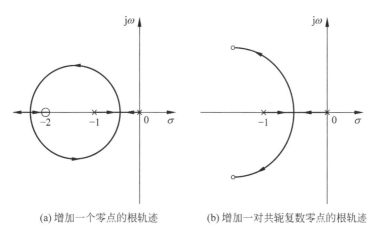

(a) 增加一个零点的根轨迹　　(b) 增加一对共轭复数零点的根轨迹

图 4-5-3　增加开环零点对根轨迹的影响

4.5.4 结论

通过上面的示例可以将用根轨迹分析自动控制系统的方法和步骤归纳如下。

(1) 根据系统的开环传递函数和绘制根轨迹的基本规则绘制出系统的根轨迹图。

(2) 由根轨迹在 s 平面上的分布情况分析系统的稳定性。①如果全部根轨迹都位于 s 平面左半部，则说明无论开环根轨迹增益为何值，系统都是稳定的；②如果根轨迹有一条（或一条以上）的分支全部位于平面的右半部，则说明无论开环根轨迹增益如何改变，系统都是不稳定的；③如果有一条（或一条以上）根轨迹从 s 平面的左半部穿过虚轴进入 s 平面的右半部（或反之），而其余的根轨迹分支位于 s 平面的左半部，则说明系统是有条件的稳定系统，即当开环根轨迹增益大于临界值时系统便由稳定变为不稳定（或反之）。此时的关键是求出开环根轨迹增益的临界值。这为分析和设计系统的稳定性提供了选择合适系统参数的依据和途径。

(3) 根据对系统的要求和系统的根轨迹图分析系统的瞬态响应指标。①对于一阶、二阶系统，很容易在它的根轨迹上确定对应参数的闭环极点；②对于三阶以上的高阶系统，通常用简单的作图法（如作等阻尼比线等）求出系统的主导极点（如果存在的话），将高阶系统近似地简化成由主导极点(通常是一对共轭复数极点)构成的二阶系统,最后求出其各项的性能指标。这种分析方法简单、方便、直观,在满足主导极点条件时,分析结果的误差很小。如果求出离虚轴较近的一对共轭复数极点不满足主导极点的条件(如它到虚轴的距离不小于其余极点到虚轴距离的 1/5 或在它的附近有闭环零点存在等),这时必须进一步考虑和分析这些闭环零、极点对系统瞬态响应性能指标的影响。

4.6 MATLAB 在本章中的应用

在 4.5 节中,我们学习了控制系统根轨迹的绘制。虽然按照传统的规则作图,可以得到系统的根轨迹,但是即使使用了所有的规则,手工获得的轨迹还是不够精确。在 MATLAB 的工作环境下,可以使用多个函数进行根轨迹分析和设计。下面以英文阅读材料的形式,分别介绍根轨迹的定义、根轨迹的绘制和根轨迹设计中的指标设置。

Reading Material

1. Closed-Loop Poles

The root locus of a/an(open-loop) transfer function $H(s)$ is a plot of the locations (locus) of all possible closed loop poles with proportional gain K and unity feedback. Fig. 4-6-1 gives a typical closed-loop control system.

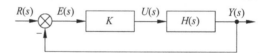

Fig. 4-6-1　A Closed-Loop system

The closed-loop transfer function is

$$\frac{Y(s)}{R(s)} = \frac{KH(s)}{1+KH(s)} \tag{4-6-1}$$

and thus the poles of the closed loop system are values of s such that $1+KH(s)=0$. If we write $H(s)=b(s)/a(s)$, then this equation has the form

$$a(s)+Kb(s)=0 \tag{4-6-2}$$

$$\frac{a(s)}{K}+b(s)=0 \tag{4-6-3}$$

Let $n=$ order of $a(s)$ and $m=$ order of $b(s)$.

We will consider all positive values of K. In the limit as $K \to 0$, the poles of the closed-loop system are $a(s)=0$ or the poles of $H(s)$. In the limit as $K \to \infty$, the poles of the closed-loop system are $b(s)=0$ or the zeros of $H(s)$.

No matter what we pick K to be, the closed-loop system must always have n poles, where n is the number of poles of $H(s)$. The root locus must have n branches, each branch starts at a pole of $H(s)$ and goes to a zero of $H(s)$. If $H(s)$ has more poles than zeros(as is often the case), $m<n$ and we say that $H(s)$ has zeros at infinity. In this case, the limit of $H(s)$ as $s \to \infty$ is zero. The number of zeros at infinity is $n-m$, the number of poles minus the number of zeros, and is the number of branches of the root locus that go to infinity(asymptotes).

Since the root locus is actually the locations of all possible closed loop poles, from the root locus we can select a gain such that our closed-loop system will perform the way we want. If any of the selected poles are on the right half plane, the closed-loop system will be unstable. The poles that are closest to the imaginary axis have the greatest influence on the closed-loop response, so even though the system has three or four poles, it may still act like a second or even first order system depending on the location(s) of the dominant pole(s).

2. Plotting the Root Locus of a Transfer Function

Example 4-6-1 Consider an open loop system which has a transfer function of

$$H(s) = \frac{Y(s)}{U(s)} = \frac{s+7}{s(s+5)(s+15)(s+20)}$$

How do we design a feed-back controller for the system by using the root locus method? Say our design criteria are 5% overshoot and 1 second rise time. Make a MATLAB file called rl.m. Enter the transfer function, and the command to plot the root locus like Fig. 4-6-2.

```
num = [1 7];
den = conv(conv([1 0],[1 5]),conv([1 15],[1 20]));
rlocus(num,den);axis([-22 3 -15 15]);
```

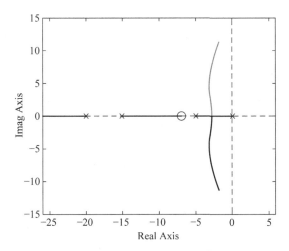

Fig. 4-6-2 Root locus of example 4-6-1

1) Choosing a value of K from the root locus

The plot above shows all possible closed-loop pole locations for a pure proportional controller. Obviously not all of those closed-loop poles will satisfy our design criteria. To determine what part of the locus is acceptable, we can use the command sgrid(Zeta, Omgn) to plot lines of constant damping ratio and natural frequency. Its two arguments are the damping ratio(Zeta) and natural frequency(ω_n: Ω). In our problem, we need an overshoot less than 5% (which means a damping ratio ξ of greater than 0.7) and a rise time of 1 second(which means a natural frequency ω_n greater than 1.8). Enter in the MATLAB command window and you will have Fig. 4-6-3.

```
Zeta = 0.7;
Omgn = 1.8;sgrid(Zeta, Omgn)
```

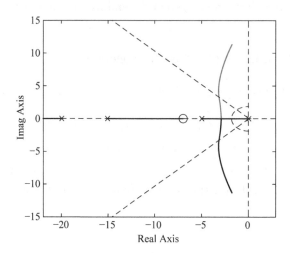

Fig. 4-6-3 Root locus with the typical parameters

On the plot above, the two white dotted lines at about a 45° degree angle indicate pole locations with Zeta = 0.7; in between these lines, the poles will have Zeta > 0.7 and outside of the lines Zeta < 0.7. The semicircle indicates pole locations with a natural frequency $\omega_n = 1.8$; inside the circle, $\omega_n < 1.8$ and outside the circle $\omega_n > 1.8$.

Going back to our problem, to make the overshoot less than 5%, the poles have to be in between the two white dotted lines, and to make the rise time shorter than 1 second, the poles have to be outside of the white dotted semicircle. So now we know only the part of the locus outside of the semicircle and in between the two lines are acceptable. All the poles in this location are in the left-half plane, so the closed-loop system will be stable.

From the plot above we see that there is part of the root locus inside the desired region. So in this case we need only a proportional controller to move the poles to the desired region. You can use rlocfind command in MATLAB to choose the desired poles on the locus.

```
[kd,poles] = rlocfind(num,den)
```

Click on the Fig. 4-6-4 the point where you want the closed-loop pole to be. You may want to select the points indicated in the plot below to satisfy the design criteria.

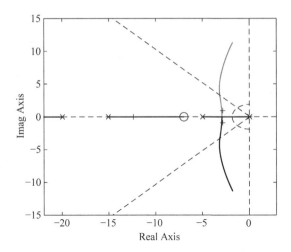

Fig. 4-6-4　Root locus with the rlocfind command

Note that since the root locus may has more than one branch, when you select a pole, you may want to find out where the other pole(poles) are. Remember they will affect the response too. From the plot above we see that all the poles selected(all the white "+") are at reasonable positions. We can go ahead and use the chosen kd as our proportional controller.

2) Closed-loop response

In order to find the step response, you need to know the closed-loop transfer function. You could compute this using the rules of block diagrams, or let MATLAB do it for you.

```
[numCL, denCL] = cloop((kd) * num, den)
```

The two arguments to the function cloop are the numerator and denominator of the open-loop system. You need to include the proportional gain that you have chosen. Unity feedback is assumed.

If you have a non-unity feedback situation, look at the help file for the MATLAB function feedback, which can find the closed-loop transfer function with a gain in the feedback loop.

Check out the step response of your closed-loop system in Fig. 4-6-5.

```
step(numCL,denCL)
```

As we expected, this response has an overshoot less than 5% and a rise time less than 1 second.

上文介绍了在 MATLAB 的工作环境下根轨迹的分析和设计。下面以英文阅读材料的形式,选择一个典型的例子,利用根轨迹法进行比例控制器和滞后控制器的设计。这些设计原理在后续章节中还会深入介绍。

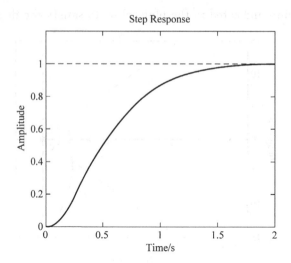

Fig. 4-6-5　Simulation result of Closed-loop response

Reading Material

Example 4-6-2　Solution to the cruise control problem using root locus method

The open-loop transfer function for this problem is:

$$\frac{Y(s)}{U(s)} = \frac{1}{ms+b}$$

where

$m = 1000$;

$b = 50$;

$U(s) = 10$;

$Y(s) =$ velocity output.

The design criteria are

$$\text{Rise time} < 5\text{s}$$
$$\text{Overshoot} < 10\%$$
$$\text{Steady state error} < 2\%$$

1. Proportional Controller

The root-locus plot shows the locations of all possible closed-loop poles when a single gain is varied from zero to infinity. Thus, only a proportional controller (K_p) will be considered to solve this problem.

Example 4-6-3　The closed-loop transfer function becomes

$$\frac{Y(s)}{U(s)} = \frac{K_p}{ms + (b + K_p)}$$

Also, from the root-locus tutorial, we know that the MATLAB command called sgrid should be used to find an acceptable region of the root-locus plot. To use the sgrid, both the damping ratio (zeta) and the natural frequency (ω_n) need to be determined first. The following two equations will be used to find the damping ratio and the natural frequency.

$$\omega_M \geq \frac{1.8}{T_r}$$

$$\xi = \sqrt{\frac{\left(\frac{\ln M_p}{\pi}\right)^2}{1+\left(\frac{\ln M_p}{\pi}\right)^2}}$$

where

ω_n = Natural frequency;

zeta = Damping ratio;

T_r = Rise time;

M_p = Maximum overshoot.

One of our design criteria is to have a rise time of less than 5 seconds. From the first equation, we see that the natural frequency must be greater than 0.36. Also using the second equation, we see that the damping ratio must be greater than 0.6, since the maximum overshoot must be less than 10%.

Now, we are ready to generate a root-locus plot and use the sgrid to find an acceptable region on the root-locus. Create a new m-file and enter the following commands with Fig. 4-6-6.

```
hold off;m = 1000;b = 50;u = 10;
numo = [1];deno = [m b];

figure
hold;axis([-0.6  0  -0.6  0.6]);
rlocus(numo,deno);sgrid(0.6, 0.36)
[Kp, poles] = rlocfind(numo,deno)

figure
hold;numc = [Kp];denc = [m (b + Kp)];
t = 0:0.1:20;
step(u * numc,denc,t)
axis([0 20 0 10])
```

Running this m-file should give you the following root-locus plot similar to Fig. 4-6-6.

The two dotted lines in an angle indicate the locations of constant damping ratio(zeta = 0.6); the damping ratio is greater than 0.6 in between these lines and less than 0.6 outside the lines. The semi-ellipse indicates the locations of constant natural frequency (ω_n = 0.36); the natural frequency is greater than 0.36 outside the semi-ellipse, and smaller than 0.36 inside.

If you look at the MATLAB command window, you should see a prompt asking you to pick a point on the root-locus plot. Since you want to pick a point in between dotted lines(zeta>0.6) and outside the semi-ellipse(ω_n>0.36), click on the real axis just outside

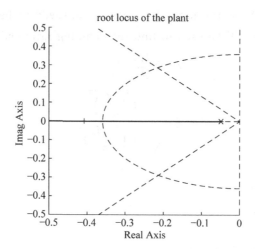

Fig. 4-6-6　Root locus of example 4-6-3

the semi-ellipse(around−0.4).

You should see the gain value(K_p) and pole locations in the MATLAB command window. Also you should see the closed-loop step response similar to the one shown in Fig. 4-6-7.

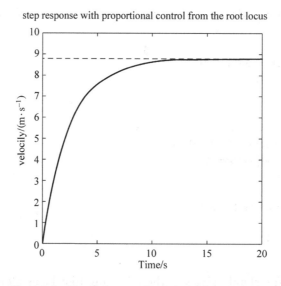

Fig. 4-6-7　Simulation result of closed-loop response

2. Lag Controller

Example 4-6-4　To reduce the steady-state error, a lag controller will be added to the system. The transfer function of the lag controller is

$$G(s) = \frac{s + z_0}{s + p_0}$$

The open-loop transfer function(not including K_p)now becomes

$$\frac{Y(s)}{U(s)} = \frac{s+z_0}{ms^2+(b+mp_0)s+bp_0}$$

The pole and the zero of a lag controller need to be placed close together. Also, it states that the steady-state error will be reduce by a factor of z_0/p_0. For these reasons, let z_0 equals—0.3 and p_0 equals—0.03.

Create a new m-file, and enter the following commands.

```
hold off;
m = 1000;b = 50;u = 10;
Zo = 0.3;Po = 0.03;
numo = [1 Zo];deno = [m b + m * Po b * Po];

figure
hold;axis([-0.6  0  -0.4  0.4])
rlocus(numo,deno)
sgrid(0.6,0.36)
[Kp, poles] = rlocfind(numo,deno)

figure
t = 0:0.1:20; numc = [Kp Kp * Zo];
denc = [m b + m * Po + Kp b * Po + Kp * Zo];
axis([0 20 0 12])
step(u * numc,denc,t)
```

Running this m-file should give you the root-locus plot similar to Fig. 4-6-8.

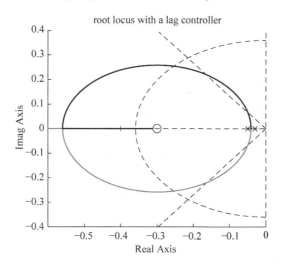

Fig. 4-6-8 Root locus of example 4-6-4

In the MATLAB command window, you should see the prompt asking you to select a point on the root-locus plot. Once again, click on the real axis around—0.4. You should have the following response in Fig. 4-6-9.

As you can see, the steady-state error has been reduced to near zero. Slight overshoot is a result of the zero added in the lag controller.

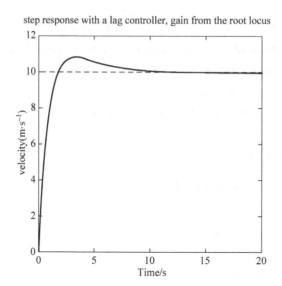

Fig. 4-6-9　Simulation result of Closed-loop response

Now all of the design criteria have been met and no further iterations will be needed. For more details, please refer to Reference[24].

本章小结

（1）根轨迹是一种图解方法，它在已知控制系统开环零点和极点的基础上，研究某一个或某些参数变化时系统闭环极点在 s 平面的分布情况。利用根轨迹法能够分析结构和参数已确定的系统的稳定性及动态响应特性，还可以根据对系统动态特性的要求确定可变参数，调整开环零、极点的位置甚至改变它们的数目，因此，根轨迹法在控制系统的分析和设计中是一种很实用的工程方法。

（2）本章介绍了根轨迹的两个基本条件（幅值条件和相角条件），以及如何利用这两个基本条件确定根轨迹上的点及相应的增益值。

（3）掌握绘制根轨迹的基本规则。对于简单的系统，能够熟练运用这些规则很快地画出根轨迹的概略图形，对于一些特殊点（如分离点或会合点等），如与分析问题无关，则不必准确求出，只要能找出它们所在的范围即可。

（4）对于结构和参数已确定的系统，能够用根轨迹法分析出主要特性。掌握闭环主导极点与动态性能之间的关系，对于主导极点以外的其他闭环极点和零点，应能定性分析出它们对动态性能的影响。

Summary and Outcome Checklist

The root locus is a powerful tool for designing and analysing feedback control systems. It provides graphical information on how the locations of the closed loop poles change as a system parameter is varied. An approximate sketch of the locus can be used to obtain qualitative information concerning the stability and performance of the system. Root locus plots, generated using MATLAB, are very useful in the design process.

Tick the box for each statement with which you agree.

☐ I can explain the meaning of root locus.

☐ I can arrange the characteristic equation in the standard form for polting the root locus by hand, and identify the "open loop ploes and zeros".

☐ I am able to state the two condition, namely the angle condition and the magnitude condition, that apply to each point of the s-plane that on the root locus of a given system.

☐ I can list the orderly procedure of twelve steps for sketching by hand the root locus of a given system.

☐ I can apply the orderly procedure to hand sketch the root locus of a given feedback control system.

☐ Given the root locus plot and a point on the plot, I am able to determine the gain that would place a closed loop pole on given point by applying the magnitude condition.

4-1 试讨论开环传递函数 $G(s)=\dfrac{K(s+z)}{(s+p_1)(s+p_2)}$ 的根轨迹特性。

4-2 设单位反馈系统的开环传递函数为
$$G(s)=\dfrac{K(s+2)}{s(s+1)}$$
试从数学上证明：复数根轨迹部分是以 $(-2,\mathrm{j}0)$ 为圆心，以 $\sqrt{2}$ 为半径的一个圆。

4-3 设单位反馈系统的开环传递函数为
$$G(s)=\dfrac{K^*}{s(s+1)(s+3.5)(s^2+6s+13)}$$
试绘制系统的概略根轨迹。

4-4 设系统的开环传递函数
$$G(s)H(s)=\dfrac{K^*}{s(s+4)(s^2+4s+20)}$$
试概略绘制闭环系统的根轨迹。

4-5 已知系统的开环传递函数
$$G(s)=\dfrac{K(s+2)}{s(s+3)(s^2+2s+2)}$$
试绘制系统的根轨迹。

4-6 负反馈系统的开环传递函数为
$$G(s)H(s)=\dfrac{K}{(s+1)(s+2)(s+4)}$$
试证明：$s_1=-1+\mathrm{j}\sqrt{3}$ 在该系统的根轨迹上，并求出相应的 K 值。

4-7 某单位反馈系统开环传递函数

$$G(s) = \frac{K^*}{s(s+1)(s+5)}$$

试概略绘制系统根轨迹。

4-8 某单位反馈系统开环传递函数

$$G(s) = \frac{K^*}{s(s+1)(s+2)}$$

试概略绘制系统根轨迹,并求临界根轨迹增益及该增益对应的三个闭环极点。

4-9 单位反馈系统开环传递函数

$$G(s) = \frac{\frac{1}{4}(s+a)}{s^2(s+1)}$$

试绘制 $a=0 \to \infty$ 时的根轨迹。

4-10 已知某单位反馈系统的开环传递函数

$$G(s)H(s) = \frac{K^*(s+1)(s+3)}{s^3}$$

试绘出 $-\infty < K^* < +\infty$ 时系统的根轨迹。

4-11 已知单位反馈系统的开环传递函数

$$G(s) = \frac{K}{s(s+1)(0.5s+1)}$$

试用根轨迹法确定系统在稳定欠阻尼状态下的开环增益 K 的范围,并计算阻尼比 $\xi=0.5$ 的 K 值以及相应的闭环极点,估算此时系统的动态性能指标。

4-12 已知某单位反馈系统的开环传递函数

$$G(s) = \frac{K^*}{s(s+3)^2}$$

(1)绘制该系统以根轨迹增益 K^* 为变量的根轨迹(求出:渐近线、分离点与虚轴的交点等)。

(2)确定使系统满足 $0 < \xi < 1$ 的开环增益 K 的取值范围。

4-13 单位反馈系统的开环传递函数

$$G(s) = \frac{K^*}{(s+1)^2(s+4)^2}$$

(1)画出根轨迹。
(2)能否通过选择 K^* 满足最大超调量 $\sigma\% \leqslant 4.32\%$ 的要求?
(3)能否通过选择 K^* 满足调节时间 $t_s \leqslant 2s$ 的要求?
(4)能否通过选择 K^* 满足误差系数 $K_p \geqslant 10$ 的要求?

4-14 (西北工业大学硕士研究生入学考试试题)系统结构图如题 4-14 图所示。

(1)分别画出常数 $a>0, a<0$ 两种情况下, $K = 0 \to \infty$ 变化时系统的根轨迹(求分离点与虚轴交点)。

(2)在保证系统单位阶跃响应稳态值 $h(\infty)=2$ 的条件下,确定使系统稳定且为欠阻尼状态的 a 值及

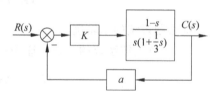

题 4-14 图

K 的取值范围。

4-15 (西北工业大学硕士研究生入学考试试题)已知某单位反馈系统的开环传递函数
$$G(s) = \frac{K_0}{s(s+3)^2}$$
(1) 绘制当 $K_0 = 0 \to \infty$ 变化时系统根轨迹(求出渐近线、分离点与虚轴的交点)。
(2) 确定开环增益 K 的取值范围,使系统同时满足以下条件:
a. 全部闭环极点均位于 s 平面中 $s = -0.5$ 左侧的区域内;
b. 阻尼比(对应闭环复极点)$\xi \geq 0.707$。

4-16 (上海交通大学硕士研究生入学考试试题)单位反馈系统的开环传递函数为
$$G(s) = \frac{K(s^2 - 2s + 5)}{(s+2)(s-1)}$$
(1) 试绘制根轨迹图。
(2) 确定系统稳定时 K 值的最大值与最小值。
(3) 确定使闭环传递函数具有阻尼比 $\xi = 0.5$ 的复数极点的 K 值。

4-17 (南京航空航天大学硕士研究生入学考试试题)某单位反馈系统的开环传递函数为
$$G(s) = \frac{\frac{1}{4}(s+a)}{s^2(s+1)}$$
(1) 试绘制参数 a 从 $0 \to \infty$ 变化的闭环根轨迹,并确定系统稳定的 a 值范围。
(2) 当系统的一对共轭复根对应的阻尼比 $\xi = 0.707$ 时,试确定系统的闭环传递函数 $\Phi(s)$,要求写出零、极点的乘积形式。

4-18 (浙江大学硕士研究生入学考试试题)设单位负反馈系统的开环传递函数为
$$G(s) = \frac{K}{s(s+6)(s^2+6s+45)}$$
(1) 试用根轨迹法画出该系统的根轨迹,并讨论本系统根轨迹的分离点情况。
(2) 求闭环系统稳定的 K 值范围。

第5章 线性系统的频域分析法

5.1 引言

在第 3 章的控制系统时域分析中,系统的输入信号一般是阶跃信号或者是斜坡信号。然而,在实际控制系统中的复杂信号可以表示成不同频率正弦信号的合成。控制系统的频率特性反映了正弦信号作用下系统响应的性能。应用频率特性研究线性系统的经典方法称为频域分析法(或频域特性法)。在经典控制理论中,它的地位等同于时域分析法和根轨迹法,具有如下特点。

(1) 频率特性具有明确的物理意义,可以将系统参数、系统结构变化和系统性能指标统一进行研究。

(2) 频率特性法的计算量较小,一般可采用近似的作图方法,简单、直观,易于在工程技术领域使用。

(3) 可以采用实验的方法求出系统或元件的频率特性,这对于机理复杂或机理不明而难以列写微分方程的系统或元件,具有重要的实用价值。

此外,频域分析法还可以用来设计某些抑制某些频段带有严重噪声的系统。因此,频域分析法是一种为广大工程技术人员所熟悉并广泛应用的有效方法。

Reading Material

In previous chapters, we examined the use of test signals such as a step and a ramp signal. In this chapter we consider the steady state response of a linear dynamic system to a sinusoidal input test signal.

An important advantage of the frequency response method is the ready availability of sinusoidal test signals for various ranges of frequencies and amplitudes. Thus the experimental determination of the frequency response of a system is easily accomplished and is the most reliable and uncomplicated method for the experimental analysis of a system. Useful performance measures based on the frequency response are the system bandwidth and the attenuation characteristics at high frequencies. These measures permit the designer to control the response of a system to undesired noise and disturbances.

The steady state response of a stable linear dynamic system to a sinusoidal input signals is an output sinusoidal signal at the same frequency as the input sinusoidal signal.

However, the magnitude and phase of the output sinusoidal signal in general differ from those of the input sinusoidal signal, and these differences are functions of the input frequency. We will explore these functional relationships in the following activity.

We will also consider different ways of presenting the frequency response data and identify some of the very useful frequency domain performance measures.

本章的主要内容：
- 频率特性的基本概念
- 典型环节的频率特性及特性曲线的绘制
- 频域稳定判据及稳定裕量
- 频率特性与控制系统性能的关系
- MATLAB 在本章中的应用

5.2 频率特性的基本概念

5.2.1 频率特性的定义

一般来讲，在正弦输入信号的作用下，系统输出的稳态分量称为频率响应。而系统的稳态输出随频率变化（ω 由 0 变到 ∞）的特性，称为该系统的频率特性。

如前所述，频率特性的研究需要将输入信号由第 3 章中的阶跃信号转换为正弦信号。考虑如图 5-2-1 所示的 RC 电路，设 RC 电路的初始条件为零，其传递函数为 $G(s)$，输入信号 $r(t)=A\sin\omega t$，则有

$$G(s) = \frac{C(s)}{R(s)} = \frac{1}{Ts+1} \tag{5-2-1}$$

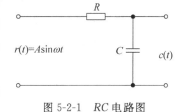

图 5-2-1　RC 电路图

式中，$T=RC$ 为时间常数。

根据正弦信号的拉普拉斯变换有

$$R(s) = \frac{A\omega}{s^2+\omega^2} = \frac{A\omega}{(s+\mathrm{j}\omega)(s-\mathrm{j}\omega)} \tag{5-2-2}$$

所以可得到输出

$$C(s) = G(s)R(s) = \frac{1}{Ts+1} \cdot \frac{A\omega}{(s+\mathrm{j}\omega)(s-\mathrm{j}\omega)} \tag{5-2-3}$$

对式(5-2-3)进行拉普拉斯反变换，可得

$$c(t) = \frac{A\omega T}{1+\omega^2 T^2}\mathrm{e}^{-\frac{t}{T}} + \frac{A}{\sqrt{1+\omega^2 T^2}}\sin(\omega t - \arctan\omega T) \tag{5-2-4}$$

由式(5-2-4)可见，式中第一项是输出的瞬态分量。当时间 $t\to\infty$ 时，瞬态分量趋于零，所以上述电路的稳态响应可以表示为

$$\begin{aligned} c(t) &= \frac{A}{\sqrt{1+\omega^2 T^2}}\sin(\omega t - \arctan\omega T) \\ &= A\left|\frac{1}{1+\mathrm{j}\omega T}\right|\sin(\omega t - \arctan\omega T) \end{aligned} \tag{5-2-5}$$

以上分析表明,当电路的输入为正弦信号时,其输出的稳态响应(频率响应)也是一个正弦信号,其频率和输入信号的频率相同,但幅值和相位发生了变化,其变化取决于频率 ω:

$$e^{j\omega t} = \cos\omega t + j\sin\omega t$$

在式(5-2-5)中,若利用数学中的欧拉公式,把输出的稳态响应和输入的正弦信号用复数表示,并求出它们的复数比,可以得

$$G(j\omega) = \frac{C'}{R'} = \frac{1}{1+j\omega T} \tag{5-2-6}$$

由式(5-2-6)可见,这个复数比不仅与电路参数 T 有关,还与输入电压的频率 ω 有关,因此称为系统的频率特性。频率特性 $G(j\omega)$ 仍是一个复数,它可以写成

$$G(j\omega) = |G(j\omega)| e^{j\varphi(\omega)} = L(\omega)e^{j\varphi(\omega)} \tag{5-2-7}$$

其中

$$\begin{cases} L(\omega) = |G(j\omega)| = \dfrac{1}{\sqrt{1+\omega^2 T^2}} & (5\text{-}2\text{-}8) \\ \varphi(\omega) = \arg G(j\omega) = -\arctan\omega T & (5\text{-}2\text{-}9) \end{cases}$$

式中,$L(\omega) = |G(j\omega)|$ 称为系统幅频特性,表示频率特性的幅值与频率的关系,是输出与输入信号的幅值之比。$\varphi(\omega)$ 称为系统相频特性,表示频率特性的相位与频率的关系,是输出与输入的相位之差。因此,可以由式(5-2-5)定义频率特性为:线性系统(或环节)在正弦函数的作用下稳态输出与输入之比。频率特性分为幅频特性和相频特性。同时,根据式(5-2-6)也可以看出,系统的频率特性 $G(j\omega)$ 与其传递函数在结构上很相似,不难证明,频率特性与传递函数之间有着确切的转换关系,即

$$G(s)\big|_{s=j\omega} = G(j\omega) = |G(j\omega)| e^{j\arg G(j\omega)} \tag{5-2-10}$$

一般在研究系统的频率特性时,只需将传递函数中 s 以 $j\omega$ 代换即可。现在来证明这种本质关系。

设闭环传递函数为 $G(s)$,则系统输出为

$$C(s) = G(s)R(s) \tag{5-2-11}$$

当 $r(t) = A\sin\omega t$ 时,则

$$R(s) = \frac{A\omega}{s^2 + \omega^2} \tag{5-2-12}$$

故

$$C(s) = \sum_{i=1}^{n} \frac{p_i}{s - s_i} + \left(\frac{M}{s+j\omega} + \frac{N}{s-j\omega}\right) \tag{5-2-13}$$

式中,s_i 为闭环特征根(设无重根);M、N 为待定系数。

$$M = -\frac{A}{2j}G(-j\omega) \tag{5-2-14}$$

$$N = \frac{A}{2j}G(j\omega) \tag{5-2-15}$$

对式(5-2-13)进行拉普拉斯反变换,系统的正弦函数响应为

$$c(t) = \sum_{i=1}^{n} p_i e^{s_i t} + (Me^{-j\omega t} + Ne^{j\omega t}) \tag{5-2-16}$$

式(5-2-16)右边后两项代表了系统的稳态响应,第一项为瞬态响应。对于稳态的系统,当 $t\to\infty$ 时,瞬态响应逐渐消失,系统最终以

$$c_{ss}(t) = Me^{-j\omega t} + Ne^{j\omega t} \tag{5-2-17}$$

做稳态运动。

将式(5-2-14)和式(5-2-15)代入式(5-2-17),则有 $G(j\omega)$ 和 $G(-j\omega)$ 是共轭复数,并利用数学中的欧拉公式,可推得

$$c_{ss}(t) = R|G(j\omega)|\sin(\omega t + \varphi) = C\sin(\omega t + \varphi) \tag{5-2-18}$$

式中,$G(j\omega)$ 就是令 $G(s)$ 中的 s 等于 $j\omega$ 所得到的复量;$|G(j\omega)|$ 为复量的模或称幅值,$\varphi=\angle G(j\omega)$ 是输出信号对于输入信号的相位移,它就等于复量 $G(j\omega)$ 的相位;$C=R|G(j\omega)|$ 是稳态响应 $c_s(t)$ 的幅值。

因此,对于稳定的线性定常系统,若传递函数为 $G(s)$,当输入量 $r(t)$ 是正弦信号时,其稳态响应是同一频率的正弦信号。此时称稳态响应的幅值 C 与输入信号的幅值 R 之比 $C/R=|G(j\omega)|$ 为系统的幅频特性,称 $c_{ss}(t)$ 与 $r(t)$ 之间的相位移 $\varphi=\angle G(j\omega)$ 为系统的相频特性,它们都是 ω 的函数。幅频特性和相频特性统称为频率特性或频率响应。

在复变函数中,复量 $G(j\omega)$ 可以写成三种形式,分别是指数式、三角式或实部与虚部相加的代数式

$$\begin{aligned}G(j\omega) &= |G(j\omega)|e^{j\varphi(\omega)} = |G(j\omega)|[\cos\varphi + j\sin\varphi] \\ &= U(\omega) + jV(\omega)\end{aligned} \tag{5-2-19}$$

式中,$U(\omega)$ 是 $G(j\omega)$ 的实部,又称实频特性;$V(\omega)$ 是 $G(j\omega)$ 的虚部,又称虚频特性。而相位角 $\varphi(\omega)$ 为

$$\varphi(\omega) = \angle G(j\omega) = \begin{cases} \arctan\dfrac{V(\omega)}{U(\omega)}, & U(\omega) > 0 \\ \pi + \arctan\dfrac{V(\omega)}{U(\omega)}, & U(\omega) < 0 \end{cases} \tag{5-2-20}$$

相位角 $\varphi(\omega)$ 本来是多值函数,为了方便起见,在计算基本环节的相位角 $\varphi(\omega)$ 时,一般取 $-180°<\varphi(\omega)<180°$。

在频率特性中,负的相位角称为相位滞后,正的相位角称为相位超前。具有负的相位角的网络称为滞后网络,具有正的相位角的网络就称为超前网络。在第 6 章自动控制系统校正中,超前网络和滞后网络将起到重要的作用。

关于频率特性这里需要说明以下几点。

(1) 频率特性不仅是对系统而言,其概念和定义对于各种控制元件也都适用。

(2) 频率特性的概念和定义只适用于线性定常系统或元件,否则不能用拉普拉斯变换求解,也不存在这种特殊的稳态对应关系。

(3) 上述理论的证明是在假定系统稳定的情况下导出的。若系统不稳定,输出响应 $c(t)$ 最终不可能趋于稳态振荡。然而,从理论上讲,$c(t)$ 中的稳态分量总是可以分离出来的,所以上述理论同样适用于不稳定的系统。

(4) 若已知系统的传递函数为 $G(s)$,令 $s=j\omega$,便可求得相应的频率特性表达式为 $G(j\omega)$。尽管频率特性是一种稳定响应,但动态过程的规律性也全部包含其中,所以它也是

控制系统或元部件的一种数学模型。

5.2.2 频率特性的表示方法

当自动控制系统较为复杂时,其频率特性的数学解析表达式较为烦琐,使用起来非常不方便。在工程上常用图形来表示频率特性,常用的表示方法有三种:幅相频率特性图、对数频率特性图和对数幅相图。

Reading Material

The frequency response method may be less intuitive than other methods you have studied previously. However, it has certain advantages, especially in real-life situations such as modeling transfer functions from physical data.

The frequency response of a system can be viewed two different ways: via the Bode plot or via the Nyquist diagram. Both methods display the same information; the difference lies in the way the information is presented. We will study both methods in this chapter.

The frequency response is a representation of the system's response to sinusoidal inputs at varying frequencies. The output of a linear system to a sinusoidal input is a sinusoid of the same frequency but with a different magnitude and phase. The frequency response is defined as the magnitude and phase differences between the input and output sinusoids. In this chapter, we will see how we can use the open-loop frequency response of a system to predict its behavior in closed-loop.

To plot the frequency response, we create a vector of frequencies (varying between zero and infinity) and compute the value of the plant transfer function at those frequencies. If $G(s)$ is the open loop transfer function of a system and ω is the frequency vector, we then plot $G(j\omega)$ vs. ω. Since $G(j\omega)$ is a complex number, we can plot both its magnitude and phase (the Bode plot) or its position in the complex plane (the Nyquist plot). More information is available on plotting the frequency response.

1. 幅相频率特性曲线图(极坐标图)

在复平面上,一个复数可以用一个点或者一个矢量来表示。幅相频率特性曲线图又称向量轨迹图、奈奎斯特图或极坐标图。绘制幅相频率特性曲线图时,把 ω 看作参变量,令 ω 由 0 变到 ∞ 时,在复平面上描绘出 $G(j\omega)$ 的轨迹,就是 $G(j\omega)$ 的幅相频率特性曲线。向量的长度表示 $G(j\omega)$ 的幅值 $|G(j\omega)|$,由正实轴方向沿逆时针方向绕原点转至向量方向的角度称为相位角,即 $\angle G(j\omega)$。一般要求在轨迹上标出 ω 值和频率变化的方向。

绘制极坐标图的根据就是式(5-2-18)。一般而言,手工绘制极坐标图较为烦琐。因此,大部分情况下不必逐点准确绘图,只要画出简图。具体做法是:找出 $\omega=0$ 及 $\omega\to\infty$ 时 $G(j\omega)$ 的位置,以及另外的一两个点或关键点(如转折频率点),再把它们连接起来并标上 ω 的变化情况,就形成极坐标简图。绘制极坐标简图的主要根据是相频特性 $\varphi(\omega)$,同时参考幅频特性 $|G(j\omega)|$。有时也要利用实频特性和虚频特性。

极坐标图的优点是在一张图上就可以较容易地得到全部频率范围内的频率特性,利用图形可以较容易地对系统进行定性分析。缺点是不能明显地表示各个环节对系统的影响和作用。后续章节中,将专门介绍基于极坐标图的稳定性判据。一般情况下,会直接给出系统的极坐标图,不要求手工绘制。

Terms and Concepts

Polar plot is a plot of the real part of $G(j\omega)$ versus the imaginary part of $G(j\omega)$.

2. 对数频率特性图

对数频率特性图又称伯德(Bode)图,其优点是易于绘制,而且容易看出参数变化和结构变化对系统性能的影响。因此,Bode图在频率特性法设计中应用非常广泛。

Bode图一般由对数幅频特性和对数相频特性两张图组成。为了能在很宽的频率范围内描绘频率特性,坐标刻度采用对数化的形式。对数幅频特性图中的纵坐标为 $L(\omega) = 20\lg|G(j\omega)|$,其单位为分贝(dB),采用线性分度,横坐标采用对数分度,表示角频率 ω。相频特性图中的纵坐标表示频率特性的相角,以度(°)为单位,采用线性分度,横坐标与对数幅频特性相同,用对数分度表示角频率 ω。横坐标用对数分度,就能在极宽的频率范围内,更好地表示系统或元件的低频特性与高频特性。但应该注意,横坐标表示的最低频率一般由我们感兴趣的频率确定。在对数频率特性中,常用到频率 ω 倍数的概念。频率变化10倍的区间称为一个十倍频程,记为 decade 或简写为 dec;ω 变化至两倍的区间叫做二倍频程,记为 octave,或简写为 oct,如图5-2-2所示。

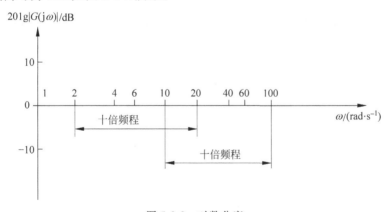

图 5-2-2 对数分度

由于幅频特性图中纵坐标是幅值的对数 $20\lg|G(j\omega)|$,如果传递函数可以写成基本环节传递函数相乘除的形式,那么它的幅频特性就可以由相应的基本环节幅频特性的代数和得到,明显简化了计算和作图过程。此外,幅频特性图中往往采用直线代替复杂的曲线,所以对数幅频特性图容易绘制。

Terms and Concepts

Bode plot: The logarithm of the magnitude of the transfer function is plotted versus the logarithm of ω, the frequency. The phase φ of the transfer function is separately plotted versus the logarithm of the frequency.

Decibel(dB): The units of the logarithmic gain.

Logarithmic magnitude: The logarithmic of the magnitude of the transfer function, usually expressed in units of 20dB.

Here, frequency response data are presented as two graphs. In the first, the magnitude ratio(in decibels)is plotted against frequency on a semilog graph sheet. In the second, the phase difference φ(in degrees)is plotted against frequency directly under the 1st graph.

Bode magnitude plot for first, and second-order transfer functions can be approximated by straight line segments in the lower and higher frequency regions. By virtue of the definition of logarithmic gain, we can convert the multiplicative factors in the transfer function into additive factors. Using asymptotic approximations, Bode diagrams can be easily hand sketched on a semi-log graph sheet, even for complex transfer functions. One can estimate the transfer function of a system from the Bode magnitude plot.

3. 对数幅相图

对数幅相图是以相角 $\varphi(\omega)$ 为横坐标,以对数幅频 $L(\omega)$(dB)为纵坐标绘出的 $G(j\omega)$ 曲线。因此它与幅相特性曲线一样,在曲线适当位置要标出 ω 的值,并且要用箭头标出 ω 增加的方向。用对数幅频特性及相频特性取得数据来绘制对数幅相图是很方便的。对数幅相图在后续章节中不再作详细介绍,下面仅以惯性环节 $G(s)=\dfrac{1}{Ts+1}$ 为例,给出对数幅相图,如图 5-2-3 所示。

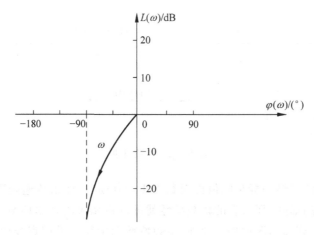

图 5-2-3 惯性环节的对数幅相图

Terms and Concepts

Nichols chart: A chart displaying the curves for the relationship between the open-loop and closed-loop frequency response.

In this case, the data set at each frequency is represented by a point in the Magnitude Ratio(in decibels)-Phase Difference plane.

5.3 典型环节的频率特性及特性曲线的绘制

在自动控制系统中,每一个典型环节的频率特性都可以用 5.2 节的三种表示方法来表示。依据前述章节中极坐标图和 Bode 图的优点,复杂自动控制系统频率特性的研究可以通过组成它的典型环节来进行。因此,各种典型环节的频率特性尤为重要。下面分别介绍各种典型环节频率特性的极坐标图和 Bode 图。

1. 比例环节

比例环节又称放大环节,其传递函数为

$$G(s) = K \tag{5-3-1}$$

其频率特性为

$$G(j\omega) = K \tag{5-3-2}$$

则幅频特性 $L(\omega)$ 和相频特性 $\varphi(\omega)$ 分别为

$$\begin{cases} L(\omega) = 20\lg K \\ \varphi(\omega) = 0° \end{cases} \tag{5-3-3}$$

显然,其极坐标图不随频率 ω 的变化而变化,只是一个定点 K,如图 5-3-1 所示。同时,其 Bode 图如图 5-3-2 所示。它的对数幅频特性为幅值等于 $20\lg K\mathrm{dB}$ 的一条水平直线,相频特性为零度线。因此,改变系统传递函数的增益,将使 Bode 图中的对数幅频特性曲线升高或降低某一常量,但不改变相频特性曲线。

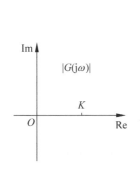

图 5-3-1 比例环节的极坐标图

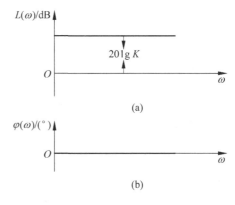

图 5-3-2 比例环节的 Bode 图

2. 积分环节

积分环节的传递函数为

$$G(s) = \frac{1}{s} \tag{5-3-4}$$

其频率特性为

$$G(j\omega) = \frac{1}{j\omega} = \frac{1}{\omega}e^{j(-90°)} \tag{5-3-5}$$

则

$$\begin{cases} L(\omega) = 20\lg\dfrac{1}{\omega} = -20\lg\omega \\ \varphi(\omega) = -90° \end{cases} \tag{5-3-6}$$

当 $\omega \to 0$ 时，$L(0) \to \infty$；当 $\omega \to \infty$ 时，$L(\infty) \to 0$，其极坐标图如图 5-3-3 所示。

同时，由式(5-3-6)可知，ω 每增加 10 倍，$L(\omega)$ 下降 20dB，所以积分环节的对数幅频特性是一条斜率为十倍频程 -20dB 的直线，此斜率记为 -20dB/dec。对数相频特性曲线则与 ω 无关，恒为 $90°$，所以其 Bode 图如图 5-3-4 所示。

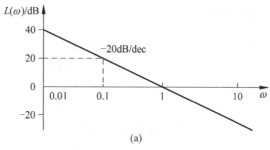

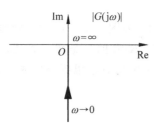

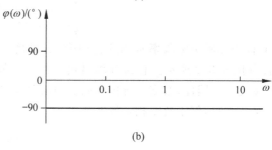

图 5-3-3　积分环节极坐标图　　　　图 5-3-4　积分环节 Bode 图

特殊情形 1：如果 n 个积分环节串联，则传递函数为

$$G(s) = \frac{1}{s^n} \tag{5-3-7}$$

对数幅频特性为

$$20\lg|G(j\omega)| = 20\lg\frac{1}{\omega^n} = -20n\lg\omega \tag{5-3-8}$$

它是一条斜率为 $\angle G(j\omega) = -n \cdot 20$dB/dec 的直线，并在 $\omega = 1$ 处穿越 0dB 线。因为

$$\angle G(j\omega) = -n \cdot 90° \tag{5-3-9}$$

所以它的相频特性是通过纵轴上 $-n \cdot 90°$ 且平行于横轴的直线。

特殊情形 2：如果一个放大环节 K 和 n 个积分环节串联，则整个环节的传递函数和频率特性分别为

$$G(s) = \frac{K}{s^n} \tag{5-3-10}$$

$$G(j\omega) = \frac{K}{j^n\omega^n} \tag{5-3-11}$$

相频特性如式(5-3-9),对数幅频特性为

$$20\lg|G(j\omega)| = 20\lg\frac{K}{\omega^n} = 20\lg K - 20n\lg\omega \tag{5-3-12}$$

这是斜率为 $-20n\text{dB/dec}$ 的直线,它在 $\omega=\sqrt[n]{K}$ 处穿越 0dB 线;它也通过 $\omega=1$、$20\lg|G(j\omega)|=20\lg K$ 这一点。

3. 惯性环节

惯性环节的传递函数为

$$G(s) = \frac{1}{Ts+1} \tag{5-3-13}$$

频率特性为

$$G(j\omega) = \frac{1}{j\omega T+1} = \frac{1}{\sqrt{1+\omega^2 T^2}} e^{j(-\arctan\omega T)} \tag{5-3-14}$$

则

$$\begin{cases} L(\omega) = 20\lg|G(j\omega)| = -20\lg\sqrt{1+\omega^2 T^2} \\ \varphi(\omega) = \angle G(j\omega) = -\arctan\omega T \end{cases} \tag{5-3-15}$$

当 $\omega \to 0$ 时,$L(0)=1$,$\varphi(0)=0°$,随着 ω 增加,$L(\omega)$ 减小,$\varphi(\omega)$ 也减小;当 $\omega \to \infty$ 时,$L(\infty) \to 0$,$\varphi(\infty) \to -90°$,其极坐标图为半圆,如图 5-3-5 所示。

当绘制其 Bode 图时,由于精确的幅频特性曲线是一条较为复杂的曲线,一般可以采取渐近线法。由式(5-3-15)可以看出:①当 $\omega \ll \frac{1}{T}$(即 $\omega T \ll 1$)时,惯性环节的对数幅频特性可近似地表示为

$$-20\lg\sqrt{1+\omega^2 T^2} \approx -20\lg 1 = 0(\text{dB}) \tag{5-3-16}$$

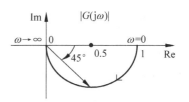

图 5-3-5 惯性环节的极坐标图

②当 $\omega \gg \frac{1}{T}$(即 $\omega T \gg 1$)时,幅频特性近似表示为

$$-20\lg\sqrt{1+\omega^2 T^2} = -20\lg\omega T(\text{dB}) \tag{5-3-17}$$

因此,惯性环节的对数幅频特性可近似成两段直线,$\omega=\frac{1}{T}$ 为转折频率,如图 5-3-6 所示。当 $\omega<\frac{1}{T}$ 时,取 0dB 水平线;当 $\omega>\frac{1}{T}$ 时,取 -20dB/dec 的直线。由式(5-3-15)逐点绘制出对数相频特性曲线,如图 5-3-6 所示。由图 5-3-6 可知,随着 ω 的增加,$\varphi(\omega)$ 由 $0°$ 逐渐趋向 $-90°$,并且曲线对 $-45°$ 点 $\left(\omega=\frac{1}{T}\right)$ 具有奇对称性质。

幅频特性曲线与渐近线图形如图 5-3-6 所示。它们在 $\omega=\frac{1}{T}$ 附近的误差较大,误差值由式(5-3-16)和式(5-3-17)来计算,典型数值列于表 5-3-1 中,最大误差发生在 $\omega=\frac{1}{T}$ 处,误差为 -3dB。渐近线容易画,误差也不大,所以绘制惯性环节的对数幅频特性曲线时,一般都绘制渐近线。绘渐近线的关键是找到转折频率 $\frac{1}{T}$。低于转折频率的频段,渐近线是

0dB 线；高于转折频率的部分，渐近线是斜率为 -20dB/dec 的直线。必要时可根据表 5-3-1 或式(5-3-16)对渐近线进行修正以得到精确的幅频特性曲线。

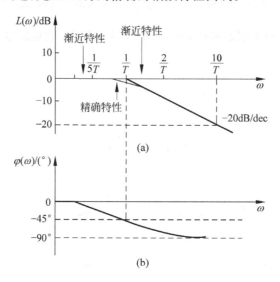

图 5-3-6 惯性环节的 Bode 图

表 5-3-1 惯性环节渐近线幅频特性误差表

ωT	0.1	0.25	0.4	0.5	1.0	2.0	2.5	4.0	10
误差/dB	-0.04	-0.26	-0.65	-1.0	-3.01	-1.0	-0.65	0.26	-0.04

由表 5-3-1 可见，最大误差为 3dB。一般情况下，渐近线即可满足控制系统的分析与设计。除特殊要求外，不需要绘制精确的频率特性曲线。

4. 一阶微分环节

一阶微分环节的传递函数为

$$G(s) = \tau s + 1 \tag{5-3-18}$$

其频率特性为

$$G(j\omega) = \tau j\omega + 1 \tag{5-3-19}$$

$$\begin{cases} L(\omega) = \sqrt{1+\omega^2\tau^2} \\ \varphi(\omega) = \arctan\omega\tau \end{cases} \tag{5-3-20}$$

当 $\omega=0$ 时，$L(0)=1$；当 $\omega\to\infty$ 时，$L(\infty)\to\infty$，其极坐标图如图 5-3-7 所示。

用绘制惯性环节 Bode 图的近似方法，绘制出一阶微分环节的对数幅频特性和对数相频特性如图 5-3-8 所示。从图 5-3-8 中可以看出一阶微分方程环节与惯性环节的对数幅频特性和相频特性以 ω 轴互为镜像。

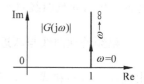

图 5-3-7 一阶微分环节的极坐标图

5. 二阶微分环节

二阶微分环节的传递函数为

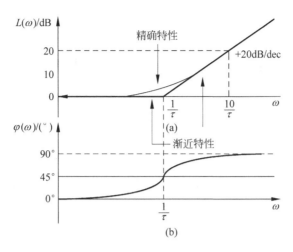

图 5-3-8 一阶微分环节的 Bode 图

$$G(s) = \tau^2 s^2 + 2\xi\tau s + 1 \quad (\xi < 1) \tag{5-3-21}$$

其频率特性为

$$G(j\omega) = 1 - \tau^2\omega^2 + j2\xi\tau\omega \tag{5-3-22}$$

则

$$\begin{cases} L(\omega) = 20\lg\sqrt{(1-\omega^2\tau^2)^2 + (2\xi\omega\tau)^2} \\ \varphi(\omega) = \begin{cases} \arctan\dfrac{2\xi\tau\omega}{1-\tau^2\omega^2}, & \omega \leqslant 1/\tau \\ 180° + \arctan\dfrac{2\xi\tau\omega}{1-\tau^2\omega^2}, & \omega > 1/\tau \end{cases} \end{cases} \tag{5-3-23}$$

当 $\omega=0$ 时,$L(0)=0$,$\varphi(0)=0°$;当 $\omega\to\infty$ 时,$\varphi(\infty)=180°$;当 $\omega=\dfrac{1}{\tau}$ 时,$G(j\dfrac{1}{\tau})=j2\xi$。图 5-3-9 所示为二阶微分环节的极坐标图。

同时,当 $\omega\ll\dfrac{1}{\tau}$ 时忽略 ωT,则

$$L(\omega) = 20\lg 1 = 0 \tag{5-3-24}$$

当 $\omega\gg\dfrac{1}{\tau}$ 时,忽略 1 和 $2\xi\tau\omega$,则

$$L(\omega) = 20\lg\tau^2\omega^2 = 40\lg\tau\omega \tag{5-3-25}$$

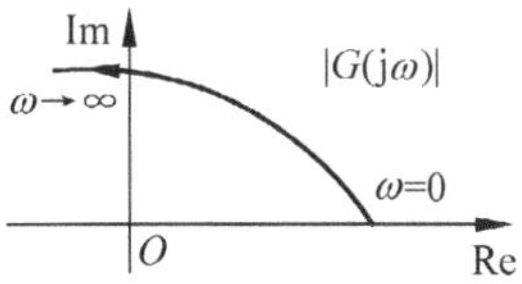

图 5-3-9 二阶微分环节的极坐标图

因此,对数幅频特性曲线可近似为一折线,在 $\omega<\dfrac{1}{\tau}$ 频率段为 0dB 线,在 $\omega>\dfrac{1}{\tau}$ 频率段为斜率 40dB/dec 的斜线。再逐点求值描绘出对数相频特性曲线,二阶微分环节的 Bode 图,如图 5-3-10 所示。

6. 振荡环节

振荡环节的传递函数为

$$G(s) = \dfrac{1}{T^2 s^2 + 2\xi Ts + 1} \tag{5-3-26}$$

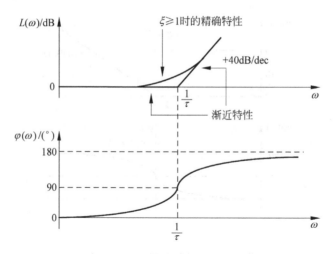

图 5-3-10 二阶微分环节的 Bode 图

式中,$\frac{1}{T}=\omega_n>0$。式(5-3-26)与欠阻尼二阶系统的传递函数相同。其频率特性为

$$G(s) = \frac{1}{T^2 s^2 + 2\xi T s + 1}$$

$$G(j\omega) = \frac{1}{(1-T^2\omega^2) + j2\xi T\omega} \tag{5-3-27}$$

则

$$\begin{cases} L(\omega) = -20\lg\sqrt{(1-T^2\omega^2)^2 + (2\xi T\omega)^2} \\ \varphi(\omega) = \begin{cases} -\arctan\dfrac{2\xi T\omega}{1-T^2\omega^2}, & \omega \leqslant \dfrac{1}{T} \\ -180° - \arctan\dfrac{2\xi T\omega}{1-T^2\omega^2}, & \omega > \dfrac{1}{T} \end{cases} \end{cases} \tag{5-3-28}$$

当 $\omega \to 0$ 时,$L(0)=1,\varphi(0)=0°$;当 $\omega \to \infty$ 时,$L(\infty) \to 0,\varphi(\infty) = -180°$;当 $\omega = \dfrac{1}{T}$ 时,$G\left(j\dfrac{1}{T}\right) = -j\dfrac{1}{2\xi}$。如图 5-3-11 所示表示不同阻尼 ξ 值的极坐标图。

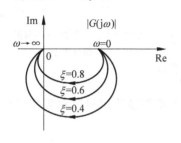

图 5-3-11 振荡环节极坐标图

同时,①当 $\omega \ll \dfrac{1}{T}$ 时,忽略 $T\omega$,则

$$L(\omega) \approx -20\lg 1 = 0 \tag{5-3-29}$$

②当 $\omega \gg \dfrac{1}{T}$ 时,忽略 1 和 $2\xi T\omega$,则

$$L(\omega) \approx -20\lg T^2\omega^2 = -40\lg T\omega \tag{5-3-30}$$

因此,对数幅频特性曲线可近似为一折线,在 $\omega < \dfrac{1}{T}$ 频率段,为 0dB 线;在 $\omega > \dfrac{1}{T}$ 频率段,是斜率为 -40dB/dec 的斜线,如图 5-3-12 所示,相频特性曲线需要逐点求值描绘。但它有两个明显的特点:①相频特性曲线对 $-90°$ 点有奇对称性质;②相频特性曲线形状与阻尼比 ξ 有关。但对任何 ξ 值

都存在：当 $\omega=0$ 时，$\varphi=0°$；当 $\omega=\dfrac{1}{T}$ 时，$\varphi=-90°$；当 $\omega\to\infty$ 时，$\varphi=-180°$，如图 5-3-12 所示。

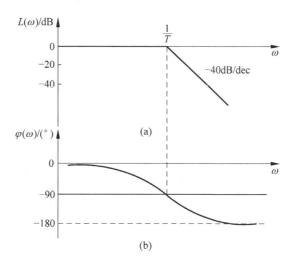

图 5-3-12　振荡环节的近似 Bode 图

在振荡环节中取不同的 ξ，对应不同的特征，如图 5-3-13 所示。正如第 3 章中讨论的，当 ξ 变化时，系统的动态特性变化主要体现在二阶系统转折频率处。当 ξ 较小时，转折频率处的峰值较高，这一点在控制系统分析与设计中需要密切关注。

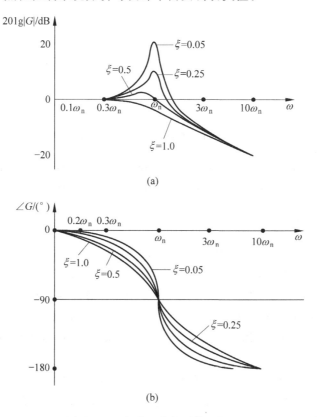

图 5-3-13　振荡环节的对数坐标图

7. 延迟环节

延迟环节的传递函数为

$$G(s) = e^{-\tau s} \tag{5-3-31}$$

其频率特性为

$$G(j\omega) = e^{-\tau j\omega} \tag{5-3-32}$$

则

$$\begin{cases} L(\omega) = -20\lg 1 = 0\text{dB} \\ \varphi(\omega) = -\tau\omega(\text{弧度}) = -57.3°\tau\omega(°) \end{cases} \tag{5-3-33}$$

可见,当 $\omega \to \infty$ 时,$\varphi(\infty) = -\infty$,而幅值恒为 1,其极坐标图为单位圆,如图 5-3-14 所示。延迟环节的对数频率特性如图 5-3-15 所示。其中对数幅频特性恒为 0dB。

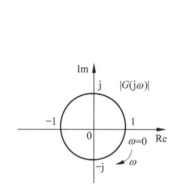

图 5-3-14 延迟环节的极坐标图

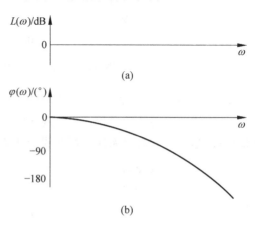

图 5-3-15 延迟环节 Bode 图

Terms and concepts

Time delay: A pure time delay, T, so that event occurring at time t at one point in the system occur at another point in the system at a later time, $t+T$.

8. 最小相位系统

如果一个环节的传递函数的极点和零点的实部全都小于或等于零,则称这个环节是最小相位环节。如果传递函数中具有正实部的零点或极点,或有延迟环节 $e^{-\tau s}$,这个环节就是非最小相位环节。对于闭环系统,如果它的开环传递函数的极点和零点的实部小于或等于零,则称它是最小相位系统。如果开环传递函数中有正实部的零点或极点,或有延迟环节 $e^{-\tau s}$,则称系统是非最小相位系统。若把 $e^{-\tau s}$ 用零点和极点的形式近似表达时,会发现它也具有正实部零点。

在一些幅频特性相同的环节之间存在着不同的相频特性,其中最小相位环节的相位移(相位角的绝对值或相位变化量)最小,也最容易控制。设系统(或环节)传递函数分母的阶次(s 的最高幂次数)是 n,分子的阶次是 m,串联积分环节的个数是 v。对于最小相位系统,

当 $\omega \to \infty$ 时,对数幅频特性的斜率为 $-20(n-m)$ dB/dec,相位等于 $-(n-m)\cdot 90°$;当 $\omega \to 0$ 时,相位等于 $-v\cdot 90°$。符合上述特征的系统一定是最小相位系统。而且,对于开环极点都在左半 s 平面上且 $n \geqslant m$ 的系统,所有具有相同开环对数幅频特性的系统,最小相位系统的相位变动是最小的。

数学上可以证明,对于最小相位系统,对数幅频特性和相频特性不是相互独立的,两者之间存在着严格确定的联系。如果已知对数幅频特性,通过公式也可以把相频特性计算出来。同样,通过公式也可以由相频特性计算出幅频特性。所以两者包含的信息内容是相同的。从建立数学模型和分析、设计系统的角度看,只要详细地画出两者中的一个就足够了。由于对数幅频特性容易画,所以对于最小相位系统,通常只绘制详细的对数幅频特性图,而对于相频特性只画简图,或者甚至不绘相频特性图。

Terms and concepts

Minimum phase: all the zeros of a transfer function lie in the left-hand side of the s-plane.
Nonminimum phase: transfer function with zeros in the right-hand s-plane.

9. 一般系统开环幅相特性曲线

对于单位负反馈系统,其开环传递函数 $G(s)$ 为回路中各串联环节传递函数之积,即

$$G(s) = G_1(s)G_2(s)\cdots G_n(s) = \prod_{i=1}^{n} G_i(s) \tag{5-3-34}$$

其频率特性为

$$G(j\omega) = G_1(j\omega)G_2(j\omega)\cdots G_n(j\omega) = \prod_{i=1}^{n} G_i(j\omega) = \prod_{i=1}^{n} M_i(\omega) e^{j\sum_{i=1}^{n}\varphi_i(\omega)} \tag{5-3-35}$$

由式(5-3-35)可得其开环幅频特性为

$$L(\omega) = \prod_{i=1}^{n} M_i(\omega) \tag{5-3-36}$$

开环相频特性为

$$\varphi(\omega) = \sum_{i=1}^{n} \varphi_i(\omega) \tag{5-3-37}$$

从上面两个式子可以看出:系统的开环幅频特性等于各串联环节的幅频特性之积;系统的开环相频特性等于各串联环节的相频特性之和。

在绘制开环幅相特性时,根据特殊的 ω 值,计算得到系统在该点的坐标值,将各点连成曲线,即为该系统的开环幅相频率特性曲线。当绘制概略开环幅相曲线时,根据稳定性判别的条件,幅相曲线应能体现系统开环频率特性的起点、终点、与负实轴的交点、与单位圆的交点以及总的变化趋势,绘制步骤如下。

(1) 开环传递函数按典型环节分解

$$G(s)H(s) = \frac{K}{s^v} \prod_{i=1}^{l} G_i(s) \tag{5-3-38}$$

其中,K 为系统的开环增益,v 为系统所含积分环节($v>0$)或微分环节($v<0$)的个数,$G_i(s)(i=1,2,\cdots,l)$ 为其他典型环节。

(2) 确定幅相曲线的起点和终点。幅相曲线的起点为 $G(j0^+)H(j0^+)$,终点为 $G(j\infty)H(j\infty)$。

(3) 确定幅相曲线与负实轴的交点。开环幅相曲线与负实轴的交点是判定系统闭环稳

定的重要因素。设 ω_g 为交点对应的频率，则有

$$\angle G(j\omega_g)H(j\omega_g) = (2k+1)\pi \quad k = 0, \pm 1, \pm 2, \cdots \quad (5\text{-}3\text{-}39)$$

或

$$\text{Im}[G(j\omega_g)H(j\omega_g)] = 0 \quad (5\text{-}3\text{-}40)$$

应该注意的是，最后一个方程的解中含有与正实轴的交点所对应的频率。由交点频率可求得与负实轴的交点为 $(\text{Re}(G(j\omega_g)H(j\omega_g)), 0)$ 或 $(G(j\omega_g)H(j\omega_g), 0)$。

(4) 根据上述确定的特征点，结合开环频率特性的变化趋势作图。若开环传递函数具有虚极点 $\pm j\omega_p$，则幅相曲线反映 ω 在 ω_p 附近变化时，开环幅相曲线的变化情况。由于 $\omega \to \omega_p$ 时，$|G(j\omega_p)H(j\omega_p)| \to \infty$，因此应确定 $\varphi(\omega_p^-)$ 和 $\varphi(\omega_p^+)$，并由此作图。

【**例 5-3-1**】 设系统的开环传递函数为

$$G(s)H(s) = \frac{10(Ts+1)}{s(s-10)} \quad T > 0$$

试绘制开环系统的大致幅相频率特性曲线。

解：系统的频率特性为

$$G(j\omega)H(j\omega) = \frac{10(j\omega T + 1)}{j\omega(j\omega - 10)}$$

幅相频率特性分别为

$$|G(j\omega)H(j\omega)| = \frac{10\sqrt{\omega^2 T^2 + 1}}{\omega\sqrt{\omega^2 + 100}}$$

$$\angle G(j\omega)H(j\omega) = \arctan\omega T - 90° - \left(180° - \arctan\frac{\omega}{10}\right)$$

计算得到特殊的 ω 点及其坐标值为

$$\omega = 0, |G(j\omega)H(j\omega)| = \infty, \angle G(j\omega)H(j\omega) = -270°$$
$$\omega = \infty, |G(j\omega)H(j\omega)| = 0, \angle G(j\omega)H(j\omega) = -90°$$

曲线与实轴的交点为

$$G(j\omega)H(j\omega) = -\frac{10\omega(10T+1) + j10(T\omega^2 - 10)}{\omega(\omega^2 + 100)} = U(\omega) + jV(\omega)$$

式中，$U(\omega)$ 与 $V(\omega)$ 分别为开环系统的实频和虚频特性。

令 $V(\omega) = 0$，解得 $\omega = \sqrt{\dfrac{10}{T}}$。将 $\omega = \sqrt{\dfrac{10}{T}}$ 代入 $U(\omega)$，得幅相频率曲线与实轴的交点为 $-T$。于是可得幅相特性曲线如图 5-3-16 所示。

下面定性地讨论系统开环频率特性曲线的一些特点。$G(j\omega)$ 可表示为

$$G(s) = \frac{K}{s^v} \frac{\prod_{k=1}^{m_1}(\tau_k s + 1) \cdot \prod_{l=1}^{m_2}(\tau_l^2 s^2 + 2\xi_l \tau_l s + 1)}{\prod_{i=1}^{n_1}(T_i s + 1) \cdot \prod_{j=1}^{n_2}(T_j^2 s^2 + 2\xi_j T_j s + 1)}$$

$$= \frac{b_m s^m + b_{m-1} s^{m-1} + \cdots + b_1 s + b_0}{a_n s^n + a_{n-1} s^{n-1} + \cdots + a_1 s + a_0} \quad (5\text{-}3\text{-}41)$$

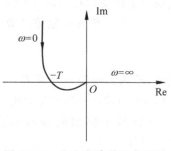

图 5-3-16 幅相频率特性曲线图

1) 极坐标曲线的起点

极坐标的起点是当 $\omega \to 0$ 时 $G(j0_+)$ 在复平面的位置。由式(5-3-41)得

$$\lim_{\omega \to 0} G(j\omega) = \lim_{\omega \to 0} \frac{K}{(j\omega)^v} = \lim_{\omega \to 0} \frac{K}{\omega^v} \angle -v \times 90° \qquad (5-3-42)$$

所以,对于不同的 v 值,特征曲线的起点将来自极坐标的不同方向,如图 5-3-17 所示。

- 0 型系统,$v=0$,$G(j\omega)$ 特性曲线起始于点 $(K,0)$ 处。
- Ⅰ 型系统,$v=1$,$G(j\omega)$ 特性曲线起始于 $-90°$ 处(负虚轴的 ∞ 处)。
- Ⅱ 型系统,$v=2$,$G(j\omega)$ 特性曲线起始于 $-180°$ 处(负实轴的 ∞ 处)。
- Ⅲ 型系统,$v=3$,$G(j\omega)$ 特性曲线起始于 $-270°$ 处(正实轴的 ∞ 处)。
- 当 $v>3$ 时,按照上面的规律重复分析。

2) 极坐标曲线的终点

因为 $\lim_{\omega \to \infty} G(j\omega) = 0 \angle -(n-m) \times 90°$,因此可以得到极坐标曲线趋向于坐标原点的规律,如图 5-3-18 所示。

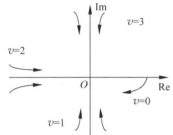

图 5-3-17 开环频率特性曲线的起始点

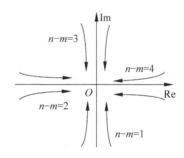

图 5-3-18 开环幅相频率特性的终点

10. 开环对数频率特性的绘制

同开环幅相频率特性一样,通过将开环传递函数分解为各个典型的环节,可方便地得到系统开环对数频率特性曲线。对于单位负反馈系统,其开环传递函数 $G(s)$ 为回路中各串联环节传递函数之积,即

$$G(s) = G_1(s)G_2(s)\cdots G_n(s) = \prod_{i=1}^{n} G_i(s) \qquad (5-3-43)$$

其频率特性为

$$G(j\omega) = G_1(j\omega)G_2(j\omega)\cdots G_n(j\omega) = \prod_{i=1}^{n} G_i(j\omega) = \prod_{i=1}^{n} M_i(\omega) e^{j \sum_{i=1}^{n} \varphi_i(\omega)} \qquad (5-3-44)$$

其开环对数频率特性为

$$L(\omega) = 20\lg |G(j\omega)| = 20\lg \left| \prod_{i=1}^{n} G_i(j\omega) \right| = 20 \sum_{i=1}^{n} \lg |G_i(j\omega)| \qquad (5-3-45)$$

$$\varphi(\omega) = \sum_{i=1}^{n} \varphi_i(\omega) \qquad (5-3-46)$$

从中可以看出:系统的开环对数频率特性等于各串联环节的幅频特性之代数和;相频特性等于组成系统的各典型环节的相频特性之代数和。

在绘制对数幅频特性图时,可以用基本环节的直线或折线渐近线代替精确幅频特性,然

后求它们的和,得到折线形式的对数幅频特性图,这样就可以明显地减少计算和绘图工作量。必要时可以对折线渐近线进行修正,以便得到足够精确的对数幅频特性。

任一段直线渐近线可看成是特殊斜率的幅频特性,则有 $G(j\omega)=k_i/s^n$,因此任一段渐近线的方程为 $20\lg|G|=-20n\lg\omega+20\lg k_i$,$\lg\omega$ 前的系数是斜率。

在求直线渐近线的和时,用到下述规则:在平面坐标图上,几条直线相加的结果仍为一条直线,和的斜率等于各直线斜率之和。

绘制开环对数幅频特性图可采用下述步骤。

(1) 将开环传递函数写成基本环节相乘的形式。

(2) 计算各基本环节的转折频率,并标在横轴上。最好同时标明各转折频率对应的基本环节渐近线的斜率。

(3) 设最低的转折频率为 ω_1,先绘 $\omega<\omega_1$ 的低频区图形,在此频段范围内只有积分(或纯微分)环节和放大环节起作用,其对数幅频特性见式(5-3-12)。

(4) 按照由低频到高频的顺序将已画好的直线或折线图形延长。每到一个转折频率,折线发生转折,直线的斜率就要在原数值之上加上对应的基本环节的斜率。在每条折线上应注明斜率。

(5) 如有必要,可对上述折线渐近线加以修正,一般在转折频率处进行修正。

相应的典型环节的种类及变化情况如表 5-3-2 所示。同样在后面的各交接频率处,渐近线频率都相应地被改变,每两个相邻交接频率间,渐近线为一直线。由此便可获得系统开环对数幅频特性曲线。

表 5-3-2 渐近线频率在交接频率处的变化

交接频率对应的典型环节	频率的变化	交接频率对应的典型环节	频率的变化
惯性环节	减小 20dB/dec	一阶微分环节	增大 20dB/dec
振荡环节	减小 40dB/dec	二阶微分环节	增大 40dB/dec

【例 5-3-2】 已知开环传递函数为

$$G(s)=\frac{7.5\left(\dfrac{s}{3}+1\right)}{s\left(\dfrac{s}{2}+1\right)\left(\dfrac{s^2}{2}+\dfrac{s}{2}+1\right)}$$

绘制系统的开环对数频率特性曲线。

解:(1) 该传递函数各基本环节的名称、转折频率和渐近线斜率,按频率由低到高的顺序排列如下:放大环节与积分环节,7.5rad/s 和 -20dB/dec;振荡环节,$\omega_1=\sqrt{2}$ rad/s、-40dB/dec;惯性环节,$\omega_2=2$rad/s、-20dB/dec;一阶微分环节,$\omega_3=3$rad/s、20dB/dec。将各基本环节的转折频率依次标在频率轴上,如图 5-3-19 所示。

(2) 最低的转折频率为 $\omega_1=\sqrt{2}$ rad/s。当 $\omega<\sqrt{2}$ rad/s 时,对数幅频特性就是 7.5rad/s 的对数幅频图。这是一条斜率为 -20dB/dec 的直线,直线位置由下述条件之一确定:当 $\omega=1$ 时,直线纵坐标为 $20\lg 7.5=17.5$dB;当 $\omega=7.5$rad/s 时直线穿过 0dB 线,如图 5-3-19 所示。

(3) 将上述直线延长至转折频率 $\omega_1=\sqrt{2}$ rad/s 处,在此位置直线斜率变为 $-20-40=-60$dB/dec。将折线延长到 $\omega_2=2$rad/s 处,在此处斜率变为 $-60-20=-80$dB/dec。将

折线延至 $\omega_3=3\text{rad/s}$ 处，斜率变为 $-80+20=-60\text{dB/dec}$。这样就得到全部开环对数幅频渐近线，如图 5-3-19 所示。如果有必要，可对渐近线进行修正。

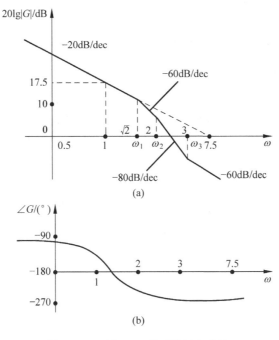

图 5-3-19 例 5-3-2 的对数频率特性

（4）求相频特性。根据频率特性函数的代数式，将分子的相位减去分母的相位就得到相频特性函数。或者将各基本环节的相频特性相加，也可求出相频特性。对于本例，有

$$\angle G(\text{j}\omega)=\arctan\frac{\omega}{3}-90°-\arctan\frac{\omega}{2}+\angle G_1(\text{j}\omega)$$

式中，$\angle G_1(\text{j}\omega)$ 表示振荡环节的相频特性，且有

$$\angle G_1(\text{j}\omega)=\begin{cases}-\arctan\dfrac{\omega}{2-\omega^2}, & \omega\leqslant\sqrt{2}\\ -180°-\arctan\dfrac{\omega}{2-\omega^2}, & \omega>\sqrt{2}\end{cases}$$

根据上两式就可以计算出各频率所对应的相位，从而画出相频特性图形，如图 5-3-19 所示。一般只绘相频特性的近似曲线。$\angle G_1(\text{j}\omega)$ 的典型数据是：当 $\omega\to 0$ 时，$\angle G_1(\text{j}\omega)\to 0°$；当 $\omega=\sqrt{2}$ 时，$\angle G_1(\text{j}\omega)\to -90°$；当 $\omega\to\infty$ 时，$\angle G_1(\text{j}\omega)\to -180°$。根据这些数据和公式就可以绘出相频特性的近似图形。

5.4 频域稳定判据及稳定裕量

第 3 章中介绍的利用闭环特征方程的系数判断系统稳定性的劳斯稳定判据，其特点是利用闭环信息来判断闭环系统的稳定性。闭环控制系统稳定的充要条件是：闭环特征方程的根均具有负的实部，或者说，全部闭环极点都位于左半 s 平面。这里要介绍的频域稳定判

据则是利用系统的开环信息——开环频率特性 $G(j\omega)$ 来判断闭环系统的稳定性。同时,时域分析法中的劳斯稳定判据只能给出自动控制系统绝对稳定性的信息,而频域分析法中还可以判断系统相对稳定性的状况。

5.4.1 奈奎斯特稳定判据

频域稳定判据是奈奎斯特(Nyquist)于 1932 年提出的,它是频率分析法的重要内容。利用奈奎斯特稳定判据,不但可以判断系统是否稳定(绝对稳定性),也可以确定系统的稳定程度(相对稳定性),还可以用于分析系统的动态性能以及指出改善系统性能指标的途径。因此,奈奎斯特稳定判据是一种重要而实用的稳定性判据,工程上应用十分广泛。

1. 辅助函数

对于图 5-4-1 所示的控制系统结构图,其开环传递函数为

$$G(s) = G_0(s)H(s) = \frac{M(s)}{N(s)} \tag{5-4-1}$$

相应的闭环传递函数为

$$\Phi(s) = \frac{G_0(s)}{1+G(s)} = \frac{G_0(s)}{1+\frac{M(s)}{N(s)}} = \frac{N(s)G_0(s)}{N(s)+M(s)} \tag{5-4-2}$$

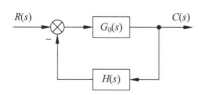

图 5-4-1 控制系统结构图

式中,$M(s)$ 为开环传递函数的 m 阶分子多项式;$N(s)$ 为开环传递函数的 n 阶分母多项式,$n \geqslant m$。由式(5-4-1)和式(5-4-2)可见,$N(s)+M(s)$ 和 $N(s)$ 分别为闭环和开环特征多项式。现以两者之比定义为辅助函数

$$F(s) = \frac{M(s)+N(s)}{N(s)} = 1+G(s) \tag{5-4-3}$$

实际系统传递函数 $G(s)$ 分母阶数 n 总是大于或等于分子阶数 m,因此辅助函数的分子分母同阶,即其零点数与极点数相等。设 $-z_1, -z_2, \cdots, -z_n$ 和 $-p_1, -p_2, \cdots, -p_n$ 分别为其零点和极点,则辅助函数 $F(s)$ 可表示为

$$F(s) = \frac{(s+z_1)(s+z_2)\cdots(s+z_n)}{(s+p_1)(s+p_2)\cdots(s+p_n)} \tag{5-4-4}$$

综上可知,辅助函数 $F(s)$ 具有以下特点。

(1) 辅助函数 $F(s)$ 是闭环特征多项式与开环特征多项式之比,其零点和极点分别为闭环极点和开环极点。

(2) $F(s)$ 的零点和极点的个数相同,均为 n 个。

(3) $F(s)$ 与开环传递函数 $G(s)$ 之间只差常量 1。$F(s)=1+G(s)$ 的几何意义为:F 平面上的坐标原点就是 G 平面上的 $(-1, j0)$ 点,如图 5-4-2 所示。

2. 幅角定理

辅助函数 $F(s)$ 是复变量 s 的单值有理复变函数。

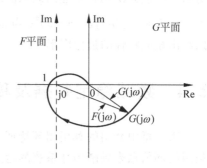

图 5-4-2 F 平面与 G 平面的关系图

由复变函数理论可知,如果函数 $F(s)$ 在 s 平面上指定域内是非奇异的,那么对于此区域内的任一点 d,都可通过 $F(s)$ 的映射关系在 $F(s)$ 平面上找到一个相应的点 d'(称 d' 为 d 的像);对于 s 平面上的任意一条不通过 $F(s)$ 任何奇异点的封闭曲线 Γ,也可通过映射关系在 $F(s)$ 平面(以下称 Γ 平面)找到一条与它相对应的封闭曲线 Γ'(Γ' 称为 Γ 的像),如图 5-4-3 所示。

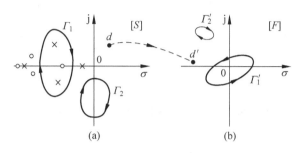

图 5-4-3 s 平面与 F 平面的映射关系图

设 s 平面上不通过 $F(s)$ 任何奇异点的某条封闭曲线 Γ,它包围了 $F(s)$ 在 s 平面上的 Z 个零点和 P 个极点。当 s 以顺时针方向沿封闭曲线 Γ 移动一周时,则在 F 平面上相对应于封闭曲线 Γ 的像 Γ' 将以顺时针的方向围绕原点旋转 R 圈。R 与 Z、P 的关系为

$$R = Z - P \tag{5-4-5}$$

3. 奈奎斯特稳定判据

为了确定辅助函数 $F(s)$ 位于右半 s 平面内的所有零点和极点数,现将封闭曲线 Γ 扩展为整个右半 s 平面。为此,设计 Γ 曲线由以下三段所组成。

(1) 正虚轴 $s=\mathrm{j}\omega$:频率 ω 由 0 变到 ∞。
(2) 半径为无限大的右半圆 $s=R\mathrm{e}^{\mathrm{j}\theta}$:$R\to\infty$,$\theta$ 由 $\pi/2$ 变化到 $-\pi/2$。
(3) 负虚轴 $s=\mathrm{j}\omega$:频率 ω 由 $-\infty$ 变化到 0。

这样,三段组成的封闭曲线 Γ(称为奈奎斯特(Nyquist)曲线)就包含了整个右半 s 平面,如图 5-4-4 所示。

在 F 平面上绘制与 Γ 相对应的像 Γ'。当 s 沿虚轴变化时,由式(5-4-3)则有

$$F(\mathrm{j}\omega) = 1 + G(\mathrm{j}\omega) \tag{5-4-6}$$

式中,$G(\mathrm{j}\omega)$ 为系统的开环频率特性。因而 Γ' 将由下面几段组成。

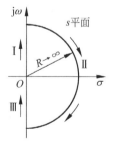

图 5-4-4 奈奎斯特曲线

(1) 和正虚轴对应的是辅助函数的频率特性 $F(\mathrm{j}\omega)$,相当于把 $G(\mathrm{j}\omega)$ 右移一个单位。

(2) 和半径为无穷大的右半圆相对应的辅助函数 $F(s)\to 1$。由于开环传递函数的分母阶数高于分子阶数,当 $s\to\infty$ 时,$G(s)\to 0$,故有 $F(s)=1+G(s)\to 1$。

(3) 和负虚轴相对应的是辅助函数频率特性 $F(\mathrm{j}\omega)$ 对称于实轴的镜像。

如图 5-4-5 所示绘出了系统开环频率特性曲线 $G(\mathrm{j}\omega)$。将曲线右移一个单位,并取镜

像,则成为 F 平面上的封闭曲线 Γ',如图 5-4-6 所示,图中用虚线表示镜像。

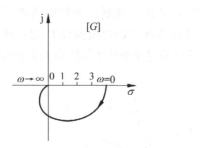
图 5-4-5　$G(j\omega)$ 开环频率特性曲线

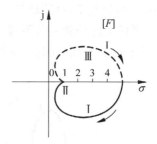
图 5-4-6　F 平面的封闭曲线

对于包含了整个右半 s 平面的奈奎斯特曲线来说,式(5-4-5)中的 Z 和 P 分别为闭环传递函数和开环传递函数在右半 s 平面上的极点数,而 R 则是 F 平面上 Γ' 曲线顺时针包围原点的圈数,也就是 G 平面上系统开环幅相特性曲线及其镜像顺时针包围 $(-1,j0)$ 点的圈数。在实际系统分析过程中,一般只绘制开环幅相特性曲线不绘制其镜像曲线,考虑到角度定义的方向性,有

$$R = -2N \tag{5-4-7}$$

式中,N 是开环幅相曲线(不包括其镜像)包围 G 平面$(-1,j0)$点的圈数(逆时针为正,顺时针为负)。将式(5-4-7)代入式(5-4-5),可得奈氏判据

$$Z = P - 2N \tag{5-4-8}$$

式中,Z 是右半 s 平面中闭环极点的个数;P 是右半 s 平面中开环极点的个数;N 是 G 平面上 $G(j\omega)$ 包围 $(-1,j0)$ 点的圈数(逆时针为正)。显然,只有当 $Z=P-2N=0$ 时,闭环系统才是稳定的。

Terms and concepts

Cauchy's theorem: (if a contour encircles Z zeros and P poles of $F(s)$ traversing clockwise, the corresponding contour in the $F(s)$-plane encircles the origin of the $F(s)$-plane $N=Z-P$ times clockwise.)

Contour map: a contour or trajectory in one plane is mapped into another plane by a relation $F(s)$.

Nyquist stability criterion: a feedback system is stable, if and only if, the contour in the $G(s)$-plane does not encircle the $(-1,j0)$ point when the number of poles of $G(s)$ in the right-hand s-plane is zero. If $G(s)$ has P poles in the right-hand plane, then the number of counterclockwise encirclements of the $(-1,j0)$ point must be equal to P for a stable system.

Conformal mapping: a contour mapping that retain the angles on the s-plane on the $F(s)$-plane.

【例 5-4-1】　系统结构图如图 5-4-7 所示,试判断系统的稳定性并讨论 K 值对系统稳定性的影响。

解:系统是一个非最小相角系统,开环不稳定。开环传递函数在右半 s 平面上有一个极点，$P=1$。幅相特性曲线如图 5-4-8 所示。当 $\omega=0$ 时，曲线从负实轴$(-K,j0)$点出发；当 $\omega=\infty$ 时，曲线以 $-90°$ 趋于坐标原点；幅相特性包围$(-1,j0)$点的圈数 N 与 K 值有关。图 5-4-8 绘出了 $K>1$ 和 $K<1$ 的两条曲线，可见，当 $K>1$ 时，曲线逆时针包围了$(-1,j0)$点的 1/2 圈，即 $N=1/2$，此时 $Z=P-2N=1-2\times(1/2)=0$，故闭环系统稳定；当 $K<1$ 时，曲线不包围$(-1,j0)$点，即 $N=0$，此时 $Z=P-2N=1-2\times0=1$，有一个闭环极点在右半 s 平面，故系统不稳定。

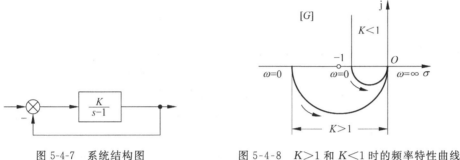

图 5-4-7　系统结构图　　　　图 5-4-8　$K>1$ 和 $K<1$ 时的频率特性曲线

Reading Material

The Nyquist diagram is basically a plot of $G(j\omega)$ where $G(s)$ is the open-loop transfer function and ω is a vector of frequencies which encloses the entire right-half plane. In drawing the Nyquist diagram, both positive and negative frequencies(from zero to infinity) are taken into account. We will represent positive frequencies with dotted line and negative frequencies with solid line. The frequency vector used in plotting the Nyquist diagram usually looks like Fig. 5-4-9(if you can imagine the plot stretching out to infinity).

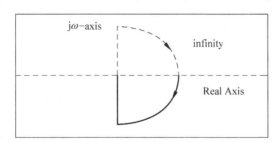

Fig. 5-4-9　Schematics of the Nyquist diagram(Case 1)

However, if we have open-loop poles or zeros on the jω axis, $G(s)$ will not be defined at those points, and we must loop around them when we are plotting the contour. Such a contour would look as follows in Fig. 5-4-10.

5.4.2　奈奎斯特稳定判据的应用

如果开环传递函数 $G(s)$ 在虚轴上有极点，则不能直接应用图 5-4-4 所示的奈奎斯特曲线，因为幅角定理要求奈奎斯特曲线不能经过 $F(s)$ 的奇点，为了在这种情况下应用奈氏判

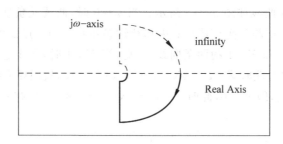

Fig. 5-4-10　Schematics of the Nyquist diagram(Case 2)

据,可以对奈奎斯特曲线略作修改。使其沿着半径为无穷小($r\to 0$)的右半圆绕过虚轴上的极点。例如,当开环传递函数中有纯积分环节时,s 平面原点有极点,相应的奈奎斯特曲线可以修改为如图 5-4-11 所示。图中的小半圆绕过了位于坐标原点的极点,使奈奎斯特曲线

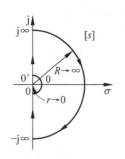

图 5-4-11　开环含有积分环节时的奈奎斯特曲线

避开了极点,又包围了整个右半 s 平面,前述的奈氏判据结论仍然适用,只是在画幅相曲线时,s 取值需要先从 j0 绕半径无限小的圆弧逆时针转 90°到 j0$^+$,然后再沿虚轴到 j∞。这样需要补充 $s=$j0$\to$j0$^+$ 小圆弧所对应的 $G(j\omega)$ 特性曲线。

设系统开环传递函数为

$$G(s) = \frac{K\prod_{i=1}^{m}(T_i s+1)}{s^v \prod_{j=1}^{n-v}(T_j s+1)} \tag{5-4-9}$$

式中,v 为系统型别。当沿着无穷小半圆逆时针方向移动时,有 $s=\lim_{r\to 0}re^{j\theta}$,映射到 G 平面的曲线可以按下式求得

$$G(s)\Big|_{s=\lim_{r\to 0}re^{j\theta}} = \frac{K\prod_{i=1}^{m}(T_i s+1)}{s^v \prod_{j=1}^{n-v}(T_j s+1)}\Bigg|_{s=\lim_{r\to 0}re^{j\theta}} = \lim_{r\to 0}\frac{K}{r^v}e^{-jv\theta} = \infty e^{-jv\theta} \tag{5-4-10}$$

由上述分析可见,当 s 沿小半圆从 $\omega=0$ 变化到 $\omega=0^+$ 时,θ 角沿逆时针方向从 0 变化到 $\pi/2$,这时 G 平面上的映射曲线将从 $\angle G(j0)$ 位置沿半径无穷大的圆弧按顺时针方向转过 $-v\pi/2$ 角度。在确定 $G(j\omega)$ 绕(-1,j0)点圈数 N 的值时,要考虑大圆弧的影响。

【例 5-4-2】　已知开环传递函数为

$$G(s) = \frac{K}{s(Ts+1)}$$

其中,$K>0$,$T>0$,绘制极坐标图并判别系统的稳定性。

解:该系统 $G(s)$ 在坐标原点处有一个极点,为 I 型系统。取奈奎斯特曲线如图 5-4-12 所示。

当 s 沿小半圆移动从 $\omega=0$ 变化到 $\omega=0^+$ 时,在 G 平面上映射曲线为半径 $R\to\infty$ 的 $\pi/2$ 圆弧。幅相曲线(包括大圆弧)如图 5-4-12 所示。此系统开环传递函数在右半

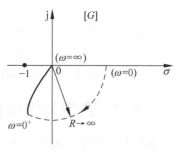

图 5-4-12　例 5-4-2 极坐标图

s 平面无极点，$P=0$；$G(s)$ 的极坐标曲线又不包围点 $(-1,j0)$，$N=0$；因此 $Z=P-2N=0$，闭环系统是稳定的。

【例 5-4-3】 已知系统开环传递函数为
$$G(s)H(s)=\frac{K(s+3)}{s(s-1)}$$
试绘制极坐标图，并分析闭环系统的稳定性。

解：由于 $G(s)H(s)$ 在右半 s 平面有一极点，故 $P=1$。当 $0<K<1$ 时，其极坐标图如图 5-4-13(a)所示，图中可见 ω 从 0 到 $+\infty$ 变化时，极坐标曲线顺时针包围 $(-1,j0)$ 点 $-1/2$ 圈，即 $N=-1/2$，$Z=P-2N=1+2(1/2)=2$，因此闭环系统不稳定。当 $K>1$ 时，其极坐标图如图 5-4-13(b)所示，当 ω 从 0 到 $+\infty$ 变化时，极坐标曲线逆时针包围 $(-1,j0)$ 点 $+1/2$ 圈，$N=+1/2$，$Z=P-2N=1-2(1/2)=0$，此时闭环系统是稳定的。

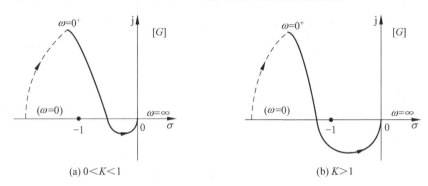

图 5-4-13　例 5-4-3 极坐标图

Reading Material

The Cauchy Criterion

The cauchy criterion(from complex analysis)states that when taking a closed contour in the complex plane, and mapping it through a complex function $G(s)$, the number of times that the plot of $G(s)$ encircles the origin is equal to the number of zeros of $G(s)$ enclosed by the frequency contour minus the number of poles of $G(s)$ enclosed by the frequency contour. Encirclements of the origin are counted as positive if they are in the same direction as the original closed contour or negative if they are in the opposite direction.

When studying feedback controls, we are not as interested in $G(s)$ as in the closed-loop transfer function.
$$G_c(s)=\frac{G(s)}{1+G(s)}$$

If $1+G(s)$ encircles the origin, then $G(s)$ will enclose the point -1. Since we are interested in the closed-loop stability, we want to know if there are any closed-loop poles (zeros of $1+G(s)$)in the right-half plane. More details on how to determine this will come later.

Therefore, the behavior of the Nyquist diagram around the -1 point in the real axis is very important; however, the axis on the standard Nyquist diagram might make it hard to see what is happening around this point. To correct this, the *nyquist.1.m* command in MATLAB plots the Nyquist diagram using a logarithmic scale and preserves the characteristics of the -1 point.

5.4.3 对数稳定判据

对于复杂的开环极坐标图,采用"包围周数"的概念判定闭环系统是否稳定比较麻烦,容易出错。为了简化判定过程,这里引用正、负穿越的概念,如图 5-4-14 所示。如果开环坐标图按逆时针方向(从上向下)穿过负实轴,称为正穿越,正穿越时相位增加;按顺时针方向(从下向上)穿过负实轴,称为负穿越,负穿越时相位减小。开环极坐标图逆时针方向包围点$(-1, j0)$的周数正好等于极坐标图在点$(-1, j0)$左方正、负穿越负实轴次数之差。因此,奈奎斯特稳定判据可叙述如下:闭环系统稳定的充要条件是,当 ω 由 $0 \to \infty$ 时,开环频率特性极坐标图在点$(-1, j0)$左方正、负穿越负实轴次数之差应为 $P/2$,P 为开环传递函数正实部极点个数。若开环极坐标图点$(-1, j0)$左方负穿越负实轴的次数大于正穿越的次数,则闭环系统一定不稳定。

由于频率特性的极坐标图较难画,所以人们希望利用开环 Bode 图来判定闭环稳定性。这里的关键问题是,极坐标中在点$(-1, j0)$左方正、负穿越负实轴的情况在对数坐标中是如何反映的。因为极坐标图上的负实轴对应于对数相频特性坐标上的$-180°$线,所以,按照正穿越相位增加、负相位减少的概念,极坐标上的正、负穿越负实轴就是 Bode 图中对数相频特性曲线正、负穿越$-180°$线,如图 5-4-15 所示。根据 Bode 图分析闭环系统稳定性的奈奎斯特稳定判据可叙述如下:闭环系统稳定的充要条件是,在开环幅频特性大于 0dB 的所有频段内,相频特性曲线对$-180°$线的正、负穿越次数之差等于 $P/2$,其中 P 为开环正实部极点个数。需注意的是,当开环系统含有积分环节时,相频特性应增补 ω 由 $0 \to 0^+$ 的部分。

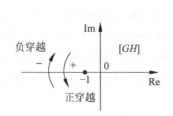
图 5-4-14 正、负穿越示意图

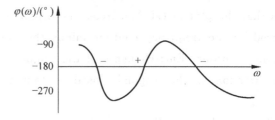
图 5-4-15 Bode 图上的正、负穿越

【例 5-4-4】 系统的开环传递函数为

$$G(s) = \frac{K}{s(T_1 s + 1)(T_2 s + 1)}$$

当 K 取最小值和最大值时的开环极坐标图如图 5-4-16 所示,判断闭环系统的稳定性。

解:开环传递函数无正实部极点。图 5-4-16(a)中开环极坐标图在点$(-1, j0)$左方没

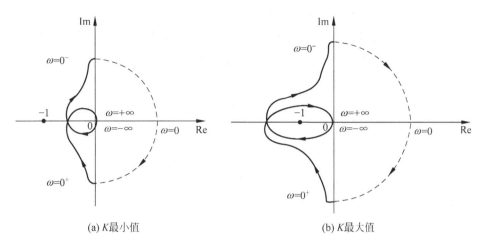

(a) K 最小值   (b) K 最大值

图 5-4-16　例 5-4-4 附图

有穿越负实轴，而图 5-4-16(b) 中开环极坐标图在点 $(-1,j0)$ 左方对负实轴有一次负穿越，所以图 5-4-16(a) 所示系统闭环稳定，而图 5-4-16(b) 的系统闭环不稳定。

【例 5-4-5】　系统开环 Bode 图和开环正实部极点个数 P 如图 5-4-17(a)～图 5-4-17(c) 所示，判定闭环系统稳定性。

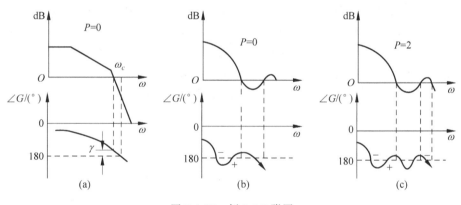

图 5-4-17　例 5-4-5 附图

解：图 5-4-17(a) 中，$P=0$，幅频特性大于 0dB 时，相频特性曲线没有穿越 $-180°$ 线，故系统稳定。

图 5-4-17(b) 中，$P=0$，幅频特性大于 0dB 的各频段内，相频特性曲线对 $-180°$ 线的正、负穿越次数之差为 $1-1=0$，故系统闭环稳定。

图 5-4-17(c) 中，$P=2$，幅频特性大于 0dB 的所有频段内，相频特性曲线对 $-180°$ 线的正、负穿越次数之差为 $1-2=-1\neq 1$，故系统闭环不稳定。

5.4.4　稳定裕量

如前所述，当控制系统结构不发生变化时，参数变化（如开环放大系数）可能导致系统的状态从稳定变为不稳定。为了使系统能很好地工作，不但要求系统稳定，而且要有一定的稳定裕量。稳定裕量是指系统的相对稳定性,它不但是可以衡量一个闭环系统稳定程度的指标,并且与系统动态性能有着密切的关系。通常在自动控制系统设计中定义了两种稳定裕量,即相角裕量 γ 和幅值裕量 h。它们实际上描述了开环极坐标曲线离点 $(-1,j0)$ 的远近程度,也是系统的重要动态性能指标。

Terms and Concepts

Gain margin: the increase in the system gain when phase=180 degree that result in a marginally stable system with intersection of the $(-1,j0)$ point on the Nyquist diagram.

Phase margin: the amount of phase shift of the $G(j\omega)$ at unity magnitude that will result in a marginally stable system with intersections of the $(-1,j0)$ point on the Nyquist diagram.

1. 相角裕量

开环频率特性幅值为1时所对应的角频率称为幅值穿越频率或剪切频率,记为 ω_c。相角裕量 γ 是指开环频率特性在幅值穿越频率 ω_c 处所对应的相角与 $-180°$ 之差。相角裕量的表达式为

$$\gamma = 180° + \angle G(j\omega_c)H(j\omega_c) = 180° + \varphi(j\omega_c) \tag{5-4-11}$$

相角裕量在极坐标图和 Bode 图上的表示如图 5-4-18(a)所示。相角裕量表示出开环奈奎斯特图在单位圆上离点 $(-1,j0)$ 的远近程度。相角裕量的几何意义是,在极坐标图上,负实轴绕原点转到 $G(j\omega_c)H(j\omega_c)$ 时所转过的角度,逆时针转向为正角,顺时针转向为负角。开环奈奎斯特图正好通过点 $(-1,j0)$ 时,称闭环系统是临界稳定。由图 5-4-18(a)可知,对于开环稳定的系统,欲使闭环稳定,其相角裕量必须为正。由于过高的相角裕量不易实现,一个性能良好的控制系统,通常要求相角裕量 $\gamma = 40° \sim 60°$。

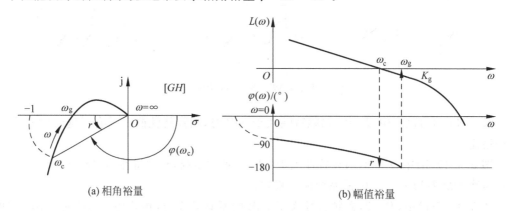

图 5-4-18 相角裕量及幅值裕量

2. 幅值裕量

开环频率特性的相位等于 $-180°$ 时所对应的角频率称为相位穿越频率,记为 ω_g,即 $\angle G(j\omega_c)H(j\omega_c) = -180°$。幅值裕量 K_g 是指开环幅相频率特性曲线与 $-180°$ 线交点 (ω_g) 处幅值的倒数,即

$$K_g = \frac{1}{|G(j\omega_g)H(j\omega_g)|} \tag{5-4-12}$$

在开环对数频率特性曲线上,相频特性曲线为$-180°$时的对数幅值离 0dB 线的距离,如图 5-4-18(b)所示。图中箭头方向为正方向。其表达式为

$$20\lg K_g = -20\lg|G(j\omega_g)H(j\omega_g)| \quad (\text{dB}) \tag{5-4-13}$$

幅值裕量的含义是指如果系统开环增益增大到原来的 K_g 倍时,系统就将处于临界稳定状态。所以说幅值裕量是系统在幅值上的稳定储备量。一个性能良好的控制系统,一般要求 $K_g = 2 \sim 3.16$ 或 $20\lg K_g = 6 \sim 10\text{dB}$。

一阶和二阶系统的相角裕量总大于零,而幅值裕量为无穷大。因此,从理论上讲,一阶、二阶系统绝对稳定。但是,某些一阶和二阶系统的数学模型本身是在忽略了一些次要因素之后建立的,实际系统常常是高阶的,其幅值裕量不可能无穷大。因此,如果开环增益太大,这些系统仍有可能不稳定。

Reading Material

1. Gain Margin

We already defined the gain margin as the change in open-loop gain expressed in decibels(dB), required at 180 degrees of phase shift to make the system unstable. Now we are going to find out where this comes from. First of all, let's say that we have a system that is stable if there are no Nyquist encirclements of -1, such as

$$G(s) = \frac{50}{s^3 + 9s^2 + 30s + 40}$$

Looking at the roots, we find that we have no open loop poles in the right half plane and therefore no closed-loop poles in the right half plane if there are no Nyquist encirclements of -1. Now, how much can we vary the gain before this system becomes unstable in closed loop? Let's look at the following Fig. 5-4-19.

The open-loop system represented by this plot will become unstable in closed loop if the gain is increased past a certain boundary. The negative real axis area between $-1/a$ (defined as the point where the 180 degree phase shift occurs…that is, where the diagram crosses the real axis) and -1 represents the amount of increase in gain that can be tolerated before closed-loop instability.

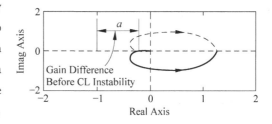

Fig. 5-4-19 Gain difference before CL instability

If we think about it, we realize that if the gain is equal to a, the diagram will touch the -1 point.

$$G(j\omega) = -1/a \Rightarrow aG(j\omega) = a(-1/a) \Rightarrow aG(j\omega) = -1$$

Therefore, we say that the gain margin is 'a' units. However, we mentioned before

that the gain margin is usually measured in decibels. Hence, the gain margin is
$$GM = 20\lg a [\text{dB}]$$

We will now find the gain margin of the stable, open-loop transfer function we viewed before. Recall that the function is
$$G(s) = \frac{50}{s^3 + 9s^2 + 30s + 40}$$

and that the Nyquist diagram can be viewed by typing

```
nyquist(50, [1 9 30 40])
```

You will have the following Fig. 5-4-20.

As we discussed before, all that we need to do to find the gain margin is find 'a', as defined in the preceding figure. To do this, we need to find the point where there is exactly 180 degrees of phase shift. This means that the transfer function at this point is real(has no imaginary part). The numerator is already real, so we just need to look at the denominator. When $s = j\omega$, the only terms in the denominator that will have imaginary parts are those which are odd powers of s. Therefore, for $G(j\omega)$ to be real, we must have:
$$-j\omega^3 + 30j\omega = 0$$

which means $\omega = 0$ (this is the rightmost point in the Nyquist diagram) or $\omega = \text{sqrt}(30)$. We can then find the value of $G(j\omega)$ at this point using polyval:

```
w = sqrt(30);
polyval(50,j*w)/polyval([1 9 30 40],j*w)
```

Our answer is: $-0.2174 + 0i$. The imaginary part is zero, so we know that our answer is correct. We can also verify by looking at the Nyquist plot again. The real part also makes sense. Now we can proceed to find the gain margin.

We found that the 180 degrees phase shift occurs at $-0.2174 + 0i$. This point was previously defined as $-1/a$. Therefore, we now have 'a', which is the gain margin. However, we need to express the gain margin in decibels.
$$-1/a = 0.2174 \Rightarrow a = 4.6 \Rightarrow GM = 20\lg 4.6 = 13.26 \text{dB}$$

We now have our gain margin. Let's see how accurate it is by using a gain of $a = 4.6$ and zooming in on the Nyquist plot in Fig. 5-4-21.

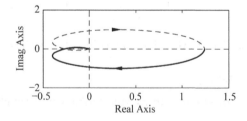

Fig. 5-4-20 Nyquist plot of the system

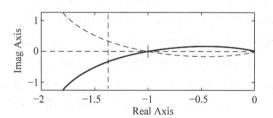

Fig. 5-4-21 Nyquist plot of the system with the different parameter

```
a = 4.6
nyquist(a * 50,[1 9 30 40])
```

The plot appears to go right through the −1 point. We will now verify the accuracy of our results by viewing the zoomed Nyquist diagrams and step responses for gains of 4.5, 4.6, and 4.7.

2. Phase Margin

We have already discussed the importance of the phase margin. Therefore, we will only talk about where this concept comes from. We have defined the phase margin as the change in open-loop phase shift required at unity gain to make a closed-loop system unstable. Let's look at the following graphical definition of this concept to get a better idea of what we are talking about. We will have Fig. 5-4-22.

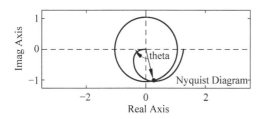

Fig. 5-4-22 Graphical definition of Phase margin

Let's analyze the previous plot and think about what is happening. From our previous example we know that this particular system will be unstable in closed loop if the Nyquist diagram encircles the −1 point. However, we must also realize that if the diagram is shifted by theta degrees, it will then touch the −1 point at the negative real axis, making the system marginally stable in closed loop. Therefore, the angle required to make this system marginally stable in closed loop is called the phase margin(measured in degrees). In order to find the point we measure this angle from, we draw a circle with radius of 1, find the point in the Nyquist diagram with a magnitude of 1(gain of zero dB), and measure the phase shift needed for this point to be at an angle of 180 deg.

本节主要介绍了频域稳定判据及其相关判定方法。一般而言,采用"包围周数"或者穿越的概念进行判断都比较复杂。下面通过英文阅读材料的方式介绍一种"−1 stability criterion"方法,又称为"阴影法",可以通过作图较为容易地进行判断,也可将此方法视为"穿越法"的一种验证方法进行使用。

Reading Material

We consider the following feedback system in Fig. 5-4-23.

Based on the above discussion, we can provide you a "−1 stability criterion", which is simply stated as the following.

The system is on the border of instability when
$|G(j\omega)| = 1$ and $\arg(G(j\omega)) = -\pi$
or $G(j\omega) = 1e^{-j\pi}$

To determine the stability of the system, you

Fig. 5-4-23 A simple control system

should follow the following steps:

(1) Draw the Nyquist plot;
(2) Shade the area at the right of the curve;
(3) Mirror the plot with respect to re-axis.

If "−1" in the shaded area, the closed loop system will be unstable. The stability of the closed system depends on the fact whether the Nyquist plot of the open system encircles −1. Fig. 5-4-24 shows the result of the specified system.

Fig. 5-4-25 gives another example with the stable result. You may check those result with the standard methods discussed above.

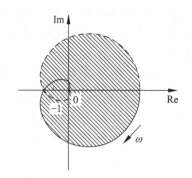

Fig. 5-4-24 Applications of "−1 stability criterion"　　Fig. 5-4-25 An example with stable result

5.5 频率特性与控制系统性能的关系

5.5.1 控制系统性能指标

控制系统性能的优劣以性能指标衡量。由于研究方法和应用领域的不同，性能指标有很多种，大体上可以归纳成两类：时间域指标和频率域指标。

时间域指标包括稳态指标和动态指标。稳态指标包括稳态误差 e_{ss}、无差度 v 以及开环放大系数 K。动态指标包括过渡过程时间 t_s、最大超调量 σ_p、上升时间 t_r、峰值时间 t_p、振荡次数 N 等，常用的是 t_s 和 σ_p。

频率域指标包括开环指标和闭环指标。开环指标有幅值穿越（剪切）频率 ω_c、相位裕度 γ、幅值裕度 K_g，常用的是 ω_c 和 γ。

系统闭环频率特性曲线，可以利用系统的开环频率特性曲线以及一些标准图线简捷方便地得到。然后，利用系统闭环频率特性曲线的一些特征量（如峰值和频带），可以进一步对系统进行分析和性能估算。频率特性是描述控制系统内在固有特性的一种工具，因而它与系统的控制性能之间有着紧密的关系。这里主要阐述用以描述控制系统性能的频域性能指标与时域性能指标之间的关系，从而揭示出从不同角度根据不同的方法分析与设计控制系统的内在联系。

设反馈系统的闭环对数幅频特性曲线如图 5-5-1 所示。

其特征量如下。

(1) 峰值 M_r：是对数幅频特性的最大值。峰值越大，意味着系统的阻尼比越小，平稳性越差，阶跃响应将有较大的超调量。

(2) 带宽频率和带宽：当闭环对数幅频特性的分贝值相对 $20\lg|\varphi(j0)|$ 的值下降 3dB 时的对应频率 ω_b，称为带宽频率。带宽频率的范围称为带宽，即 $0<\omega\leqslant\omega_b$。带宽频率范围越大，表明系统复现快速变化信号的能力越强，失真小，系统快速性好，阶跃响应上升时间和调节时间短。但另一方面系统抑制输入端高频噪声的能力相应地被削弱。

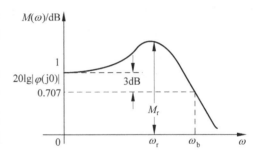

图 5-5-1 典型闭环对数幅频特性曲线

经验表明，闭环对数幅频特性曲线在带宽频率附近的斜率越小（即对数幅频特性衰减越快），系统从噪声中区别有用信号的特性越好。但是，一般呈现较大的峰值 M_r，因而系统平稳性较差，稳定程度也差。

(3) 零频值：是指频率等于零时的闭环对数幅值，即 $20\lg|\varphi(j0)|$。

当 $20\lg|\varphi(j0)|=0$（即 $|\varphi(j0)|=1$）时，系统阶跃响应的终值等于输入信号的幅值，稳态误差为零。当 $20\lg|\varphi(j0)|\neq0$ 时，系统存在稳态误差，故零频幅值反映了系统的稳态精度。

对于一阶、二阶系统，闭环频率特性的特征量与时域指标之间的关系是很明确的。二阶系统（即振荡环节）表明 M_r 与阻尼比 ξ、阻尼比与超调量 $\sigma\%$ 以及 M_r 与 $\sigma\%$ 都存在确定关系。带宽 ω_b 与调节时间 t_s 之间的关系也很明确。

例如一阶系统，其闭环传递函数为

$$\Phi(s) = \frac{1}{Ts+1} \tag{5-5-1}$$

可求出系统带宽

$$\omega_b = \frac{1}{T} \tag{5-5-2}$$

在第 3 章时域分析中可知，当取 5％误差带时

$$t_s = 3T, \quad t_r = 2.2T$$

因此与频域中带宽之间的关系为

$$t_s = \frac{3}{\omega_b}, \quad t_r = \frac{2.2}{\omega_b} \tag{5-5-3}$$

显然，带宽频率和调节时间成反比，和上升时间成反比。

对于欠阻尼的二阶系统，在时域分析中可知

$$t_s = \frac{3.5}{\xi\omega_n}, \quad t_r = \frac{\pi-\beta}{\omega_n\sqrt{1-\xi^2}}, \quad t_p = \frac{\pi}{\omega_n\sqrt{1-\xi^2}} \tag{5-5-4}$$

可见 t_s、t_r 和 t_p 都和自然频率 ω_n 成反比。而带宽 ω_b 与自然频率 ω_n 成正比。因此，t_s、t_r 和 t_p 也都和 ω_b 成反比。根据闭环幅频特性及带宽频率定义可得 ω_b 与 ω_n 的确定关系式为

$$\omega_b = \omega_n\left[(1-2\xi^2) + \sqrt{(1-2\xi^2)+1}\right]^{1/2} \tag{5-5-5}$$

最后应指出，系统的截止频率 ω_c 和闭环的带宽频率 ω_b 之间有密切关系。如果两个系统的稳定程度相仿，则 ω_c 大的系统 ω_b 也大；ω_c 小的系统 ω_b 也小。即 ω_c 和系统响应速度存

在着反比关系。因此,也常用 ω_c 来衡量系统的响应速度。

对于高阶系统的 M_r、ω_b 与时域指标之间没有精确的关系式。一般情况下采用下面的经验公式近似地表示高阶系统性能指标间的关系。实际性能指标一般比计算结果偏好。若闭环系统的幅频特性曲线如图 5-5-2 所示,时域性能指标的估算公式为

$$M_r = \frac{1}{\sin\gamma°} \tag{5-5-6}$$

$$\sigma\% \approx \left\{ 41\ln\left[\frac{M_r M\left(\frac{\omega_1}{4}\right)}{M_0^2} \frac{\omega_b}{\omega_{0.5}}\right] + 17 \right\}\% \tag{5-5-7}$$

$$t_s = \left(13.57 \frac{M_r \omega_b}{M_0 \omega_{0.5}} - 2.51\right)\frac{1}{\omega_{0.5}}(s) \tag{5-5-8}$$

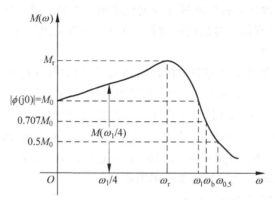

图 5-5-2 闭环系统的幅频特性曲线

由于系统闭环频率特性曲线不如开环频率特性曲线那么容易得到。另外,既然闭环系统稳定性及稳定程度均可以通过开环频率特性得到,自然会想到能否根据开环频率特性来估算闭环系统的时域指标。

Reading Material

Performance Specifications in the Frequency Domain

Frequency response(magnitude)plot of a typical second order system with the transfer function

$$G(s) = \frac{1}{1 + 2\xi\frac{s}{\omega_n} + \frac{s^2}{\omega_n^2}}$$

is shown in the following Fig. 5-5-3.

ω_r is the resonance frequency.

$M_{P\omega}$ is the maximum magnitude(or resonance).

ω_B is the bandwidth.

Resonance frequency, **$M_{P\omega}$**, is the frequency at which the maximum value of the frequency response is attained.

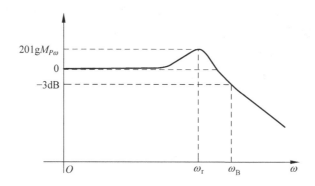

Fig. 5-5-3　Frequency response(magnitude)plot

Bandwidth, ω_B, is a measure of system's ability to faithfully reproduce an input signal.

It is the frequency, ω_B, at which frequency response has declined by 3dB from its low-frequency value.

In the case of the second order transfer function, $M_{P\omega}$ is a function of ξ only, and hence related to the overshoot to a step input.

Resonant frequency, ω_r, and bandwidth, can be related to the speed of response(rise time). In the case of the second order transfer functions, $\omega_B \approx \omega_n$, for values of ξ around 0.7.

5.5.2　开环对数幅频特性与性能指标间的关系

对于常见的系统,特别是最小相位系统,主要是利用开环对数幅频特性分析和设计系统。按照频率的范围,一般可以从低频段、中频段和高频段来分析控制系统的频率特性,以提出控制系统设计的性能要求。

1. 低频段

如果开环幅频特性中最低的转折频率是 ω_1,则低于 ω_1 的频段称为低频段。在低频部分,系统的开环频率特性为

$$G(j\omega)H(j\omega) = \frac{K}{j^v \omega^v} \tag{5-5-9}$$

低频部分的对数幅频特性是

$$20\lg|G(j\omega)H(j\omega)| = 20\lg K - 20v\lg\omega \tag{5-5-10}$$

式(5-5-10)是直线方程,斜率为 $-20v$dB/dec,直线通过 $\omega=1, 20\lg K$ 点;同时直线或其延长线在 $\omega = \sqrt[v]{K}$ 处通过 0dB 线。所以,由低频部分的斜率和直线位置可求出系统型别或串联积分环节个数 v 和开环放大系数 K。因此开环对数幅频特性的低频部分反映出系统的稳态性能,或者说,系统的稳态性能指标取决于开环幅频特性的低频部分。

2. 中频段

幅值穿越频率 ω_c 属于中频段。在相位相角裕量一定的情况下,ω_c 的大小取决于系统的响应速度的大小,参见式(5-5-6)~式(5-5-8)。

经验表明，为了使闭环系统稳定并具有足够的相位裕度，开环对数幅频特性最好以 $-20\text{dB}/\text{dec}$ 的斜率通过 0dB 线，如图 5-5-4 所示。如果以 $-40\text{dB}/\text{dec}$ 的斜率通过 0dB 线，则闭环系统可能不稳定，即使稳定，相位裕量往往也比较小。如果以 $-60\text{dB}/\text{dec}$ 或更负的斜率通过 0dB 线，则闭环系统肯定不稳定，如图 5-5-4 所示，设

$$h = \frac{\omega_3}{\omega_2} \tag{5-5-11}$$

或

$$h = \lg\omega_3 - \lg\omega_2 \tag{5-5-12}$$

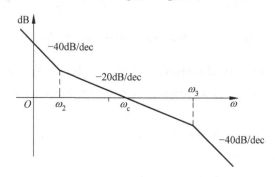

图 5-5-4 理想幅频特性曲线中频段示意图

建议按照下述公式选取 ω_2 和 ω_3

$$\omega_2 \leqslant \omega_c \frac{M_r - 1}{M_r} \tag{5-5-13}$$

$$\omega_3 \geqslant \omega_c \frac{M_r + 1}{M_r} \tag{5-5-14}$$

其中的关系可用如下经验公式表示

$$h \geqslant \frac{M_r + 1}{M_r - 1} \tag{5-5-15}$$

3. 高频段

比幅值穿越频率 ω_c 高出许多倍的频率范围称为高频段。系统开环幅频特性的高频部分对系统性能指标影响不大，一般只要求高频部分有比较负的斜率，幅值衰减得快一些。

频率特性设计法中，以系统的开环对数频率特性（Bode 图）为设计对象，主要是使开环对数幅频特性的低频、中频和高频部分满足要求。对低频段的要求是要具有足够高的放大系数。有时也要求加入积分环节以提高系统型别。对中频段的要求是要有足够宽的幅值穿越频率 ω_c，并确保足够的相位裕度。对数幅频渐近线以 $-20\text{dB}/\text{dec}$ 的斜率通过 0dB 线，并能保持足够的长度，就能达到要求的相位裕度。高频段一般不再特殊设计，依靠控制对象自身的特性实现高频衰减。

Reading Material

The bandwidth frequency is defined as the frequency at which the closed-loop magnitude response is equal to -3dB. However, when we design via frequency response, we are interested in predicting the closed-loop behavior from the open-loop response. Therefore, we will use a second-order system approximation and say that the bandwidth frequency equals the frequency at which the open-loop magnitude response is between -6 and -7.5dB, assuming the open loop phase response is between -135 deg and -225 deg.

In order to illustrate the importance of the bandwidth frequency, we will show how the output changes with different input frequencies. We will find that sinusoidal inputs with frequency less than ω_b (the bandwidth frequency) are tracked "reasonably well" by the system. Sinusoidal inputs with frequency greater than ω_b are attenuated (in magnitude) by a factor of 0.707 or greater (and are also shifted in phase).

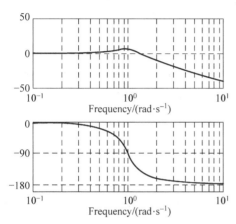

Fig. 5-5-5 Bode plot of the system

Let's say that we have the following closed-loop transfer function representing a system.

$$G(s) = \frac{1}{s^2 + 0.5s + 1}$$

First of all, let's find the bandwidth frequency by looking at the Bode plot in Fig. 5-5-5.

```
bode(1, [1 0.5 1])
```

Since this is the closed-loop transfer function, our bandwidth frequency will be the frequency corresponding to a gain of -3dB. Looking at the plot, we find that it is approximately 1.4rad/s. We can also read off the plot that for an input frequency of 0.3 radians, the output sinusoid should have a magnitude about one and the phase should be shifted by perhaps a few degrees (behind the input). For an input frequency of 3 rad/s, the output magnitude should be about -20dB (or 1/10 as large as the input) and the phase should be nearly -180 (almost exactly out-of-phase). We can use the *lsim* command to simulate the response of the system to sinusoidal inputs.

First, consider a sinusoidal input with a frequency lower than ω_b. We must also keep in mind that we want to view the steady state response. Therefore, we will modify the axes in order to see the steady state response clearly (ignoring the transient response). We will have Fig. 5-5-6 as the following.

```
w = 0.3;
num = 1;
den = [1 0.5 1 ];
t = 0:0.1:100;
u = sin(w*t);
[y,x] = lsim(num,den,u,t);
plot(t,y,t,u)
axis([50,100,-2,2])
```

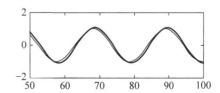

Fig. 5-5-6 Simulation result of the system with lsim command

Note that the output tracks the input fairly well; it is perhaps a few degrees behind the input as expected.

However, if we set the frequency of the input higher than the bandwidth frequency for the system, we get a very distorted response(with respect to the input)in Fig. 5-5-7.

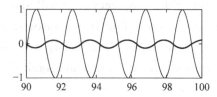

```
w = 3;
num = 1;
den = [1 0.5 1];
t = 0:0.1:100;
u = sin(w*t);
[y,x] = lsim(num,den,u,t);
plot(t,y,t,u)
axis([90, 100, -1, 1])
```

Fig. 5-5-7 A very distorted response of the system

Again, note that the magnitude is about 1/10 that of the input, as predicted, and that it is almost exactly out of phase(180 degrees behind)the input. Feel free to experiment and view the response for several different frequencies ω, and see if they match the Bode plot.

5.6　MATLAB 在本章中的应用

在本章中，主要介绍了控制系统频域表示方法，例如极坐标图、Bode 图等。这些频域特性图的手工绘制一般较为烦琐，得到的图形也不精确。在 MATLAB 的工作环境中，有专门的频域分析函数及工具。下面将以英文阅读材料的形式，首先复习频域分析法中的基本概念，进而介绍 MATLAB 中的相关函数及得到的频域分析曲线。最后，通过实例介绍如何利用频率特性进行控制系统设计。

Reading Material

1. Bode Plots

As noted above, a Bode plot is the representation of the magnitude and phase of $G(j\omega)$ (where the frequency vector w contains only positive frequencies). To see the Bode plot of a transfer function, you can use the MATLAB bode command. For example

bode(50,[1 9 30 40])

displays the Bode plots for the transfer function.

$$G(s) = \frac{50}{s^3 + 9s^2 + 30s + 40}$$

You will have the Bode plot as Fig. 5-6-1.

Please note the axes of the figure. The

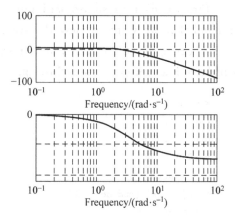

Fig. 5-6-1 A typical Bode plot in MATLAB

frequency is on a logarithmic scale, the phase is given in degrees, and the magnitude is given as the gain in decibels.

2. Gain and Phase Margin

Let's say that we have the following system in Fig. 5-6-2.

where K is a variable(constant) gain and $G(s)$ is the plant under consideration. The gain margin is defined as the change in open loop gain required to make the system unstable. Systems with greater gain margins can withstand greater changes in system parameters before becoming unstable in closed loop. The phase margin is defined as the change in open loop phase shift required to make a closed loop system unstable.

The phase margin also measures the system's tolerance to time delay. If there is a time delay greater than π/ω_{pc} in the loop(where ω_{pc} is the frequency where the phase shift is 180 deg), the system will become unstable in closed loop. The time delay can be thought of as an extra block in the forward path of the block diagram that adds phase to the system but has no effect on the gain. That is, a time delay can be represented as a block with magnitude of 1 and phase $\omega \times$ time_delay(in rad/s).

For now, we won't worry about where all this comes from and will concentrate on identifying the gain and phase margins on a Bode plot in Fig. 5-6-3.

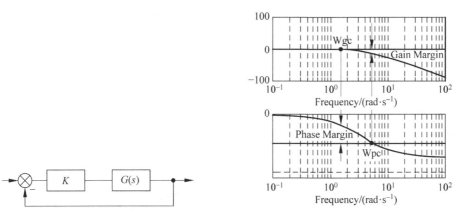

Fig. 5-6-2 A typical control system   Fig. 5-6-3 Gain and phase margins on a Bode plot

The phase margin is the difference in phase between the phase curve and -180 deg at the point corresponding to the frequency that gives us a gain of 0dB(the gain cross over frequency, ω_{gc}). Likewise, the gain margin is the difference between the magnitude curve and 0dB at the point corresponding to the frequency that gives us a phase of -180 deg(the phase cross over frequency, ω_{pc}).

One nice thing about the phase margin is that you don't need to replot the Bode in order to find the new phase margin when changing the gains. If you recall, adding gain only shifts the magnitude plot up. This is the equivalent of changing the y-axis on the magnitude plot. Finding the phase margin is simply the matter of finding the new cross-over frequency and reading off the phase margin. For example, suppose you entered the command

```
bode(50,[1 9 30 40])
```

You will get the following bode plot like Fig. 5-6-4.

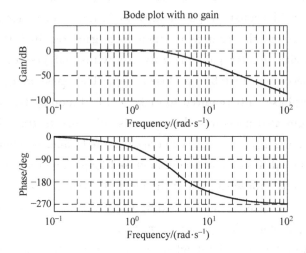

Fig. 5-6-4 A Bode plot with no gain

You should see that the phase margin is about 100 degrees. Now suppose you added a gain of 100, by entering the command

```
bode(100 * 50,[1 9 30 40])
```

You should get the following plot(note I changed the axis so the scale would be the same as the plot above, your bode plot (Fig. 5-6-5) may not be exactly the same shape, depending on the scale used).

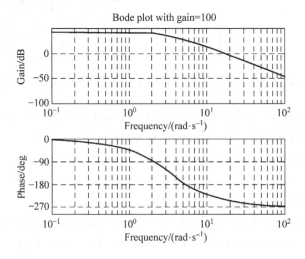

Fig. 5-6-5 A Bode plot with a gain

As you can see the phase plot is exactly the same as before, and the magnitude plot is shifted up by 40dB(gain of 100). The phase margin is now about −60 degrees. This same result could be achieved if the y-axis of the magnitude plot was shifted down −40dB. Try this, look at the first Bode plot, find where the curve crosses the 40dB line, and read off the phase margin. It should be about −60 degrees, the same as the second Bode plot.

We can find the gain and phase margins for a system directly, by using MATLAB. Just enter the *margin* command. This command returns the gain and phase margins, the gain and phase cross over frequencies, and a graphical representation of these on the Bode plot like Fig. 5-6-6. Let's check it out.

```
margin(50,[1  9  30  40])
```

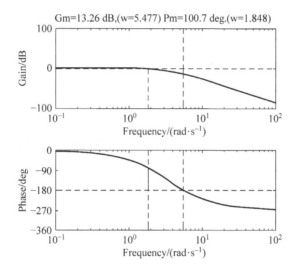

Fig. 5-6-6 A Bode plot with gain and phase margins

3. The Nyquist Diagram

The Nyquist plot allows us to predict the stability and performance of a closed-loop system by observing its open-loop behavior. The Nyquist criterion can be used for design purposes regardless of open-loop stability(remember that the Bode design methods assume that the system is stable in open loop). Therefore, we use this criterion to determine closed-loop stability when the Bode plots display confusing information.

To view a simple Nyquist plot using MATLAB, we will define the following transfer function and view the Nyquist plot.

$$G(s) = \frac{0.5}{s - 0.5}$$

```
nyquist(0.5,[1 -0.5])
```

You can easily get the result in MATLAB.

4. Closed-loop Performance

In order to predict closed-loop performance from open-loop frequency response, we need to have several concepts clear.

(1) The system must be stable in open loop if we are going to design via Bode plots.

(2) If the gain cross over frequency is less than the phase cross over frequency (i.e. $\omega_{gc} < \omega_{pc}$), then the closed-loop system will be stable.

(3) For second-order systems, the closed-loop damping ratio is approximately equal to the phase margin divided by 100 if the phase margin is between 0 and 60 deg. We can use this concept with caution if the phase margin is greater than 60 deg.

(4) For second-order systems, a relationship between damping ratio, bandwidth frequency and settling time is given by an equation described on the bandwidth page. A very rough estimate that you can use is that the bandwidth is approximately equal to the natural frequency.

Let's use these concepts to design a controller for the following system.

5. Case Studies

Suppose we have a open-loop control system with the following transfer function.

$$G(s) = \frac{10}{1.25s + 1}$$

The design must meet the following specifications.

☐ Zero steady state error.

☐ Maximum overshoot must be less than 40%.

☐ Settling time must be less than 2 seconds.

There are two ways of solving this problem: one is graphical and the other is numerical. Within MATLAB, the graphical approach is best, so that is the approach we will use. First, let's look at the Bode plot in Fig. 5-6-7. Create an m-file with the following code.

```
num = 10;
den = [1.25,1];bode(num, den)
```

There are several characteristics of the system that can be read directly from this Bode plot. First of all, we can see that the bandwidth frequency is around 10 rad/s. Since the bandwidth frequency is roughly the same as the natural frequency (for a second order system of this type), the rise time is 1.8/BW=1.8/10=1.8 seconds. This is a rough estimate, so we will say the rise time is about 2 seconds.

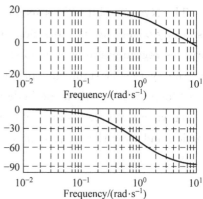

Fig. 5-6-7 A Bode plot of the specific control system

The phase margin for this system is approximately 95 degrees. This corresponds to a damping of PM/100 = 95/100 = 0.95. Plugging in this value into the equation relating overshoot and the damping ratio (or consulting a plot of this relation), we find that the damping ratio corresponding to this overshoot is approximately 1%. The system will be close to being over damped.

The last major point of interest is steady-state error. The steady-state error can be read directly off the Bode plot as well. The constant (K_p, K_v, or K_a) are located at the intersection of the low frequency asymptote with the $\omega = 1$ line. Just extend the low frequency line to the $\omega = 1$ line. The magnitude at this point is the constant. Since the Bode plot of this system is a horizontal line at low frequencies (slope = 0), we know this system is of type zero. Therefore, the intersection is easy to find. The gain is 20dB (magnitude 10). What this means is that the constant for the error function is 10. Click here to see the table of system types and error functions. The steady-state error is $1/(1+K_p) = 1/(1+10) = 0.091$. If our system was type one instead of type zero, the constant for the steady-state error would be found in a manner similar to the above.

Let's check our predictions by looking at a step response plot. This can be done by adding the following two lines of code into the MATLAB command window. Then, you will have Fig. 5-6-8.

[numc,denc] = cloop(num,den,-1);step(numc,denc)

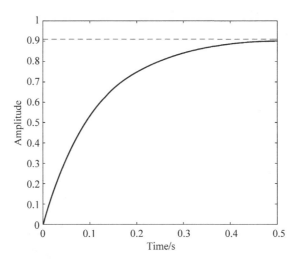

Fig. 5-6-8 Closed-loop step response without controller

As you can see, our predictions were very good. The system has a rise time of about 2 seconds, is overdamped, and has a steady-state error of about 9%. Now we need to choose a controller that will allow us to meet the design criteria. We choose a PI controller because it will yield zero steady state error for a step input. Also, the PI controller has a zero, which we can place. This gives us additional design flexibility to help us meet our criteria. Recall that a PI controller is given by

$$G_c(s) = \frac{K(s+a)}{s}$$

The first thing we need to find is the damping ratio corresponding to a percent overshoot of 40%. Plugging in this value into the equation relating overshoot and damping ratio(or consulting a plot of this relation), we find that the damping ratio corresponding to this overshoot is approximately 0.28. Therefore, our phase margin should be approximately 30 degrees. We must have a bandwidth frequency greater than or equal to 12 if we want our settling time to be less than 2 seconds which meets the design specs.

Now that we know our desired phase margin and bandwidth frequency, we can start our design. Remember that we are looking at the open-loop Bode plots. Therefore, our bandwidth frequency will be the frequency corresponding to a gain of approximately -7dB.

Let's see how the integrator portion of the PI or affects our response. Change your m-file to look like the following(this adds an integral term but no proportional term)Fig. 5-6-9.

```
num = [10];den = [1.25, 1];
numPI = [1];denPI = [1 0];
newnum = conv(num,numPI); newden = conv(den,denPI);
bode(newnum, newden, logspace(0,2))
```

Our phase margin and bandwidth frequency are too small. We will add gain and phase with a zero. Let's place the zero at 1 for now and see what happens. Change your m-file to look like the following in Fig. 5-6-10.

```
num = [10];den = [1.25, 1];
numPI = [1 1];denPI = [1 0];
newnum = conv(num,numPI);newden = conv(den,denPI);
bode(newnum, newden, logspace(0,2))
```

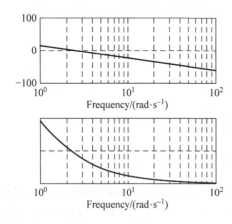
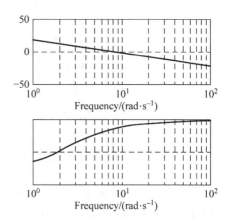

Fig. 5-6-9 A Bode plot with a PI controller(Ⅰ)   Fig. 5-6-10 A Bode plot with a PI controller(Ⅱ)

It turns out that the zero at 1 with a unit gain gives us a satisfactory answer. Our phase margin is greater than 60 degrees(even less overshoot than expected) and our bandwidth frequency is approximately 11 rad/s, which will give us a satisfactory response. Although satisfactory, the response is not quite as good as we would like. Therefore, let's try to get a higher bandwidth frequency without changing the phase margin too much. Let's try to increase the gain to 5 and see what happens. This will make the gain shift and the phase will remain the same in Fig. 5-6-11.

```
num = [10];den = [1.25, 1];
numPI = 5 * [1 1];denPI = [1 0];
newnum = conv(num,numPI);newden = conv(den,denPI);
bode(newnum, newden, logspace(0,2))
```

That looks really good. Let's look at our step response and verify our results. Add the following two lines to your m-file.

```
[clnum,clden] = cloop(newnum,newden,-1); step(clnum,clden)
```

As you can see in Fig. 5-6-12, our response is better than we had hoped for. However, we are not always quite as lucky and usually have to play around with the gain and the position of the poles and/or zeros in order to achieve our design requirements. For more details, refer to Reference [24].

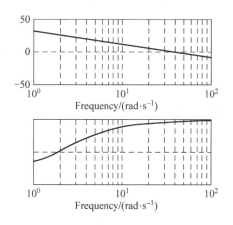

Fig. 5-6-11 A Bode plot with a PI controller(Ⅲ)

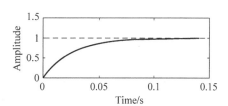

Fig. 5-6-12 Final step responses after design

本章小结

（1）频率特性表示线性定常系统在正弦信号作用下，其输出稳态值与输入量之比。它是传递函数的一种特殊形式，即 $G(j\omega) = G(s)|_{s=j\omega}$，因而频率特性与传递函数、线性定常微分方程一样，也是线性定常系统的一种数学模型。由于频率特性可用图形表示，相应的计算机仿真也简单，故频率特性法在控制工程中得到了广泛的应用。

（2）频率特性曲线主要包括幅相频率特性曲线和对数频率特性曲线。幅相频率特性曲线又称奈奎斯特曲线或极坐标图，对数频率特性曲线又称伯德图。频率特性法的重要特点之一，是可以根据系统的开环频率特性曲线分析系统的闭环性能。频率特性法的另一个优点是,可用实验的方法来获取线性系统的 Bode 图,进而得到系统的传递函数。

(3) 开环传递函数的极点和零点均在 s 左半平面的系统称为最小相位系统。由于这类系统的幅频和相频特性之间有着唯一的对应关系,因而只要根据它的对数幅频特性曲线就能确定其数学模型及相应的性能。

(4) 奈奎斯特稳定判据是根据开环频率特性曲线(−1,j0)点的情况和 s 右半平面上的极点数来判别对应闭环系统的稳定性的。相应地,在对数频率特性曲线上,可采用对数频率稳定判据。

(5) 考虑到系统的内部参数和外界环境的变化对系统稳定性的影响,要求控制系统不仅能稳定地工作,而且还有足够的稳定裕量。稳定裕量一般用相位裕量和幅值裕量来表征。在控制工程中,通常要求系统的相位裕量在 $\gamma = 40° \sim 60°$ 范围内,这是十分必要的。

(6) 为了方便地绘制对数频率特性曲线并用它来定性分析系统的性能,常将开环频率特性曲线分成低频段、中频段和高频段三个频段。低频段反映了系统的稳态精度;中频段主要表征系统的动态性能,它反映了系统的动态响应的平稳性和快速性;高频段则反映了系统高频干扰的能力。

(7) 二阶系统的开环频域指标与幅值穿越(剪切)频率 ω_c、相位裕度 γ、幅值裕度 K_g 有确定的关系;闭环频域指标与谐振频率 ω_r、谐振峰值 M_r 有着确定的关系;进而与时域指标相关。

(8) 若某线性系统(或部件)是稳定的,就可以用实验的方法来估计数学模型,这是频率特性法的一大优点。

Summary and Outcome Checklist

The frequency response of a stable system has been defined as the steady state response of the system to a sinusoidal input signal. Several alternative forms of frequency response plots have been considered. One can evaluate the frequency response of a system from the transfer function model of the system, or determine it by experiment. The stability of a closed-loop system can be ascertained from the open loop frequency response plot. Powerful control system design techniques exist that are based on the frequency response of the open loop system.

Frequency response methods can also be used to investigate stability of feedback control systems. Powerful design techniques based on frequency response exist for the design of control systems. These techniques are particularly useful in those cases where a transfer function model of control system is not available, and experimental data are used to design the controller. Moreover, even unstable systems can be studied and improved by determining open loop frequency response of the control system.

Tick the box for each statement with which you agree.

☐ I am able to explain the meaning of frequency response of a dynamic system.

☐ I can state the relationship that exists between the frequency transfer function of

a stable linear dynamic system and the ratio of the magnitudes of the output and input sinusoidal signals.
- [ ] I can state the relationship that exists between the frequency transfer function of a linear dynamic system and the phase difference between the output and input sinusoidal signals.
- [ ] I am able to present the frequency response data as a polar plot.
- [ ] I am able to present the frequency response data as a Bode diagrams.
- [ ] I am able to present the frequency response data as a log magnitude and phase diagram.
- [ ] I can explain how the frequency response of a stable linear dynamic system can be determined by experiment.
- [ ] I can explain how the transfer function of a linear dynamic system can be estimated from the frequency response data.
- [ ] I am able to list and explain the various frequency domain performance measures of feedback control systems.
- [ ] For a system that has a dominant pair of complex poles, I can state the relationship between the closed loop frequency domain performance measures and the time domain performance measures.
- [ ] I am able to use MATLAB bode and logspace commands. And obtain the frequency plots for a given transfer function.
- [ ] In the particular context of feedback control systems, I am able to distinguish between the open loop frequency response and the closed loop frequency response.

习题

5-1 系统的闭环传递函数为 $G(s) = \dfrac{C(s)}{R(s)} = \dfrac{K(T_2 s + 1)}{T_1 s + 1}$，输入信号为 $r(t) = R\sin\omega t$，求系统的稳态输出。

5-2 已知系统如题 5-2 图所示，$G_1(s) = s + 2$，$G_2(s) = \dfrac{1}{s(s+1)(s+2)+K}$，试用奈氏判据确定使系统稳定的 K 值。

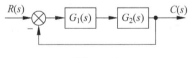

题 5-2 图

5-3 绘制下列传递函数的对数幅频特性图。

(1) $G(s) = \dfrac{1}{s(s+1)(2s+1)}$ (2) $G(s) = \dfrac{250}{s(s+5)(s+15)}$

(3) $G(s) = \dfrac{250(s+1)}{s^2(s+5)(s+15)}$ (4) $G(s) = \dfrac{500(s+2)}{s(s+10)}$

(5) $G(s) = \dfrac{2000(s-6)}{s(s^2+4s+20)}$ (6) $G(s) = \dfrac{2000(s+6)}{s(s^2+4s+20)}$

(7) $G(s) = \dfrac{2}{s(0.1s+1)(0.5s+1)}$ (8) $G(s) = \dfrac{2s^2}{(0.04s+1)(0.4s+1)}$

5-4 已知最小相位开环系统对数幅频特性如题 5-4 图所示,求开环传递函数。

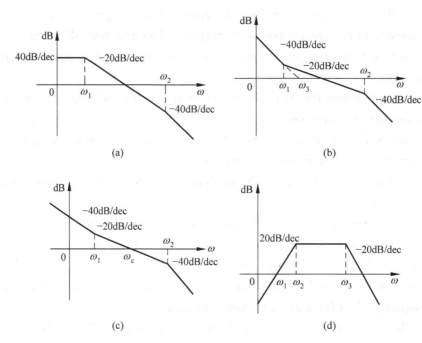

题 5-4 图

5-5 已知单位反馈系统其开环传递函数如下,试绘制其伯德图并判断其稳定性。

(1) $G(s) = \dfrac{100}{(0.25s+1)(0.0625s+1)} \cdot \dfrac{0.2s^2}{0.8s+1}$

(2) $G(s) = \dfrac{5(1-0.5s)}{s(0.1s+1)(1-0.2s)}$

(3) $G(s) = \dfrac{1000}{s(s^2+25)(0.2s+1)}$

5-6 设系统开环幅相频率特性如题 5-6 图所示,试判别闭环系统的稳定性。(图中,r 为原点的开环极点数,n 为位于右半 s 平面的开环极点数。)

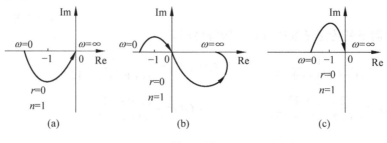

题 5-6 图

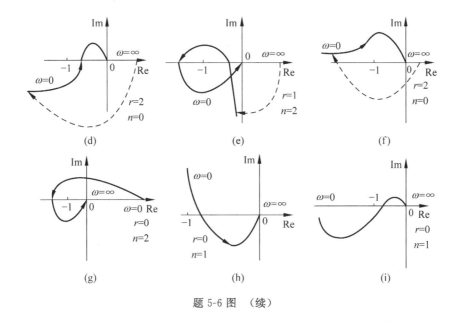

题 5-6 图 （续）

5-7 题 5-7 图表示开环奈奎斯特图的负频部分，P 为开环正实部极点个数，判断闭环系统是否稳定。

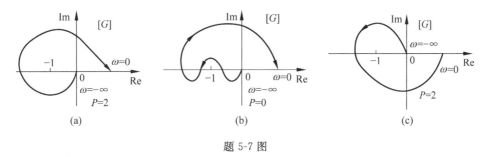

题 5-7 图

5-8 已知单位反馈系统的开环传递函数如下，其相应的幅相曲线如题 5-8 图所示，试用奈氏判据判别闭环系统的稳定性（$\xi, \omega_n, K, T_i(i=1,2,3,4,5,6)$ 皆大于零）。

5-9 某单位反馈系统的前向通道的开环传递函数为

$$G(s) = \frac{2s^4 + 8s^3 + 12s^2 + 8s + 2}{s^6 + 5s^5 + 10s^4 + 10s^3 + 5s^2 + s}$$

（1）求系统的开环增益。

（2）$G(s)$ 是否是最小相位系统？

（3）闭环系统是否稳定？

（4）求系统的开环对数幅频特性的高频段 $\omega \to +\infty$ 的渐近线的斜率及相频特性自 $\omega \to 0$ 至 $\omega \to +\infty$ 的相角变化量。

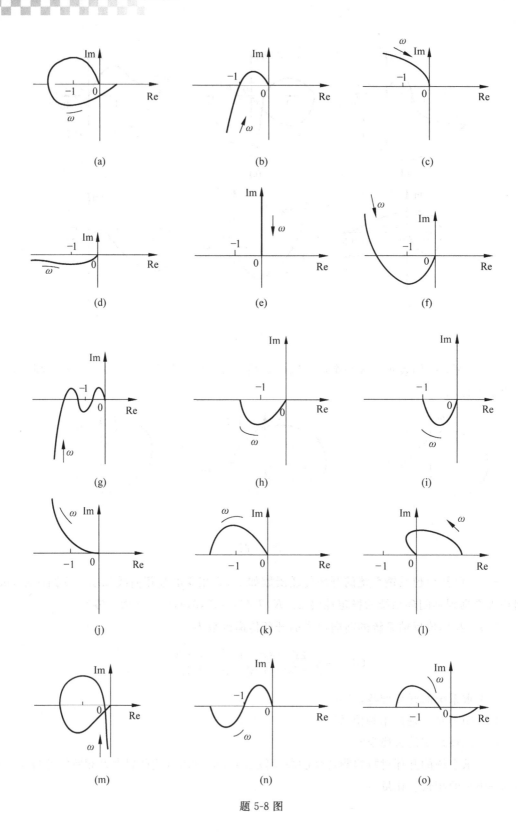

题 5-8 图

5-10 已知单位反馈系统的开环传递函数为

$$G(s) = \frac{K\prod_{j=1}^{m}(\tau_j s+1)}{s(Ts-1)\prod_{j=1}^{m}(\tau_j s+1)}$$

其中除了一个位于原点和一个位于正实轴的极点外,其余开环零极点均位于负实轴上。当开环放大系数为 $K=1$ 时,系统的开环对数频率特性曲线 $L(\omega)$ 和 $\varphi(\omega)$ 如题 5-10 图所示。讨论欲使闭环系统稳定的开环放大系数 K 的取值范围 $K>0$。

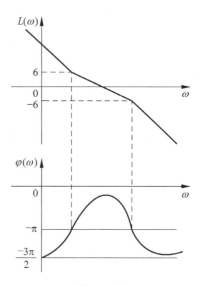

题 5-10 图

5-11 已知系统开环传递函数为

$$G(s)H(s) = \frac{K(T_1 s+1)}{s^2(T_2 s+1)} \quad T_1 > T_2 > 0$$

试求 K 变化下相角裕度 γ 的最大值。

5-12 (西安交通大学硕士研究生入学考试试题)应用奈奎斯特准则研究系统稳定性。已知闭环系统的开环传递函数:

$$G(s) = \frac{K}{s(0.2s+1)(2s+1)}$$

求解下列问题:

(1) 写出幅频特性、相频特性、实频特性和虚频特性的表达式。
(2) 当 $K=10$ 时,画出该系统的奈奎斯特图,并说明闭环系统的稳定性。
(3) 求出系统处于临界状态时的 K 值。

5-13 (华南理工大学硕士研究生入学考试试题)某最小相角系统的开环对数幅频特性如题 5-13 图所示。试求:

(1) 写出系统开环传递函数。
(2) 利用相位裕量判断系统稳定性。
(3) 将其对数幅频特性向右平移十倍频程,试讨论对系统性能的影响。

5-14 (燕山大学硕士研究生入学考试试题)系统结构图如题 5-14 图所示。其中 a, K_1 均大于零。当输入为 $r(t)=\sin t$ 时,系统的稳态响应 $C(t)=\sin(t-45°)$;问系统的相角裕量是多少?

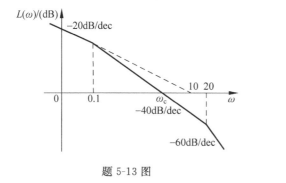

题 5-13 图

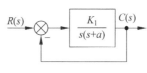

题 5-14 图

5-15　（武汉科技大学硕士研究生入学考试试题）设单位负反馈系统开环传递函数为：

$$G_K(s) = \frac{10}{s(s^2 + 4s + 100)}$$

试求：

（1）写出系统开环幅频、相频特性表达式。

（2）绘制系统开环对数幅频特性（渐近线）和相频特性。

（3）确定该系统的相位穿越频率，并判断系统的稳定性。

5-16　（南京航空航天大学硕士研究生入学考试试题）已知单位反馈系统开环传递函数为

$$G(s) = \frac{K}{(s+1)(0.1s+1)(0.5s+1)}$$

（1）试用奈氏判据确定闭环系统稳定的 K 值范围。

（2）若希望系统的闭环极点全部位于 $s=-1$ 垂线左侧，试用奈氏判据确定此时 K 的取值范围（其他方法无效）。

5-17　（浙江大学硕士研究生入学考试试题）单位反馈系统的开环传递函数为

$$G(s) = \frac{K(s+1)}{s^3 + 0.8s^2 + 2s + 1}$$

（1）确定引起闭环系统产生等幅振荡的 K 值。

（2）求相应的振荡频率。

5-18　（电子科技大学硕士研究生入学考试试题）某单位负反馈系统的开环传递函数为

$$G(s) = \frac{K(s+1)}{s(0.1s+1)(10s+1)}$$

（1）绘制系统的对数渐近幅频特性曲线。

（2）绘制系统的对数相频特性曲线。

5-19　（哈尔滨工业大学硕士研究生入学考试试题）已知单位负反馈系统的开环传递函数为

$$G(s) = \frac{K(T_1 s + 1)}{s^2 (T_2 s + 1)} \quad (T_1 > T_2 > 0)$$

$T_1 = 2, T_2 = 1$ 时，求使系统相角裕量最大的 K 值。

第6章 线性系统的校正方法

6.1 引言

在前面几章里,我们学习了控制系统的三种工程分析方法,即时域方法(Time Domain Methods)、根轨迹方法(Root Locus Methods)和频域方法(Frequency Domain Methods),利用这些方法能够在系统结构和参数已经确定的条件下,计算或估算出它们的性能。但在工程实际中往往需要设计一个自动控制系统,这种设计过程一般经过以下三步。

(1) 根据任务要求,选定控制对象。

(2) 根据性能指标的要求,确定系统的控制规律,并设计出满足这个控制规律的控制器,初步选定构成控制器的元器件。

(3) 将选定的控制器与被控对象组成一个控制系统,如果构成的系统不能满足或不能全部满足设计要求的性能指标,还必须增加合适的元件,按一定的方式连接到原系统中,使重新组合起来的系统全面满足设计要求。

在控制理论中,这种添加新环节的过程称为控制系统的校正,附加的环节称为校正装置或校正元件。工业过程控制中所用的比例-积分-微分控制器(Proportional Integral Derivative, PID)就属于校正装置。同时,在自动控制原理中这类问题称为系统的综合。综合的目的就是在系统中引入各种附加装置,使系统的缺点得到校正,从而满足一定的性能指标。

校正装置的设计是自动控制系统全局设计中的重要组成部分。设计者的任务是在不改变系统基本部分的情况下,选择合适的校正装置,并计算、确定其参数,以使系统满足各项性能指标的要求。

Reading Material

In this chapter, we address the central issue of improving the performance of feedback control system by making minor modification to the structure of the feedback control system. Additional hardware, referred to as the compensator, is incorporated into the control system structure that enables us to achieve the desired system performance. A compensator is an additional component or circuit that is inserted into a control system to compensate for a deficient performance. Commonly a two port electric network can serve as a compensator in many control systems. Several types of compensation schemes are available of which the lead and lag networks are the most widely used. Root locus method can be used for the synthesis of the compensator transfer function.

本章主要内容：
- □ 系统校正的基本概念
- □ 系统校正装置
- □ 反馈校正
- □ MATLAB 在本章中的应用

6.2 系统校正的基本概念

6.2.1 受控对象

一般而言，将受控对象和控制装置同时进行设计是比较合理的，这样就能充分发挥控制的作用，使受控对象获得特殊要求下良好的技术性能。然而，在很多实际生产过程中，一般先给定受控对象，然后才进行系统设计。这就要求不论在何种情形下，都要详细了解受控对象的工作原理和特点，准确掌握受控对象的动态数学模型，以及受控对象性能的要求，作为系统设计的主要依据。

Terms and Concepts

Design is the process of conceiving or inventing the forms, parts, and details of a system to achieve a specified purpose.

Specifications: Statement that explicitly state what the device or product is to be and to do. It is a set of prescribed performance criteria.

Synthesis: The process by which new physical configurations are created. The combining of separate elements or devices to form a coherent whole.

Trade-off: The result of making a judgment about how much compromise must be between conflicting criteria.

6.2.2 性能指标概述

性能指标是衡量控制系统性能优劣的尺度，也是综合和校正控制系统的主要技术依据。校正装置的设计通常是针对某些具体性能指标来进行的。结合前述章节的讨论，从稳态性能和动态性能方面考虑，有以下两种常用的性能指标范畴。

(1) 稳态性能指标。主要包括静态位置误差系数 K_p、静态速度误差系数 K_v 和静态加速度系数 K_a。

(2) 动态性能指标。常用的时域性能指标有最大超调量 $\sigma_p\%$、峰值时间 t_p、调节时间 t_s、稳态误差 e_{ss}、相角裕度 γ、幅值裕度 K_g、谐振峰值 M_p、谐振频率 ω_p、系统带宽 ω_b 等。

将时域分析指标和频域分析指标统一分类，常用的时域性能指标有最大超调量 $\sigma_p\%$、调节时间 t_s、静态位置误差系数 K_p、静态速度误差系数 K_v 和静态加速度系数 K_a。一般应用根轨迹法进行综合与校正会比较方便，即使用根轨迹法确定合适的校正装置的传递函数，使最终的根轨迹具有期望的闭环特征根分布。常用的频域性能指标有相角裕度 γ、幅值裕度 K_g、谐振峰值 M_p、谐振频率 ω_p、系统带宽 ω_b，通常应用频率特性法进行综合与校正，即校正装置的设计根据绘制在极坐标图、Bode 图或尼柯尔斯图上的频率响应曲线确定。

6.2.3 系统校正连接方式

按照校正装置在系统中的连接方式，控制系统的校正方式有串联校正方式（Cascade Compensation）、并联校正方式（Parallel Compensation）和复合校正方式（Compound Compensation）三种。

Terms and Concepts

Cascade compensation network: A compensator network placed in cascade or series with the system process.

Compensation: The alteration or adjustment of a control system in order to provide a suitable performance.

Compensator: An additional component or circuit that is inserted into the system to compensate for a performance deficiency.

1. 串联校正方式

串联校正装置作为一种重要的控制手段，其接入位置应视校正装置本身的物理特性和原系统的结构而定，一般接在系统误差测量之后和放大器之前，如图 6-2-1 所示。其中，$G_c(s)$ 为校正装置传递函数，$G(s)$ 为原系统前向通道的传递函数，$H(s)$ 为原系统反馈通道的传递函数。对于体积小、质量轻、容量小的校正装置（电器装置居多），常加在系统信号容量不大的地方，即比较靠近输入信号的前向通道中；相反，对于体积、质量、容量较大的校正装置（如无源网络、机械、液压、气动装置等），常串接在容量较大的部位，即比较靠近输出信号的前向通道中。

2. 并联校正方式

如果校正装置接在系统局部反馈通道中，这种校正方式称为并联校正，或称为反馈校正（Feedback Compensation），如图 6-2-2 所示。其中，$G_c(s)$ 的反馈作用除了使系统的性能得到改善之外，还能抑制系统参数的波动并减小非线性因素对系统性能的影响。由于并联校正装置的输入端信号取自于原系统的输出端或原系统前向通道中某个环节的输出端，信号功率一般都比较大，因此，在校正装置中不需要设置放大电路，有利于校正装置的简化。但由于输入信号功率比较大，校正装置的容量和体积相应要大一些。

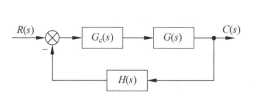

图 6-2-1 系统的串联校正方式

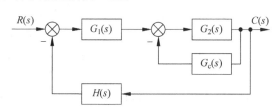

图 6-2-2 系统的并联校正方式

3. 复合校正方式

复合校正方式是在系统中除反馈回路之外采用的校正方法，如图 6-2-3(a) 和图 6-2-3(b) 所示。其中，图 6-2-3(a) 为输入补偿的复合控制形式，图 6-2-3(b) 为按扰动补偿的复合控制形式。复合控制不但可以在保持系统稳定的前提下极大地减小稳态误差，甚至可以消除稳态误差，而且几乎可以抑制所有可测量的扰动。因此在高精度的控制系统中，复合控制得到了广泛的应用。

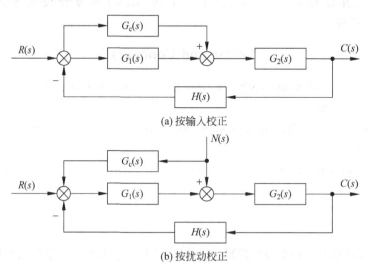

图 6-2-3　系统的复合校正方式

通过结构图的变换，一种连接方式可以等效地转换成另一种连接方式，它们之间的等效性决定了系统的综合与校正的非唯一性。在工程应用中，究竟采用哪一种连接方式，这要视具体情况而定。一般来说，要考虑的因素有原系统的物理结构、性能指标的要求、信号是否便于取出和加入、信号的性质、系统中各点功率的大小、可供选用的元件、环境使用条件，还有设计者的经验和经济条件等。串联校正简单且较易实现，也比较容易对信号进行各种必要形式的变换。

Reading Material

A suitable control system should have some of the following properties：

(1) It should be stable and present acceptable response to input command, i. e., the controlled variable should follow the changes in the input at a suitable speed without unduly large oscillations or overshoots；

(2) It should operate with as little error as possible；

(3) It should be able to mitigate the effect of undesirable disturbances.

All in all, a feedback control system must：

(1) be stable；

(2) have an acceptable transient response to input commands；

(3) have minimum steady state error for steady input commands;

(4) be sufficiently insensitive to system parameter changes;

(5) reduce the effect of undesirable disturbance inputs.

The preceding chapters have shown that it is often possible to adjust the system parameters in order to provide the desired system response. When the achievement of a simple performance requirement may be met by selecting a particular value of K, the process is called gain compensation. However, we often find that it is not sufficient to adjust a system parameter and thus obtain the desired performance. In order to obtain a suitable system, we must reconsider the structure of the system and redesign the system. The alteration or adjustment of a control system in order to provider a suitable performance is called compensation.

6.2.4 基本控制规律

通常自动化控制系统是由被控对象和控制器两大部分组成的。控制器按实际要求向被控对象以某种规律发出控制信号,以达到控制要求。包含校正装置在内的控制器,常常采用比例、微分、积分等基本控制规律,或者采用这些基本控制规律的某些组合,如比例-微分、比例-积分、比例-积分-微分等组合控制规律,以实现对被控对象的有效控制。

Reading Material

In redesigning a control system, an additional component or device is inserted within the structure of the feedback system to compensate for the performance deficiency. The compensating device may be electric, mechanical, hydraulic, pneumatic, or some other type of devices or networks and is often called a compensator. Commonly an electric network serves as a compensator in many control systems.

1. 比例(P)控制

控制器的输出量与输入量成正比,这种控制器为比例控制器,简称 P 控制器,如图 6-2-4 所示。

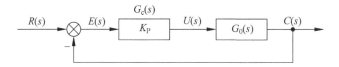

图 6-2-4 比例控制系统结构图

由定义可知,比例控制器的时域方程为
$$u(t) = K_P e(t) \tag{6-2-1}$$
对其进行拉普拉斯变换,则比例控制器的传递函数为
$$G_c(s) = K_P \tag{6-2-2}$$
比例控制器实质上是一个具有可调增益的放大器。它的作用是调整系统的开环比例系

数,以提高系统的稳态精度,同时 P 控制器只改变信号的增益而不影响其相位,是一个纯增益。但在串联校正中,它会降低系统的相对稳定性,甚至可能造成闭环系统不稳定。因此在系统校正中,很少单独采用比例控制器,而是与其他控制环节共同使用。

2. 积分(I)控制

控制器的输出量是输入量对时间的积分,则称这种控制器为积分控制器,简称 I 控制器,如图 6-2-5 所示。

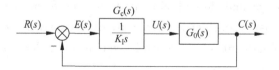

图 6-2-5　积分控制系统结构图

由定义可知,积分控制器的时域方程为

$$u(t) = \frac{1}{K_I} \int e(t) \mathrm{d}t \tag{6-2-3}$$

对其进行拉普拉斯变换,则积分控制器的传递函数为

$$G_c(s) = \frac{1}{K_I s} \tag{6-2-4}$$

式中,K_I 为可调比例系数,由于 I 控制器的积分作用,当输入 $e(t)$ 消失后,输出信号 $u(t)$ 有可能是一个不为零的常量。

加入 I 控制器可以提高系统的型别,0 型系统变为 I 型系统,并且由静差系统变为一阶无静差系统,有利于系统稳定性能的提高,但积分控制使系统增加了一个位于原点的开环极点,使信号产生 90°的滞后,从而导致系统可能不稳定。因此在控制系统的校正设计中,通常不宜采用单一的积分控制器。

3. 比例-积分(PI)控制

控制器的输出量与输入量成正比,又与输入量对时间的积分成正比,则称这种控制器为比例-积分控制器,简称 PI 控制器,如图 6-2-6 所示。

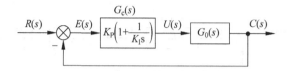

图 6-2-6　比例-积分控制系统结构图

由定义可知,比例-积分控制器的时域方程为

$$u(t) = K_P \left[e(t) + \frac{1}{K_I} \int e(t) \mathrm{d}t \right] \tag{6-2-5}$$

对其进行拉普拉斯变换,则比例-积分控制器的传递函数为

$$G_c(s) = K_P \left(1 + \frac{1}{K_I s} \right) \tag{6-2-6}$$

其中，K_P 和 K_I 都为可调系数，调节 K_I 只能影响积分控制分量，而调节 K_P 则对比例控制分量及积分控制分量均有影响。为了研究方便，令 $K_P=1$，则

$$G_c(s) = 1 + \frac{1}{K_I s} \tag{6-2-7}$$

其中，比例控制器的作用是使系统趋于稳定，积分控制器的作用是使原来的 0 型系统变为 Ⅰ 型系统，Ⅰ 型系统变为 Ⅱ 型系统，提高了系统的型别，趋于消除或减小对各种输入响应中的稳态误差。在串联校正中，PI 控制器相当于在系统中增加了一个位于原点的开环极点，同时也增加了一个位于 s 左半平面的开环零点。在控制工程实践中，PI 控制器主要用来改善控制系统的稳态性能。

4. 比例-微分(PD)控制

控制器的输出量与输入量成正比，又与输入量的一阶导数成正比，则称这种控制器为比例-微分控制器，简称 PD 控制器，如图 6-2-7 所示。

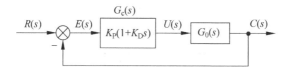

图 6-2-7　比例-微分控制系统结构图

由定义可知，比例-微分控制器的时域方程为

$$u(t) = K_P \left[e(t) + K_D \frac{\mathrm{d}e(t)}{\mathrm{d}t} \right] \tag{6-2-8}$$

对其进行拉普拉斯变换，则比例-微分控制器的传递函数为

$$G_c(s) = K_P(1 + K_D s) \tag{6-2-9}$$

其中，K_P 和 K_D 都为可调系数，调节 K_D 只能影响微分控制分量，而调节 K_P 则对比例控制分量及微分控制分量均有影响。

微分控制能够反映信号的变化速度，并且在作用误差的值变得很大之前会产生一个有效的修正，以增加系统的阻尼程度，从而改善系统的稳定性。但是微分控制只对动态过程起作用，而对稳态过程没有影响，且对噪声非常敏感，所以微分作用不能单独使用，它总是与比例控制或比例-积分控制组合应用于实际的控制系统。

由于微分控制可以预见输入量变化，并能及时采取措施以控制系统，这就是 PD 控制器的"预见性"。但需要注意的是，微分控制虽然有"预见"信号变化趋势的优点，但同时存在着将高频干扰信号放大的缺点，所以在设计控制系统时，应注意这个问题。

5. 比例-积分-微分(PID)控制

控制器的输出量与输入量成正比，又与输入量对时间的积分成正比，还与输入量的一阶导数成正比，则称这种控制器为比例-积分-微分控制器，简称 PID 控制器，如图 6-2-8 所示。PID 控制器用于瞬态响应和稳态响应都需要改善的控制系统。

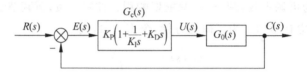

图 6-2-8　比例-积分-微分控制系统结构图

由定义可知，PID 控制器的时域方程为

$$u(t) = K_\text{P}\left[e(t) + \frac{1}{K_\text{I}}\int e(t)\,\mathrm{d}t + K_\text{D}\frac{\mathrm{d}e(t)}{\mathrm{d}t}\right] \tag{6-2-10}$$

对其进行拉普拉斯变换，则比例-积分-微分控制器的传递函数为

$$G_\text{c}(s) = K_\text{P}\left(1 + \frac{1}{K_\text{I}s} + K_\text{D}s\right) \tag{6-2-11}$$

当利用 PID 控制器进行串联校正时，除可使系统的型别提高一级外，还将提供两个负实部零点。与 PI 控制器相比，PID 控制器除了同样具有提高系统的稳态性能的优点外，还多提供了一个负实部零点，使在提高系统动态性能方面，具有更大的优越性。因此在工业过程控制系统中，PID 控制器得到广泛应用。PID 控制器各部分参数的选择，在系统现场调试中最后确定。通常 I 部分发生在系统频率特性的低频段，以提高系统的稳态性能；D 部分发生在系统频率特性的中频段，以提高系统的动态性能。

实际中的自动控制系统传递函数往往和标准形式略有不同并更为复杂，因此需要对 PID 控制器进行许多修正，以使它们具有更优良的性能。下面的英文阅读材料给出 PID 控制的基本原理，并针对参数调节作深入介绍。

Reading Material

In fact, the PID is probably the most commonly used compensator in feedback control systems.

Control effort produced by a PID controller is given by

$$u_\text{c} = K\left[e + \frac{1}{T_\text{r}}\int e\,\mathrm{d}t + T_\text{d}\frac{\mathrm{d}e}{\mathrm{d}t}\right] = u_\text{p} + u_\text{r} + u_\text{d} \tag{6-2-12}$$

Integral action u_r looks after long persistent errors. Its magnitude continues to grow with the passage of time as long as error e persists. Proportional action u_p responds immediately to changes in deviation but it is insensitive to how rapidly the deviation is changing. Derivative action u_d adds in an extra term which is proportional to the rate of change of deviation. The factor, T_d, called the derivative time, is used to weight the amount of derivative action relative to the proportional action. Similarly, the amount of integral action is weighted relative to the proportional action by a factor, T_r, called the reset time.

The idea behind a general purpose PID process controller is to employ the controller in a flexible manner so that the three settings, K, T_d and T_r, can be adjusted in situ to produce the desired response. The values of the parameters can often be determined by trial and error, if the plant is not known exactly. If the parameters of the plant are subject to large variation, the parameters of PID controller can be adjusted to improve the performance.

To get the best performance from a three-term controller, the amount of each action has to be selected carefully. If a perfect model of the plant is available, then the selection process could be done through simulation or other analytical techniques. With process plant the model is only vague and liable to change, and so selection procedure has to be done in situ and this is termed control-loop tuning. Two method widely used in industry are both based on an article published by Ziegler and Nichols dating from 1942 which established a set of empirical rules for tuning controllers. We give A typical PID control system like Fig. 6-2-9.

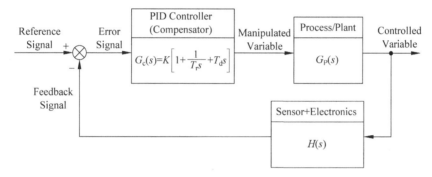

Fig. 6-2-9　A typical PID control system

The manipulated variable is computed as a sum of the error signal, integral of the error signal, and the derivative of error signal. The weighing factors, $K = \dfrac{100}{\%PB}$ (where $\%PB$ is referred to as the proportional band in the process industries), T_r (reset time) and T_d (derivative time) will have to be selected to suit the particular system. Integral component increases the type number of the system by one, and hence improves steady state error. Derivative component improves stability and transient response of the system. Pure derivative action is not practical and it would amplify high frequency noise. Therefore the derivative component is usually used in conjunction with a low pass filter, i. e. $\dfrac{T_d s}{1 + \dfrac{T_d}{N} s}$, instead of $T_d s$. N is conveniently chosen to suit the particular system.

A number of general-purpose PID controllers are used in the process industries. These controllers are incorporated into the closed loop and tuned in situ by carrying out a few basic experiments on the actual system (Ziegler-Nichols methods of tuning PID controllers). The design problem then is to choose the correct type of compensator in a given situation, and to determine the optimum values of the compensator parameters. Root locus method can be used for the synthesis of the controller parameters. MATLAB's *rltool* can be used for designing the compensator.

6.3 系统校正装置

根据校正装置的特性，常分为无源及有源校正网络。无源校正又可分为超前校正装置、滞后校正装置和滞后-超前校正装置三种。其中，滞后-超前校正装置是在滞后校正后面紧接着超前校正，因此它是二阶的。比起单独的滞后校正或者超前校正，滞后-超前校正的灵活性要高得多。

Terms and Concepts

Phase-lag network: A network that provides a negative phase angle and a significant attenuation over the frequency range of interest.

Phase-lead network: A network that provides a positive phase angle over the frequency range of interest. Thus phase lead can be used to cause a system to have an adequate phase margin.

Lead-lag network: A network with the characteristics of both a lead network and a lag network.

6.3.1 超前校正装置

校正装置输出信号在相位上超前于输入信号，即校正装置具有正的相角特性，这种校正装置称为超前校正装置，对系统的校正称为超前校正(Lead Compensation)。超前校正的基本原理是利用超前校正网络的相位超前去增大系统的相角裕度，以改善系统的暂态响应。从原理上讲，6.2节中介绍的PD控制器就属于超前补偿网络。

1. RC 超前网络的特性

如图 6-3-1 所示为一个无源超前网络的电路图。

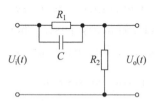

图 6-3-1 无源超前网络的电路图

其传递函数为

$$G(s) = \frac{U_o(s)}{U_i(s)}$$
$$= \frac{R_1}{R_1 + R_2} \cdot \frac{R_1 Cs + 1}{\frac{R_2}{R_1 + R_2} R_1 Cs + 1} \quad (6\text{-}3\text{-}1)$$

令 $T = \frac{R_1 R_2}{R_1 + R_2} C, \alpha = \frac{R_1 + R_2}{R_2} > 1$，则传递函数可写成

$$G(s) = \frac{1}{\alpha} \cdot \frac{\alpha Ts + 1}{Ts + 1} \quad (6\text{-}3\text{-}2)$$

由式(6-3-2)可知，无源超前网络具有幅值衰减作用，衰减系数为 $1/\alpha$，如果给超前无源网络串接一个放大系数为 α 的比例放大器，就可补偿幅值。此时，超前网络传递函数可写成

$$\alpha G(s) = \frac{\alpha Ts + 1}{Ts + 1} \quad (6\text{-}3\text{-}3)$$

根据式(6-3-3),作出无源超前网络的 Bode 图,如图 6-3-2 所示。显然超前网络对数频率特性在 $1/(\alpha T) \sim 1/T$ 之间的输入信号有明显的微分作用,在该频率范围内,输出信号的相位比输入信号的相位超前,超前网络的名称由此而来。

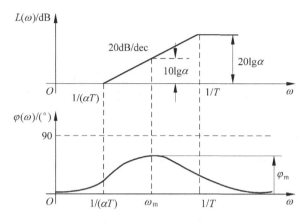

图 6-3-2　无源超前网络的 Bode 图

图 6-3-2 表明,在最大超前角频率 ω_m 处,具有最大超前角 φ_m,且 ω_m 正好处于两转折频率 $1/(\alpha T)$ 和 $1/T$ 的几何中心。其中最大超前角 φ_m 和对应的最大超前角频率 ω_m 分别为

$$\omega_m = \frac{1}{T\sqrt{\alpha}}$$
$$\varphi_m = \arcsin\frac{\alpha+1}{\alpha-1} \tag{6-3-4}$$

或写成

$$\alpha = \frac{1+\sin\varphi_m}{1-\sin\varphi_m} \tag{6-3-5}$$

式(6-3-5)表明,最大超前角 φ_m 仅与分度系数 α 有关。φ_m 随 α 的增大而增加,但当 $\alpha>15$ 以后,φ_m 几乎不增加,故一般很少取 $\alpha>15$。此外,φ_m 处的对数幅值为

$$L_c(\omega) = 10\lg\alpha \tag{6-3-6}$$

2. 串联超前校正

这种简单的超前网络若在系统的前向通路上(一般是串接于两级放大器之间),就构成了串联超前校正。给系统串入串联超前校正,减小对数幅频特性在幅值穿越频率上的负斜率,可以有效地改善原系统的平稳性和稳定性,并可以提高系统的频带宽度,对快速性也将产生有利的影响,但是超前校正很难使原系统的低频段特性得到改善。如果采取进一步开环增益的办法,使低频段上移,则系统的平稳性有所下降;幅频段过分上移,还会大大削弱系统干扰的能力。故超前校正对提高系统稳态精度的作用是很小的。

3. 利用 Bode 图设计超前校正网络

超前校正的基本原理是利用超前网络的相位特性去增大系统的相位裕度,以改善系统

的瞬态响应，具体设计步骤如下。

(1) 求出满足稳态指标的开环放大系数 K 值。

(2) 根据求得的 K 值，画出未校正系统的 Bode 图，并计算出其幅值穿越频率 ω_c、相位裕度 γ、幅度裕度 K_g。

(3) 确定需要对系统增加的相位超前 φ_m，φ_m 可表示为

$$\varphi_m = \gamma' - \gamma - \Delta \tag{6-3-7}$$

式中，γ' 和 γ 分别表示期望的相位裕度和未校正系统（原系统）的相位裕度，Δ 为增加超前网络后使幅值穿越频率向右方移动所带来的原系统相位的滞后量，一般该滞后量为 $5°\sim12°$。

(4) 确定 α 值。

(5) 确定校正后系统的幅值穿越频率 ω_c'。

为了最大限度利用超前网络的相位超前量。ω_c' 应与 ω_m 相重合，即 ω_c' 应选在未校正系统的 $L(\omega) = -10\lg\alpha$ 处。

(6) 确定校正装置的传递函数。

令 $\omega_m = \omega_c' = 1/(T\sqrt{\alpha})$，从而求出超前校正网络的两个转折频率

$$\begin{cases} \omega_1 = \dfrac{1}{\alpha T} \\ \omega_2 = \dfrac{1}{T} \end{cases} \tag{6-3-8}$$

由此得出校正装置具有的传递函数为

$$G(s) = \frac{\dfrac{s}{\omega_1}+1}{\dfrac{s}{\omega_2}+1} = \frac{\alpha T s + 1}{T s + 1} \tag{6-3-9}$$

(7) 验证校正后系统的相位裕度 γ。

如果不满足，则需要从步骤(3)重复上述设计过程，直到获得满意的结果为止。

【例 6-3-1】 考虑开环传递函数为

$$G_o(s) = \frac{K}{s(0.1s+1)(0.01s+1)}$$

的系统，希望稳态误差系统 $K_v \geqslant 100\text{s}^{-1}$，相角裕度 $\gamma \geqslant 30°$，且穿越频率 $\omega_c \geqslant 45\text{rad/s}$。

解：由于该系统为 I 型，因此选择

$$K = K_v = 100\text{s}^{-1}$$

具体步骤如下：

(1) 如图 6-3-3 所示，以绘制未校正系统的 Bode 图。

(2) 在本例中，校正后的穿越频率应当满足响应速度的要求。如果选择 $\omega_c = 50\text{rad/s}$，则 $G_o(j\omega)$ 在该频率处的相角为

$$\angle G_o(j\omega)|_{\omega=50} = -90° - \arctan(0.1\times50) - \arctan(0.01\times50) = -195°$$

(3) 因为希望相角裕度 $\gamma \geqslant 30°$，所以，所需的超前相角至少为

$$30° - (180° - 195°) = 45°$$

在留有余地的情况下，选择由校正装置提供的超前相角为 $\varphi_m = 55°$。

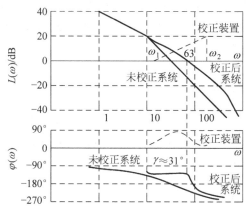

图 6-3-3 例 6-3-1 的 Bode 图

(4) 由式(6-3-5)可求 α 值为

$$\alpha = \frac{1+\sin 55°}{1-\sin 55°} = 10$$

由式(6-3-8)得超前校正网络的两个转折频率为

$$\omega_1 = \frac{1}{\alpha T} = 15.8$$

$$\omega_2 = \frac{1}{T} = 158$$

因此,校正装置的传递函数为

$$G_c(s) = \frac{\alpha Ts+1}{Ts+1} = \frac{0.063s+1}{0.0063s+1}$$

(5) 校正装置和校正后系统的 Bode 图如图 6-3-3 所示。

(6) 由图 6-3-2 有

$$20\lg|G_c(\omega_m)| = 10\lg\alpha = 10\text{dB}$$

而由图 6-3-3 则有

$$L_o(\omega_m) = 20\lg G_o|(j\omega)|_{\omega=\omega_m} = -8\text{dB}$$

这样校正后实际的穿越频率大于 $\omega_m = 50$,由图 6-3-3 可看到合适的校正后穿越频率为

$$\omega_c' = 63 \text{rad/s}$$

因此,最终的相角裕度为

$$\gamma = 180° + \angle G_o(j\omega_c') + \angle G_c(j\omega_c') = 180° - 203.2° + 54.3° = 31.1°$$

最终的相角裕度大于相角裕度的期望值。

4. 结论

超前校正具有一些其他校正形式所不具备的优点,同时使用时也可能有些不太方便。通过上述的例子,可以归纳出以下三点结论。

(1) 超前校正提供了超前相位,能使系统的超调量限制在希望的范围内。

(2) 开环(以及一些情况下的闭环)带宽增大。通常这是有利的,因为系统的响应包括较高的频率导致较快的响应速度。但是在较高频率处存在噪声,这也可能带来麻烦。

(3) 当 φ_m 附近相频特性曲线比较陡峭时有可能发生麻烦。这是由于新的增益穿越点向右移动时,要求校正装置提供越来越大的相位超前,因此需要很大的 α 值。校正装置用物理元部件实现时就很难做到。

6.3.2 滞后校正装置

校正装置输出信号在相位上滞后于输入信号,即校正装置具有负的相角特性,这种校正装置称为滞后校正装置,对系统的校正称为滞后校正(Lag Compensation)。I 控制器和 PI 控制器就属于滞后补偿网络。

1. RC 滞后网络的特性

如图 6-3-4 所示为一个无源滞后网络的电路图,其传递函数为

$$G(s) = \frac{U_o(s)}{U_i(s)} = \frac{R_2 Cs + 1}{(R_1 + R_2)Cs + 1}$$

$$T = (R_1 + R_2)C, \beta = \frac{R_2}{R_1 + R_2} < 1 \qquad (6\text{-}3\text{-}10)$$

则传递函数可写成

$$G(s) = \frac{\beta Ts + 1}{Ts + 1} \qquad (6\text{-}3\text{-}11)$$

根据式(6-3-11),作出无源滞后网络的 Bode 图,如图 6-3-5 所示。滞后网络在频率 $1/T \sim 1/(\beta T)$ 呈积分效应,而对数相频呈滞后特性。与超前网络类似,最大滞后角 φ_m 发生在最大滞后角频率 ω_m 处,且 ω_m 正好处于 $1/T \sim 1/(\beta T)$ 的几何中心。

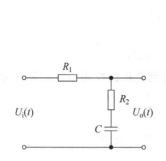

图 6-3-4 无源滞后网络的电路图

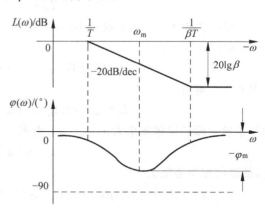

图 6-3-5 无源滞后网络的 Bode 图

2. 串联滞后校正

给系统串入串联滞后校正装置,主要作用是在高频段上造成显著的幅值衰减,其最大衰减量与滞后网络传递函数中的参数成反比。当在控制系统中采用串联滞后校正时,其高频衰减特性可以保证系统在有较大开环放大系数的情况下获得满意的相角裕度或稳态性能。

3. 利用 Bode 图设计滞后校正网络

串联滞后的作用主要有两条：其一是提高系统低频响应的增益，减少系统的稳态误差，同时基本保证系统的暂态性能不变；其二是滞后校正装置的低通滤波器(Low-Pass Filter)的特性，将使系统高频响应的增益衰减，降低系统的截止频率，改善系统的稳定性和某些暂态性能。具体设计步骤如下。

(1) 求出满足稳态指标的开环放大系数 K 值。

(2) 根据求得的 K 值，画出未校正系统的 Bode 图，并计算其幅值穿越频率 ω_c、相位裕度 γ、幅度裕度 K_g。

(3) 确定校正后系统的幅值穿越频率 ω_c'。

(4) 选择频率 ω_c'，使 $\omega_c = \omega_c'$ 时，未校正系统的相位为

$$\varphi(\omega_c') = -180 - \gamma' - \Delta \qquad (6\text{-}3\text{-}12)$$

式中，γ' 表示期望的相位裕度，Δ 为相位滞后校正网络在 $\omega_c = \omega_c'$ 处所引起的相位滞后量，一般该滞后量为 $5°\sim12°$。

(5) 确定滞后网络参数 β 和 T 值。

(6) 验证校正后系统的相位裕度 γ' 和幅度裕度 K_g'。

如果不满足，则需要从步骤(3)重复上述设计过程，直到获得满意的结果为止。

【例 6-3-2】 考虑开环传递函数为

$$G_o(s) = \frac{K}{s(s+1)(0.5s+1)}$$

的系统，为该系统设计一个串联校正装置，使得稳态误差系统 $K_v \geqslant 5s^{-1}$，相角裕度 $\gamma \geqslant 40°$，而且增益裕度 $L(K_g) \geqslant 10\text{dB}$。

解：为了满足稳态性能，因此选择

$$K = 5$$

步骤如下：

(1) 未校正系统的穿越频率为 $\omega_{c0} \approx 2\text{rad/s}$，相位裕度为 $\gamma_0 \approx -18°$，而增益裕度 $L(K_{g0}) \approx -7\text{dB}$，未校正系统不稳定。由于相频特性曲线在 ω_{c0} 附近具有较陡的频率，可能相位滞后校正较为合适。

(2) 确定新的穿越频率 ω_c。考虑到 β、T 取值的物理实现以及允许校正装置 $12°$ 的相位滞后，放置滞后校正装置时使新的穿越频率处有

$$\angle G_o(\text{j}\omega) = -180° + 40° + 12° = -128°$$

由相频特性图可有

$$\omega_c = 0.5\text{rad/s}$$

(3) 确定转折频率。由于已取 $12°$ 的安全裕度，则转折频率

$$\omega_1 = \frac{1}{\beta T} = 0.2\omega_c = 0.1\text{rad/s}$$

由幅频特性图可知，使 $\omega_c = 0.5\text{rad/s}$ 成为新的穿越频率所需的衰减等于 20dB。因而有 $20\lg\beta = -20\text{dB}$，即 $\beta = 0.1$。

这样,另一个转折频率为

$$\omega_1 = \frac{1}{T} = 0.01 \text{rad/s}$$

(4) 确定 $G_c(s)$。校正装置的传递函数为

$$G_c(s) = \frac{10s+1}{100s+1}$$

(5) 检验校正后系统的特性。由 Bode 图可知,对于校正后系统有 $K_v \geq 5s^{-1}$,$\omega_c = 0.5 \text{rad/s}$,$\gamma = 40°$ 和 $L(K_g) = 11 \text{dB}$。所有设计指标都已满足。

4. 结论

(1) 滞后校正提供必要的阻尼比以使超调限制在要求的范围内。
(2) 与超前校正相比,转折频率的选择不是十分苛刻,因此校正过程比较简单。
(3) 由校正后系统可见,相位滞后法减小了开环系统的带宽,并且也减小了闭环系统的带宽,从而导致响应速度变慢。
(4) 与超前校正不同,理论上相位滞后校正可以使相位裕度的变化超过 90°。

6.3.3 滞后-超前校正装置

单纯采用超前校正或滞后校正均只能改善系统暂态或稳态一个方面的性能。利用超前校正改善系统的暂态响应,利用滞后校正提高系统的稳态精度。更具体地说,超前网络串入系统,可增加频带宽度以提高快速性,并且可使稳定裕度加大改善平稳性,但会降低稳态精度。滞后网络串入系统,可以提高平稳性和稳态精度,而降低了快速性。故同时采用超前校正和滞后校正,可全面提高系统的控制性能,因此在某些情况下,将滞后和超前校正网络组合在一起。若校正装置在某一频率范围内具有负的相角特性,而在另一频率范围内具有正的相角特性,这种校正装置称为滞后-超前校正装置,对系统的校正称为滞后-超前(Lag-Lead Compensation)校正。PID 控制器就属于滞后-超前校正网络。

1. RC 滞后-超前网络的特性

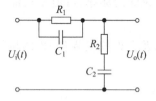

图 6-3-6 无源滞后-超前网络的电路图

如图 6-3-6 所示为一个无源滞后网络的电路图,其传递函数为

$$G(s) = \frac{U_o(s)}{U_i(s)}$$

$$= \frac{(R_1 C_1 s + 1)(R_2 C_2 s + 1)}{R_1 R_2 C_1 C_2 s^2 + (R_1 C_1 + R_2 C_2 + R_1 C_2)s + 1} \quad (6\text{-}3\text{-}13)$$

令 $\tau_1 = R_1 C_1$,$\tau_2 = R_2 C_2$,$\tau_{12} = R_1 C_2$,则传递函数可写成

$$G(s) = \frac{(\tau_1 s + 1)(\tau_2 s + 1)}{\tau_1 \tau_2 s^2 + (\tau_1 + \tau_2 + \tau_{12})s + 1} \quad (6\text{-}3\text{-}14)$$

若适当选择参数,使式(6-3-14)具有两个不相等的负实数极点,可以改写为

$$G(s) = \frac{(\tau_1 s + 1)(\tau_2 s + 1)}{(T_1 s + 1)(T_2 s + 1)} \quad (6\text{-}3\text{-}15)$$

曲线的低频部分具有负频率、负相移,起滞后校正作用;后一段具有正频率、正相移,起超前校正作用,如图 6-3-7 所示。

2. 串联滞后-超前校正

串联超前校正主要是利用超前网络的相角超前特性来提高系统的相角裕量或相对稳定性,而串联滞后校正是利用滞后网络在高频段的幅值衰减特性来提高系统的开环放大系数,从而改善系统的稳态性能。串联滞后校正主要用来校正开环频率的低频区特性,超前校正主要用于改变中频区特性的形状和参数。在确定参数时,两者基本上可独立进行。可按照前面的知识分别确定超前和滞后装置的参数。一般地,先根据动

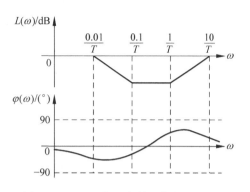

图 6-3-7　无源滞后-超前网络的 Bode 图

态性能指标的要求确定超前校正装置的参数,在此基础上,再根据稳态性能指标的要求确定滞后装置的参数。应注意的是,在确定滞后校正装置时,尽量不影响已由超前装置校正好了的系统的动态指标,在确定超前校正装置时,要考虑到滞后装置加入对系统动态性能的影响,对参数的选择应留有裕量。

3. 利用 Bode 图设计滞后-超前校正网络

这种校正方法兼有滞后和超前校正的优点,即校正后系统响应速度较快、超调量小,抑制高频噪声的性能也较好。它的基本原理是,利用滞后-超前校正网络的超前部分来增大系统的相位裕度,同时利用滞后部分来改善系统的稳态性能。具体设计步骤如下:

(1)求出满足稳态指标的开环放大系数 K 值。

(2)根据求得的 K 值,画出未校正系统的 Bode 图,并计算其幅值穿越频率 ω_c、相位裕度 γ、幅度裕度 K_g。

(3)确定校正后系统的幅值穿越频率 ω_c'。

选择频率 ω_c',使得 $\omega_c = \omega_c'$ 时能通过校正网络超前环节所提供的相位超前量,使系统满足相位裕度要求,又能通过滞后环节的作用把此点原幅频特性衰减到零分贝。

(4)验证校正后系统中滞后环节的角频率 $1/T_2$ 和超前环节的角频率 $1/T_1$。

(5)验证校正后系统的相位裕度 γ' 和幅度裕度 K_g'。

如果不满足,则需要从步骤(3)重复上述设计过程,直到获得满意的结果为止。

【例 6-3-3】 开环传递函数为

$$G_o(s) = \frac{K}{s(0.1s+1)(0.01s+1)}$$

设计一个串联校正系统,使得闭环系统的稳态误差系统 $K_v \geqslant 100s^{-1}$,相角裕度 $\gamma \geqslant 40°$,而且穿越频率 $\omega_c = 20s^{-1}$。

解:满足稳态误差的开环增益为 $K = K_v = 100$,绘制未校正系统 Bode 图,如图 6-3-8 所示。

由 Bode 图可知,未校正系统的穿越频率为 $\omega_{c0} = 30 \text{rad/s}$,而未校正系统的相角裕度为

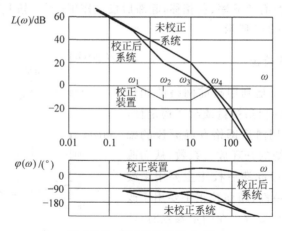

图 6-3-8 例 6-3-3 的 Bode 图

$\gamma_o \approx 2°$，这说明需要某种形式的校正。由于期望的穿越频率低于未校正系统的穿越频率，因此需要相位滞后校正。在 $\omega_{co}=20\text{rad/s}$ 处可有

$$\angle G_o(j\omega) = -90° - 63.4° - 11.3° \approx -165°$$

显然，为了确保相角裕度的要求，还需要相位超前校正。

滞后部分考虑 5° 的补偿，在 $\omega_m = \omega_c = 20\text{rad/s}$ 处要增加的最大相位超前至少为

$$\varphi_m = 40° - (180° - 165°) + 5° \approx 30°$$

由式(6-3-5)得

$$\alpha \geqslant \frac{1+\sin 30°}{1-\sin 30°} = 3$$

超前部分的两个转折频率应当满足

$$\omega_3 = \frac{1}{\alpha T_2} \leqslant 11.5\text{rad/s}$$

$$\omega_4 = \frac{1}{T_2} \geqslant 35\text{rad/s}$$

考虑到未校正系统在 $\omega = 10\text{rad/s}$ 处的开环极点，取 $\omega_3 = 10\text{rad/s}$ 和 $\omega_4 = 35\text{rad/s}$。在 $\omega_c = 20\text{rad/s}$ 处进行估测，可以看到需要 8dB 的衰减，即 $20\lg G_c|(j\omega)|_{\omega=20} = -8\text{dB}$。由此，超前部分的 Bode 图如图 6-3-8 所示，在 ω_3 和 ω_4 之间幅频特性斜率为 $+20\text{dB/dec}$。

至于滞后部分，也考虑取 5° 的补偿，转折频率定位于

$$\omega_2 = \frac{1}{\beta T_1} = 0.1\omega_c = 2\text{rad/s}$$

$$\omega_4 = \frac{1}{T_1} = 0.35\text{rad/s}$$

因此，滞后-超前校正装置的传递函数为

$$G_c(s) = \frac{0.5s+1}{2.86s+1} \cdot \frac{0.1s+1}{0.029s+1}$$

校正后系统的 Bode 图如图 6-3-8 所示。最终的相角裕度为

$$\gamma = 180° - 165° + \angle G_c(j\omega_c) = 15° - 4.7° + 33.3° = 43.6°$$

要求的所有性能指标都得到了满足。

6.3.4 超前校正、滞后校正和滞后-超前校正的比较

（1）超前校正通过其相位超前特性获得所需要的结果，滞后校正则是通过其高频衰减特性获得所需要的结果。

（2）超前校正通常用来改善稳定裕度。超前校正比滞后校正可能提供更高的幅值穿越频率。如果需要有大的带宽，或者说具有快速的响应特性，应当采用超前校正。

（3）超前校正需要一个附加的放大器，以补偿超前网络本身的衰减。这表明超前校正比滞后校正需要更大的放大系数。在大多数情况下，放大系数越大，意味着系统的体积和质量越大，成本也越高。

（4）滞后校正降低了系统在高频段的放大系数，但是并不降低系统在低频段的放大系数。

如果需要获得快速的响应特性，又需要获得良好的稳态精度，则可以采用滞后-超前校正装置。通过应用滞后-超前校正装置，增大低频放大系数（改善了稳态精度），同时也增大系统的带宽和稳定裕度。

6.4 反馈校正

在控制系统的校正中，除了采用串联校正方案外，反馈校正也是常用的校正方式之一。反馈校正除了与串联校正一样，可改善系统的性能以外，还可抑制反馈环内不利因素对系统的影响。常见的有被控量的速度、加速度反馈，执行机构的输出及其速度的反馈，复杂对象的中间变量反馈等，如图 6-4-1 所示。

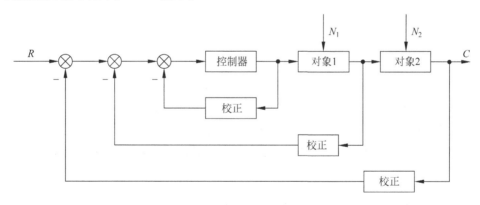

图 6-4-1 反馈校正的连接方式

在随动系统和调速系统中，转速、加速度、电枢电流等都可以作为反馈信号，而具体的反馈元件实际上是一些传感器，如测速电机、电压加速度传感器、电流互感器等。

6.4.1 反馈校正的特点

一般情况下，如图 6-4-2 所示的反馈校正中，校正环节会对原系统的某些环节进行包围，因此，不同形式的反馈环节对系统有不同的影响。例如，比例反馈包围惯性环节，校正后

的系统时间常数下降,惯性减弱,过渡过程时间缩短,系统增益下降,可以改善稳态性能,但增加了系统的稳态误差。比例反馈包围积分环节由积分性质转变为惯性环节,可以提高系统稳定性,但是降低了系统的稳态精度。微分反馈包围惯性环节,增大时间常数可以使系统各环节的时间常数拉开,从而改善系统的动态平稳性。

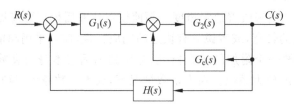

图 6-4-2　反馈校正

6.4.2　反馈校正系统的设计

在设计反馈系统时,经常利用反馈校正取代局部结构,以改造不希望有的某些环节,消除非线性、变参数的影响,抑制干扰。如图 6-4-2 所示的局部反馈回路,前向通道传递函数为 $G_1(s)$,反馈为 $H(s)$,则回路的传递函数为

$$G(s) = \frac{G_1(s)}{1 + G_1(s)H(s)} \tag{6-4-1}$$

在一定的频率范围内,如果选择结构参数,使

$$|G_1(j\omega)H(j\omega)| \approx 1 \tag{6-4-2}$$

则

$$G(j\omega) \approx \frac{1}{H(j\omega)} \tag{6-4-3}$$

此时反馈回路的传递函数等效为

$$G(s) \approx \frac{1}{H(s)} \tag{6-4-4}$$

可见,被包围环节与传递函数无关。这样,在感兴趣的频段,只要满足 $|G_1(j\omega)H(j\omega)| \approx 1$,就可以利用反馈环节取代原来的特性。我们称感兴趣的频段为接受频段。当 $|G_1(j\omega)H(j\omega)| \not\approx 1$ 时,等效传递函数为原有环节,反馈环节不起作用,称此频段为不接受校正频段。

6.4.3　串联校正与反馈校正比较

(1) 串联校正比反馈校正简单,但串联校正对系统元件特性的稳定性有较高的要求。反馈校正对系统元件特性的稳定性要求较低,因为其减弱了元件特性变化对整个系统特性的影响。

(2) 反馈校正常由一些昂贵而庞大的部件所构成,对某些系统可能难以应用。

(3) 反馈校正可以在需要的频段内,消除不需要的特性,抑制参数变化对系统性能的影响,串联校正无此特性。

所以在需要的控制系统结构简单、成本低而无特殊要求时,可采用串联校正。若有特殊要求,特别是被控对象参数不稳定时,可以考虑采用反馈校正。但是当系统低频扰动比较大时,反馈校正的作用不明显,可以引入误差补偿通路,与原来的反馈控制一起进行复合控制。

复合控制通过在系统中引入输入或扰动作用的开环误差补偿通路(顺馈或前馈通路),与原来的反馈控制回路一起实现系统的高精度控制。关于复合控制的基本原理本书不再讲述,可以参考有关控制系统方面的书籍。

【例 6-4-1】 控制系统的结构图如图 6-4-3 所示,其中

$$G_1(s) = \frac{238}{0.06s+1}, \quad G_2(s) = \frac{238}{0.36s+1}, \quad G_3(s) = \frac{0.0208}{s}$$

试设计反馈校正装置,使系统的性能指标 $\sigma \leqslant 25\%, t_s \leqslant 0.8$s。

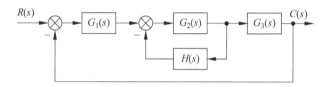

图 6-4-3　例 6-4-1 控制系统结构图

解：校正前系统的开环传递函数为

$$G(s) = G_1(s)G_2(s)G_3(s) = \frac{1130}{s(0.06s+1)(0.36s+1)}$$

(1) 绘制原系统的对数幅频特性 L_0,如图 6-4-4 所示。

(2) 绘制系统的期望对数幅频特性 L_g,如图 6-4-4 所示。

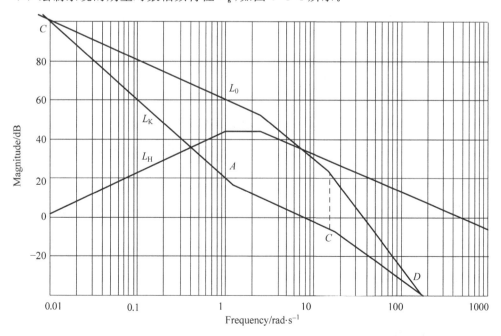

图 6-4-4　控制系统的对数幅频特性

(3) 将 $L_o - L_g$ 得到 $20\lg|G_2(j\omega)H(j\omega)|$，如图中 L_H 所示，其传递函数为

$$G_2(j\omega)H(j\omega) = \frac{K_H s}{(T_1 s+1)(T_2 s+1)}$$

其中，$T_1 = 1/1.1 = 0.9$，$T_2 = 1/2.78 = 0.36$，$K_H = 1/0.009 = 111$，得

$$H(s) = \frac{G_2(s)H(s)}{G_2(s)} = \frac{0.487s}{0.9s+1}$$

6.5　MATLAB 在本章中的应用

本节讨论 MATLAB 在自动控制系统校正问题中的应用。利用 MATLAB 进行系统设计可以快速而精确地得到最终结果，在 Bode 图上进行校正设计的一般步骤如下。

（1）以 Bode 图绘制满足稳态误差要求的原系统 Bode 图，以 margin 函数求取原函数的 ω_c、ω_g、γ、K_g。

（2）根据所要求的性能指标，分析采用何种校正方式。

（3）大致计算参数取值范围。

（4）以运算结果为依据不断进行调整，直到满意为止。

【例 6-5-1】 已知单位反馈系统开环传递函数为

$$G_k(s) = \frac{K_0}{s(s+2)}$$

试设计系统的相位超前校正，使系统：

（1）在斜坡信号 $r(t) = t$ 作用下，系统的稳态误差 $e_{ss} \leqslant 0.001v_0$；

（2）校正系统的相位稳定裕度 γ 满足 $43° < \gamma < 48°$。

解：（1）求 K_0。

在斜坡信号作用下，系统的稳态误差

$$e_{ss} = \frac{v_0}{K_v} = \frac{v_0}{K} = \frac{v_0}{K_0} \leqslant 0.001 v_0$$

可得 $K_v = K = K_0 \geqslant 1000 s^{-1}$，取 $K_0 = 1000 s^{-1}$，即被控对象的传递函数为

$$G_k(s) = 1000 \cdot \frac{1}{s(s+2)}$$

（2）作原系统的 Bode 图与阶跃响应曲线，如图 6-5-1 检查是否满足题目要求。

```
MATLAB Program
K0 = 1000;n1 = 1;d1 = conv([1 0],[1 2]);
[mag,phase,w] = bode(K0 * n1,d1);
figure(1);margin(mag,phase,w);
hold on
figure(2);s1 = tf(K0 * n1,d1);
sys = feedback(s1,1);step(sys)
```

运行结果：

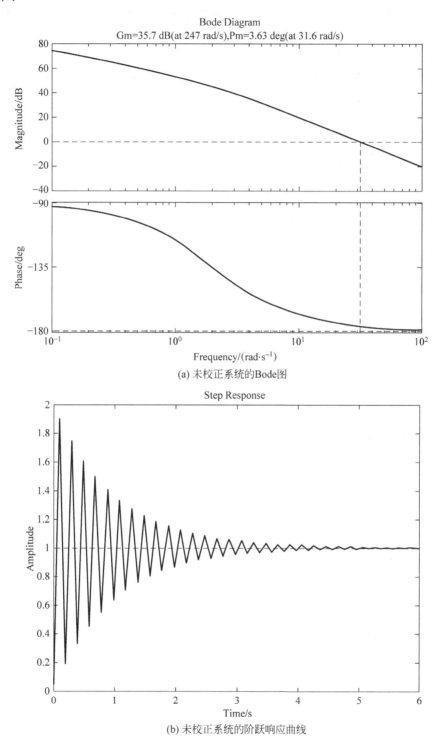

(a) 未校正系统的Bode图

(b) 未校正系统的阶跃响应曲线

图 6-5-1　未校正系统的 Bode 图及阶跃响应曲线

由图 6-5-1 可知，系统的模稳定裕度量 $G_m=35.7\text{dB}$，相稳定裕度 $P_m=3.63\text{deg}$，为满足题目中 $43°<\gamma<48°$ 的要求，此外，系统阶跃响应曲线虽然衰减，但振荡较剧烈，同样说明系统不符合要求。

① 求超前校正器的传递函数。

根据相位稳定裕度 $43°<\gamma<48°$ 的要求，取 $\gamma=45°$。根据以下程序，计算超前校正器的传递函数：

```
MATLAB Program
K0 = 1000;n1 = 1;d1 = conv([1 0],[1 2]);
sope = tf(K0 * n1,d1);
[mag,phase,w] = bode(sope);
gama = 45;[mu,pu] = bode(sope,w);
gam = gama * pi/180;
alfa = (1-sin(gam))/(1 + sin(gam));
adb = 20 * log10(mu);am = 10 * log10(alfa);
ca = adb + am;wc = spline(adb,w,am);
T = 1/(wc * sqrt(alfa));
alfat = alfa * T;Gc = tf([T 1],[alfat 1])
```

运行结果：

```
Transfer function:
0.04916s + 1
---------
0.008434s + 1
```

即校正装置传递函数为

$$G(s)=\frac{0.049\,16s+1}{0.008\,434s+1}$$

② 检验系统校正后是否满足要求。

根据校正后系统的结构与参数，给出以下程序：

```
MATLAB Program
K0 = 1000;n1 = 1;d1 = conv([1 0],[1 2]);
s1 = tf(K0 * n1,d1);
n2 = [0.04916 1];d2 = [0.008434 1];
s2 = tf(n2,d2);sope = s1 * s2;
[mag,phase,w] = bode(sope);
margin(mag,phase,w);
```

如图 6-5-2 所示，稳定裕度为 $G_m=97.9\text{dB}$，$P_m=47.2\text{deg}$。

③ 计算系统校正后阶跃响应及其性能指标。

校正系统响应曲线及性能指标的程序：

```
MATLAB Program
global y t;
K0 = 1000;n1 = 1;d1 = conv([1 0],[1 2]);
s1 = tf(K0 * n1,d1);
n2 = [0.04916 1];d2 = [0.008434 1];
s2 = tf(n2,d2);
scope = s1 * s2;
sys = feedback(scope,1);
step(sys)
[y,t] = step(sys)
```

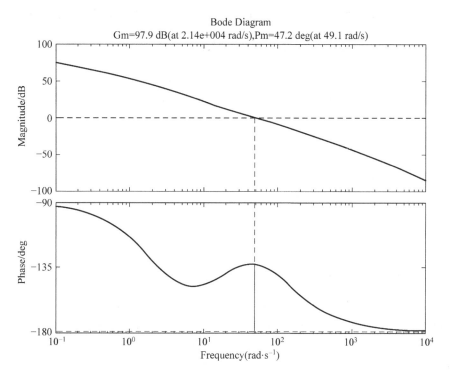

图 6-5-2 校正后系统的 Bode 图

校正后系统阶跃响应曲线如图 6-5-3 所示,系统性能指标:

$$\text{Sigma}=0.2957, t_p=0.0587, t_s=0.1136$$

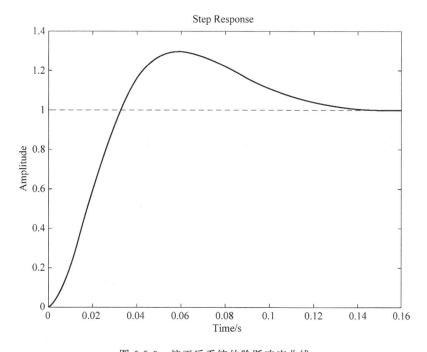

图 6-5-3 校正后系统的阶跃响应曲线

下面的英文阅读材料讲述针对一个开环的控制系统,利用频域分析方法进行设计。

Reading Material

Example 6-5-2: Solution to the Cruise Control Problem Using Frequency Response. The open-loop transfer function for this problem is

$$\frac{Y(s)}{U(s)} = \frac{1}{ms+b} \qquad (6\text{-}5\text{-}1)$$

$m=1000$, $b=50$, $U(s)=10$, $Y(s)=$ Velocity output.

The design criteria are:

$$\text{Rise time} < 5 \text{ s}$$
$$\text{Overshoot} < 10\%$$
$$\text{Steady state error} < 2\%$$

1. Bode Plot and Open-Loop Response

The first step in solving this problem using frequency response is to determine what open-loop transfer function to use. Just like for the Root-Locus design method, we will only use a proportional controller to solve the problem. The block diagram and the open-loop transfer function are shown below in Fig. 6-5-4.

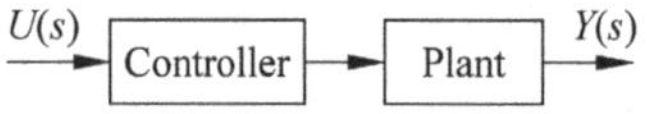

$$\frac{Y(s)}{U(s)} = \frac{K_P}{ms+b} \qquad (6\text{-}5\text{-}2)$$

Fig. 6-5-4 A open-loop control system

In order to use a Bode plot, the open-loop response must be stable. Let K_p equals 1 for now and see how the open-loop response looks like. Create a new m-file and enter the following commands.

```
m = 1000;b = 50;
u = 500;Kp = 1;
numo = [Kp];deno = [m b];
step(u * numo,deno)
```

Running this m-file in the MATLAB command window should give you the following plot like Fig. 6-5-5.

As you can see, the open-loop system is stable; thus, we can go ahead and generate the Bode plot. Change the above m-file by deleting step command and add in the following command.

```
bode(numo,deno)
```

Running this new m-file should give you the following Bode plot like Fig. 6-5-6.

2. Proportional Controller

Let's see what system characteristics we can determine from the above Bode plot. Recall from the Root-Locus part, the bandwidth frequency(BW)(the frequency at the gain

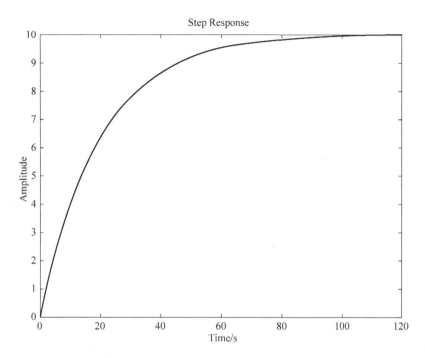

Fig. 6-5-5　The step response of the system

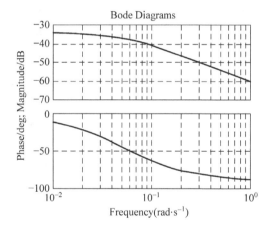

Fig. 6-5-6　The Bode diagrams of the system

$M(\mathrm{dB}) = -6 \sim 7.5\mathrm{dB})$ is roughly equals to the natural frequency(ω_n). Using the equation

$$T_r = \frac{1.8}{\omega_n} = \frac{1.8}{\mathrm{BW}} \tag{6-5-3}$$

the rise time(T_r) for our system can be determined to be extremely long since the gain shown above do not reach $-6 \sim 7.5\mathrm{dB}$. Moreover, we see from the Root-Locus Part that the damping ratio is roughly equals to the phase margin(in degrees) divided by 100.

$$\xi = \frac{\mathrm{PM(deg)}}{100} \tag{6-5-4}$$

Since our phase margin is approximately 155 degrees, the damping ratio will be 1.55.

Thus, we know that the system is overdamped. Finally, the steady-state error can be found from the following equation

$$ss_{error} = \frac{1}{1 + M_{\omega \to 0}} \times 100\% \tag{6-5-5}$$

For our system, since the low frequency gain $M(dB)$ is approximately $-35dB$, the steady-state error should be 98%. We can confirm these by generating a closed-loop step response in Fig. 6-5-7.

In terms of the Bode plot, if we can shift the gain upward so that both the bandwidth frequency and the low frequency gain increase, then both the rise time and the steady-state error will improve. We can do that by increasing the proportional gain(K_p). Let's increase the proportional gain(K_p) to, say, 100 and see what happens. Change the m-file to the following.

```
m = 1000;b = 50;
u = 10;Kp = 100;
numo = [Kp];deno = [m b];
bode(numo,deno)
```

Running this m-file in the MATLAB command window should give you the following Bode plot like Fig. 6-5-8.

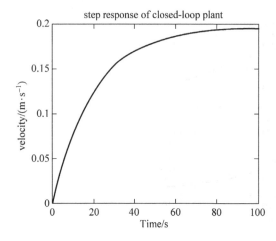

Fig. 6-5-7 The step response of the close-loop system

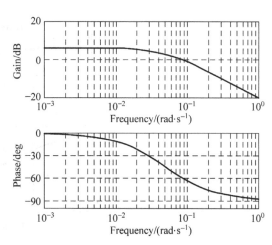

Fig. 6-5-8 The Bode diagrams of the system

Now, the low frequency gain is about $-6dB$(magnitude 2) which predicts the steady-state error of 33%. The bandwidth frequency is about 0.1 rad/sec so that the rise time should be around 18 seconds. Let's take a look at the closed-loop step response and confirm these in Fig. 6-5-9.

As we predicted, both the steady-state error and the rise time have improved. Let's increase the proportional gain even higher and see what happens. Change the m-file to the following and rerun it. You should get the following Bode plot in Fig. 6-5-10.

```
m = 1000;b = 50;
u = 10;Kp = 100;
numo = [Kp/b];deno = [m/b 1];
bode(numo,deno)
```

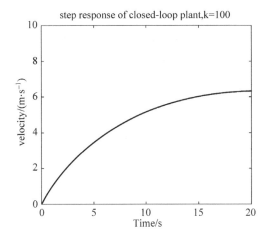

Fig. 6-5-9   The step response of the close-loop system

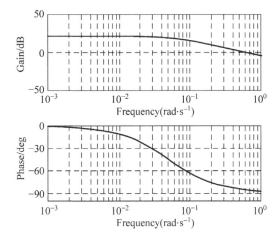

Fig. 6-5-10   The Bode diagrams of the system

Now, the low frequency gain is approximately 20dB(magnitude 10) that predicts the steady-state error of 9%, and the bandwidth frequency is around 0.6 that predicts the rise time of 3 seconds(within the desired value). Thus, both the steady-state error and the rise time should have been improved. Again, let's confirm these by generating the closed-loop step response like Fig. 6-5-11.

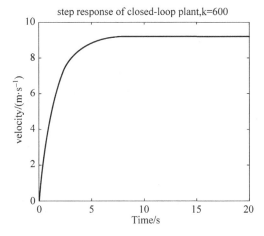

Fig. 6-5-11   The step response of the close-loop system

If you noticed, the steady-state error will eventually reach the desired value by increasing the proportional gain even higher. However, by the time the steady-state error reaches the desired value, the rise time becomes too fast(unrealistic for the real physical system). Thus, let's leave the K_p as it is and implement a lag controller to handle the steady-state error problem.

3. Lag Controller

The lag controller adds gain at the low frequency while keeping the bandwidth frequency at the same place. This is actually what we need: Larger low frequency gain to reduce the steady-state error and keep the same bandwidth frequency to maintain the desired rise time. The transfer function of the lag controller is shown below.

$$G(s) = \frac{1}{\alpha}\left(\frac{1+\alpha Ts}{1+Ts}\right) \quad \alpha < 1 \tag{6-5-6}$$

Now, we need to choose a value for α and T. The steady-state error will decrease by a factor of α. The value T should be chosen so that two corner frequencies will not be placed close together because transient response gets worse. So let α equals 0.05 and T equals 700 and see what happens. Copy the following m-file and run it in the MATLAB command window. You should see the following Bode plot in Fig. 6-5-12.

```
m = 1000;b = 50;
u = 10;Kp = 600;
numo = [Kp/b];deno = [m/b 1];
a = 0.05;T = 700;
numlag = [a * T 1];denlag = a * [T 1];
newnum = conv(numo,numlag);
newden = conv(deno,denlag);
bode(newnum,newden)
figure
[numc,denc] = cloop(newnum,newden);
step(u * numc,denc)
```

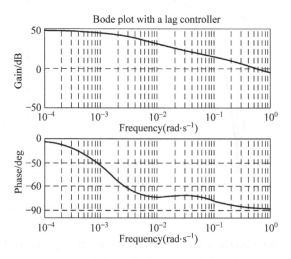

Fig. 6-5-12 The Bode diagrams of the system

Since the low frequency gain has increased while the bandwidth frequency stayed the same, the steady-state error should be reduced and the rise time should stay the same. Let's confirm this by generating a closed-loop step response like Fig. 6-5-13.

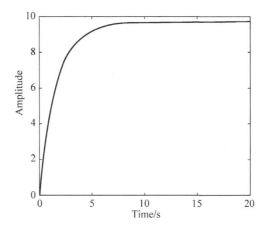

Fig. 6-5-13　The step response with a lag controller

It may be hard to see, but there should be a green, dotted line across just below 10. This line shows the steady-state value of the step, and we can see that the steady-state error has been met. However, the settling time is too long. To fix this, raise the proportional gain to $K_P = 1500$. This gain was chosen from trial-and-error that will not be described here in the interest of length. With this change made, the following Bode and step response plots can be generated in Fig. 6-5-14 and Fig. 6-5-15.

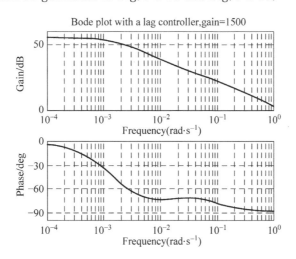

Fig. 6-5-14　The Bode diagrams of the system in $K_P = 1500$

As you can see, the overshoot is in fact zero, the steady state error is close to zero, the rise time is about 2 seconds, and the settling time is less than 3.5 seconds. The system has now met all of the design requirements. No more iteration is needed.

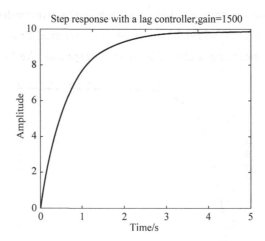

Fig. 6-5-15 The simulation results of the system

上述内容中,以频率特性法的方式讨论了控制系统的分析与校正。下面利用经典的 PID 控制方法以英文方式给出设计实例。

This tutorial will show you the characteristics of the each of proportional (P), the integral (I), and the derivative (D) controls, and how to use them to obtain a desired response. In this tutorial, we will consider the following unity feedback system (Fig. 6-5-16):

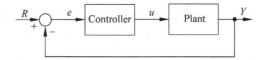

Fig. 6-5-16 A typical control system

Plant: A system to be controlled.

Controller: Provides the excitation for the plant; Designed to control the overall system behavior.

1. The three-term controller

The transfer function of the PID controller looks like the following:

$$K_P + \frac{K_I}{s} + K_D s = \frac{K_D s^2 + K_P s + K_I}{s} \tag{6-5-7}$$

K_P = Proportional gain

K_I = Integral gain

K_D = Derivative gain

First, let's take a look at how the PID controller works in a closed-loop system using the schematic shown above. The variable (e) represents the tracking error, the difference between the desired input value (R) and the actual output (Y). This error signal (e) will be sent to the PID controller, and the controller computes both the derivative and the integral of this error signal. The signal (u) just past the controller is now equal to the proportional gain (K_P) times the magnitude of the error plus the integral gain (K_I) times the integral of the error plus the derivative gain (K_D) times the derivative of the error.

$$u = K_P e + K_I \int e \mathrm{d}t + K_D \frac{\mathrm{d}e}{\mathrm{d}t} \tag{6-5-8}$$

This signal (u) will be sent to the plant, and the new output (Y) will be obtained. This new output (Y) will be sent back to the sensor again to find the new error signal (e). The controller takes this new error signal and computes its derivative and its integral again. This process goes on and on.

2. The characteristics of P, I, and D controllers

A proportional controller (K_P) will have the effect of reducing the rise time and will reduce, but never eliminate, the steady-state error. An integral control (K_I) will have the effect of eliminating the steady-state error, but it may make the transient response worse. A derivative control (K_D) will have the effect of increasing the stability of the system, reducing the overshoot, and improving the transient response. Effects of each of controllers K_P, K_D, and K_I on a closed-loop system are summarized in the Tab. 6-5-1 shown below.

Tab. 6-5-1 The characteristics of PID controllers

CL RESPONSE	RISE TIME	OVERSHOOT	SETTLING TIME	S-S ERROR
K_P	Decrease	Increase	Small Change	Decrease
K_I	Decrease	Increase	Increase	Eliminate
K_D	Small Change	Decrease	Decrease	Small Change

Note that these correlations may not be exactly accurate, because K_P, K_I, and K_D are dependent of each other. In fact, changing one of these variables can change the effect of the other two. For this reason, the table should only be used as a reference when you are determining the values for K_I, K_P and K_D.

3. Example Problem

Suppose we have a simple mass, spring, and damper problem, For example Fig. 6-5-17.

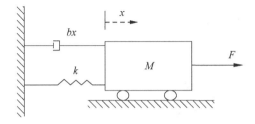

Fig. 6-5-17 A typical mechanical system

The modeling equation of this system is

$$M\ddot{x} + b\dot{x} + kx = F \tag{6-5-9}$$

Taking the Laplace transform of the modeling equation(6-5-9)

$$Ms^2X(s) + bsX(s) + kX(s) = F(s) \qquad (6\text{-}5\text{-}10)$$

The transfer function between the displacement $X(s)$ and the input $F(s)$ then becomes

$$\frac{X(s)}{F(s)} = \frac{1}{Ms^2 + bs + k} \qquad (6\text{-}5\text{-}11)$$

Let
- $M = 1\text{kg}$
- $b = 10\text{N} \cdot \text{s/m}$
- $k = 10\text{N/m}$
- $F(s) = 1$

Plug these values into the above transfer function

$$\frac{X(s)}{F(s)} = \frac{1}{s^2 + 10s + 20} \qquad (6\text{-}5\text{-}12)$$

The goal of this problem is to show you how each of K_P, K_I, and K_D contributes to obtain
- Fast rise time
- Minimum overshoot
- No steady-state error

1) Open-loop step response

Let's first view the open-loop step response. Create a new m-file and add in the following code:

```
num = 1; den = [1 10 20];
step(num, den);
```

Running this m-file in the Matlab command window should give you the plot shown below(Fig. 6-5-18).

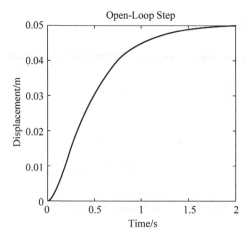

Fig. 6-5-18 The result of open-loop control

The DC gain of the plant transfer function is 1/20, so 0.05 is the final value of the output to an unit step input. This corresponds to the steady-state error of 0.95, quite large indeed. Furthermore, the rise time is about one second, and the settling time is about 1.5 seconds. Let's design a controller that will reduce the rise time, reduce the settling time, and eliminates the steady-state error.

2) Proportional control

From the table shown above, we see that the proportional controller (K_p) reduces the rise time, increases the overshoot, and reduces the steady-state error. The closed-loop transfer function of the above system with a proportional controller is:

$$\frac{X(s)}{F(s)} = \frac{K_P}{s^2 + 10s + (20 + K_P)} \qquad (6\text{-}5\text{-}13)$$

Let the proportional gain (K_P) equals 300 and change the m-file to the following:

```
Kp = 300;num = [Kp];den = [1 10 20 + Kp];
t = 0:0.01:2;step(num,den,t)
```

Running this m-file in the Matlab command window should gives you the following plot(Fig. 6-5-19).

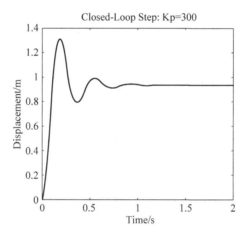

Fig. 6-5-19 The result of closed-loop Proportional control

Note: The Matlab function called cloop can be used to obtain a closed-loop transfer function directly from the open-loop transfer function (instead of obtaining closed-loop transfer function by hand). The following m-file uses the cloop command that should give you the identical plot as the one shown above.

```
num = 1;den = [1 10 20];Kp = 300;
[numCL,denCL] = cloop(Kp * num,den);
t = 0:0.01:2;step(numCL, denCL,t)
```

The above plot shows that the proportional controller reduced both the rise time and the steady-state error, increased the overshoot, and decreased the settling time by small amount.

3) Proportional-Derivative control

Now, let's take a look at a PD control. From the table shown above, we see that the derivative controller (K_D) reduces both the overshoot and the settling time. The closed-loop transfer function of the given system with a PD controller is:

$$\frac{X(s)}{F(s)} = \frac{K_D s + K_P}{s^2 + (10 + K_D)s + (20 + K_P)} \qquad (6\text{-}5\text{-}14)$$

Let K_P equals to 300 as before and let K_D equals 10. Enter the following commands into an m-file and run it in the Matlab command window.

```
Kp = 300;Kd = 10;num = [Kd Kp];den = [1 10 + Kd 20 + Kp];
t = 0:0.01:2;step(num,den,t)
```

Fig. 6-5-20 shows that the derivative controller reduced both the overshoot and the settling time, and had small effect on the rise time and the steady-state error.

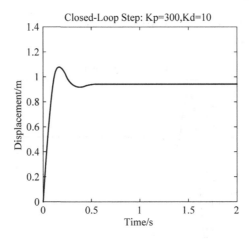

Fig. 6-5-20　The result of closed-loop PD control

4) Proportional-Integral control

Before going into a PID control, let's take a look at a PI control. From the table, we see that an integral controller (K_I) decreases the rise time, increases both the overshoot and the settling time, and eliminates the steady-state error. For the given system, the closed-loop transfer function with a PI control is:

$$\frac{X(s)}{F(s)} = \frac{K_P s + K_I}{s^3 + 10s^2 + (20 + K_P)s + K_I} \qquad (6\text{-}5\text{-}15)$$

Let's reduce the K_P to 30, and let K_I equals to 70. Create a new m-file and enter the following commands.

```
Kp = 30;Ki = 70;num = [Kp Ki];den = [1 10 20 + Kp Ki];
t = 0:0.01:2;step(num,den,t)
```

Run this m-file in the Matlab command window, and you should get the following plot(Fig. 6-5-21).

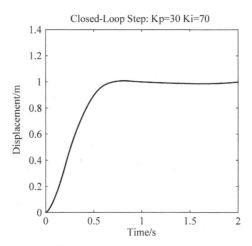

Fig. 6-5-21 The result of closed-loop PI control

We have reduced the proportional gain (K_P) because the integral controller also reduces the rise time and increases the overshoot as the proportional controller does (double effect). The above response shows that the integral controller eliminated the steady-state error.

5) Proportional-Integral-Derivative control

Now, let's take a look at a PID controller. The closed-loop transfer function of the given system with a PID controller is:

$$\frac{X(s)}{F(s)} = \frac{K_D s^2 + K_P s + K_I}{s^3 + (10 + K_D)s^2 + (20 + K_P)s + K_I} \qquad (6-5-16)$$

After several trial and error runs, the gains $K_P = 350$, $K_I = 300$, and $K_D = 50$ provided the desired response. To confirm, enter the following commands to an m-file and run it in the command window. You should get the following step response and plot(Fig. 6-5-22).

```
Kp = 350;Ki = 300;Kd = 50;
num = [Kd Kp Ki];den = [1 10 + Kd 20 + Kp Ki];
t = 0:0.01:2;step(num,den,t)
```

Now, we have obtained the system with no overshoot, fast rise time, and no steady-state error. For more details, please refer to Reference [24].

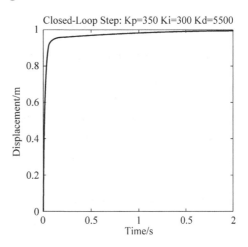

Fig. 6-5-22 The result of closed-loop PID control

本章小结

（1）在自动控制系统设计中，为了满足系统性能要求，常需在系统中附加一些装置，通过改变系统的结构和参数，来改变系统的性能。这种措施称为系统的校正，所引入的装置称为校正装置。系统中除校正装置外的部分称为系统的固有部分。控制系统的校正也就是根据系统的固有部分和对性能指标的要求，确定校正装置的结构和参数。

（2）按照校正装置在系统中的连接方式，可将校正的方法分为串联校正、并联校正和复合校正。根据校正原理的不同，又可将串联校正分为超前校正、滞后校正和滞后-超前校正。另外，根据校正装置所采用的器件的不同，又可分为无源校正和有源校正。

串联超前校正是利用校正装置的超前相位，增加系统的相位裕量，改善系统的稳定性。由于校正之后系统对数幅频曲线斜率的改变，使得其穿越频率增大，从而提高了系统的快速性，但降低了系统抗高频干扰的能力。若原系统需补偿的相角太大，则一般的超前校正装置难以收到较好的效果。PD调节器就是一种超前校正装置。

（3）串联滞后校正是利用校正装置的高频幅值衰减特性，以减少系统穿越频率为代价使系统的相位裕量增加，系统抗高频干扰的能力也得到加强；另外，滞后校正没有改变原系统最低频段的特性，往往还允许增加开环增益，从而可改善系统的稳态精度。PI调节器也是一种滞后校正装置，但由于积分环节的引入，可提高系统的无静差度，改善系统的稳态性能。滞后-超前校正综合了滞后和超前的优点，利用校正装置的超前部分，改善系统的动态性能；利用校正装置的滞后部分，改善系统的稳态精读。PID调节器也是一种滞后-超前校正装置。

（4）控制系统工程设计方法的主要思路是，根据对系统性能指标的要求，选择确定预期的典型数学模型。根据系统固有部分的数学模型与预期典型数学模型的对照，选择校正装置的结构和部分参数，以使系统校正为典型系统的结构形式；然后再进一步选择校正装置的参数，以满足动态性能指标的要求。

Summary and Outcome Checklist

Several alternative approaches have been considered to improve the performance of feedback control systems. The effect of introducing cascade compensating networks within the feedback loops of control systems have been studied using the root locus method. It was noted that the phase-lead compensator improves the stability. When large error constants are specified for a feedback system, it is usually easier to compensate the system by using phase lag networks. Operational amplifier circuits are widely used in industrial practice to implement the compensator.

Tick the box for each statement with which you agree.

☐ I can list the desirable features that the control system designer would be striving to achieve when designing a feedback control system.

☐ I am able to explain what a compensator is in the context of feedback control systems, and state the various ways in which a compensator can be placed in a

- ☐ I am able to explain what a PID controller is, and what effect it usually has on the performance of a closed loop system.
- ☐ I am able to explain what a lead compensator is, and what effect it usually has on the performance of a closed loop system.
- ☐ I am able to explain what a lag compensator is, and what effect it usually has on the performance of a closed loop system.
- ☐ I can use MATLAB's rltool and synthesise suitable compensator for a given control system.
- ☐ I am able to design a lead compensator to improve the transient response and stability of a given feedback control system.
- ☐ I am able to design a lag compensator to achieve a reduced steady state error for a given feedback control system.
- ☐ I am able to design a rate feedback compensator to improve the stability and transient response of a given feedback control system.

习题

6-1 设单位反馈系统的开环传递函数

$$G(s) = \frac{200}{s(0.1s+1)}$$

试设计无源校正网络,使校正系统的相角裕度不小于45°,截止频率不低于50rad/s。

6-2 设Ⅰ型单位反馈系统原有部分的开环传递函数为 $G_0(s) = \dfrac{K}{s(s+1)}$,要求设计串联校正装置,使系统具有 $K=12$ 及 $\gamma=40°$ 的性能指标。

6-3 某单位反馈系统的开环传递函数

$$G(s) = \frac{K}{s(s+1)(0.25s+1)}$$

(1) 设计一个串联校正装置,使得速度误差系数 $K_v \geqslant 5\text{rad/s}$,相角裕度 $\gamma \geqslant 45°$。
(2) 如果还要求增益穿越频率 $\omega_c \geqslant 2\text{rad/s}$,重新设计一个串联校正装置。

6-4 某单位反馈系统的开环传递函数

$$G(s) = \frac{K}{s(0.2s+1)(0.05s+1)}$$

设计一个串联校正装置,使得速度误差系数 $K_v \geqslant 2\text{rad/s}$,最大超调量 $\sigma_p \leqslant 25\%$,调整时间 $t_s \leqslant 1\text{s}$。

6-5 某单位反馈系统的开环传递函数为

$$G(s) = \frac{40}{s(0.2s+1)(0.0625s+1)}$$

(1) 设计相位超前校正装置,使得相角裕度 γ 为30°,增益裕度为10~12dB。
(2) 设计相位滞后校正装置,使得相角裕度 γ 为50°,增益裕度为30~40dB。

6-6 某单位反馈系统被控对象的传递函数为

$$G(s) = \frac{40}{s(s+2)}$$

要求对于斜坡输入 $r(t)=At$ 的稳态误差小于 $0.05A$,相角裕度为 $30°$,同时要求穿越频率 ω_c 为 10rad/s,确定是否需要一个超前或滞后校正装置。

6-7 考虑单位反馈系统,开环传递函数为

$$G(s) = \frac{20}{s(s+1)(s+3)}$$

我们希望加入一个滞后-超前校正装置

$$G_c(s) = \frac{(s+0.15)(s+0.7)}{(s+0.015)(s+7)}$$

证明校正后系统的相角裕度为 $75°$ 且增益裕度为 24dB。

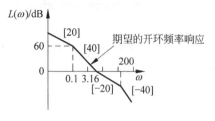

题 6-8 图

6-8 反馈系统开环传递函数为

$$G(s) = \frac{10(0.316s+1)}{(0.1s+1)(0.01s+1)}$$

确定使系统满足题 6-8 图所示期望开环频率响应时所需要的串联校正装置 $G_c(s)$。

6-9 (武汉大学硕士研究生入学考试试题)简答题:试简述系统串联滞后校正网络的校正原理。

6-10 (西北工业大学硕士研究生入学考试试题)已知单位反馈的典型二阶系统在 $r(t)=\sin 2t$ 作用下的稳态输出响应为

$$C_s(t) = 2\sin(2t - 90°)$$

欲采用串联校正,使校正后系统仍为典型二阶系统,并且同时满足条件:在 $r(t)=t$ 作用时,系统的稳态误差 $e_{ss}=0.25$,超调量 $\sigma\%=16.3\%$。

(1)试确定校正前系统的开环传递函数 $G_0(s)$。

(2)确定校正后系统的开环传递函数 $G(s)$,求校正后系统的截止频率 ω_c 和相角裕量 r_0。

(3)确定校正装置的传递函数 $G_c(s)$。

6-11 (南京航空航天大学硕士研究生入学考试试题)简答题:

(1)何谓系统的截止频率 ω_c?何谓系统的带宽频率 ω_b?带宽频率通常取大好还是取小好?为什么?

(2)简要说明比例积分微分 PID 控制规律中 P、I 和 D 的作用。

(3)串联超前校正的目的是利用其超前角使系统相角裕量 r 增加,串联滞后校正的相角为负值,为何也会使系统的相角裕量增加?请说明原因。

6-12 (南京航空航天大学硕士研究生入学考试试题)设单位反馈系统的开环传递函数为

$$G(s) = \frac{20}{s(s+1)}$$

试设计串联校正网络使校正后的系统相角裕量 $r \geqslant 45°$。

6-13 (天津大学硕士研究生入学考试试题)单位反馈系统的开环传递函数为:

$$G(s) = \frac{100}{s(Ts+1)}$$

(1) 当 $T=0.001$,试用频域法设计比例积分串联控制器 $G_c(s) = K_p(1 + \frac{1}{T_1 s})$ 的参数,使系统的幅值穿越频率 $\omega_c = 100 \text{rad/s}$,相角裕量 $r=60°$,并绘制校正前和校正后系统开环传递函数的对数幅频特性曲线和相频特性曲线。

(2) 串联上述比例积分控制器后,为使系统稳定,参数 T 的变化范围为多少? 系统可以做到对速度输入信号无静差吗? 可以做到对加速度输入信号无静差吗?

6-14 (浙江大学硕士研究生入学考试试题)设单位负反馈系统如题 6-14 图所示。

设计有源串联校正装置 $G_c(s)$,使校正后系统满足:

(1) 跟踪输入信号 $r(t)=t^2$ 时的稳态误差为 0.2。

(2) 相角裕量 $r=30°$。

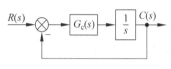

题 6-14 图

第 7 章 计算机控制系统概述

7.1 引言

计算机控制系统(Computer Control System,CCS)是应用计算机参与控制并借助一些辅助部件与被控对象相联系,从而达到一定控制目的的系统。近年来,随着计算机制造成本的降低、功能的增强,它在工业上的应用也越来越广泛。各种各样的计算机控制系统、智能仪器仪表已逐渐占据了工业控制装置的主导地位。本章主要讨论信号的采样和复现;介绍离散系统的数学模型;进行采样控制系统的稳定性分析。考虑到其他课程中已讲述 z 变换,故在此不再赘述。需要补充这部分知识的读者,请参照本书附录 B。

Terms and Concepts

A distributed control system is composed of many digital computers that perform different control functions such as feedback control, logical, and sequencing. The different computers communicate over a network. There is also a database where all system variables are stored.

Digital control system is a control system using digital signals and a digital computer to control a process.

Digital computer compensator is a system that uses a digital computer as the compensator element.

本章主要内容:
- [ ] 计算机控制系统概述
- [ ] A/D 转换采样过程与采样定理
- [ ] 采样信号的复现
- [ ] 离散控制系统的数学模型
- [ ] 采样控制系统的稳定性分析
- [ ] 其他控制系统简介
- [ ] MATLAB 在本章中的应用

7.2 计算机控制系统概述

计算机控制系统又称为数字控制系统,是采用数字技术实现各种控制功能的自动控制系统。它的主要特点是整个系统中一处或几处的信号具有数字代码形式。在计算机控制系统中,计算机的作用主要有三个方面:①信息处理,对于复杂的控制系统,输入信号和根据控制规律的要求实现的输出信号的计算工作量很大,采用模拟计算装置不能满足精度要求,需要采用数字计算机;②用数字计算机的软件程序实现对控制系统的校正以保证控制系统具有所要求的动态特性;③由于数字计算机具有快速完成复杂的工程计算的能力,因而可以实现对系统的最优控制、自适应控制等高级控制功能及多功能计算调节。

控制系统引入计算机后,仍可以按模拟系统的构成原则和信息处理流程构成系统。所不同的是要考虑计算机接收信号的形式和计算本身的特点。计算机只能处理数值信号,不能处理模拟量,因此,输入计算机的所有信息均要转化为数字量。计算机接收数字量后,按预定的程序(计算方法)对控制变量进行计算,所得的也是数字形式的控制变量。由于大部分的工业被控对象是按模拟信号工作的,所以这些数字式的控制量还要转换成模拟量,然后由模拟形式的控制变量去控制被控对象。

计算机控制系统需要有专门的电子器件把输入的模拟量转换成计算机所能接收的数字量,这种转换称为模/数转换,用 A/D 表示。而计算机本身的数据处理功能则用来作为补偿器(运算器),其输出的控制变量仍然是数字量,然后再用一种专门的电子器件把数字量转换成模拟量,即数/模转换,用 D/A 表示。经 D/A 转换后的模拟量用来控制被控对象。A/D 和 D/A 转换器都是由大规模集成电路组成的,它们能在很短的时间内完成信号转换。

典型的计算机控制系统如图 7-2-1 所示。

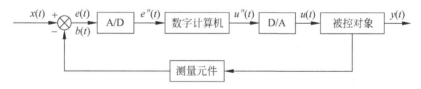

图 7-2-1 计算机控制系统

图中 A/D 转换器对连续误差信号 $e(t)$ 进行定时采样并转换成数字信号 $e''(t)$ 送入计算机。计算机输出的控制信号 $u''(t)$ 也是数字信号,通过 D/A 转换器将其恢复为连续的控制信号 $u(t)$,然后再去控制被控对象。

Reading Material

Over the years, digital computers have become more powerful, smaller, faster, reliable, and cheaper. From about 1980s digital control has been the standard technique for implementing control systems. This applies to dedicated, simple, single-loop controllers as well as to large distributed control systems. Digital control offers several important advantages over analogue control.

Functions of the control computer

Feedback control is only one of the functions of the control computer. Other functions may include:

- ☐ Reference signal generation;
- ☐ Process monitoring and date logging;
- ☐ Alarming and taking appropriate actions when variables exceed permissible limits;
- ☐ Sequencing of multiple parallel actuators;
- ☐ Process start up and shutdown;
- ☐ Sensing the status of contacts;
- ☐ Indicating whether on-off valves are closed or open;
- ☐ Switching on a motor or opening a valve.

Digital computer operates sequentially in time and each operation takes some time. To enable the computer to meet the variety of demands imposed on it, it is time-shard among its tasks.

7.3 A/D 转换采样过程与采样定理

7.3.1 采样过程

把连续信号转换成脉冲或数字序列的过程,称为采样过程。实现采样的装置,称为采样开关或采样器,用 K 表示。如果采样开关 K 以周期 T 时间闭合,并且闭合的时间为 τ,这样就把一个连续的函数 $e(t)$ 变成了一个断续的脉冲序列 $e^*(t)$ $(t=0,T,2T,\cdots)$,如图 7-3-1 所示。

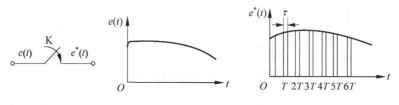

图 7-3-1 采样过程

由于采样开关 K 闭合持续时间很短,即 $\tau \ll T$,因此在分析时可以近似地认为 τ 趋于 0。

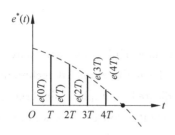

这样可看出,当采样器输入为连续信号 $e(t)$ 时,输出采样信号就是一串理想脉冲,采样瞬时 $e^*(t)$ 的脉冲等于相应瞬时 $e(t)$ 的幅值,即 $e(0T), e(T), e(2T), e(3T), \cdots, e(kT)$,如图 7-3-2 所示。

根据图 7-3-2 可以写出采样过程的数学描述为

$$e^*(t) = e(0T)\delta(t) + e(T)\delta(t-T) + \cdots + e(kT)\delta(t-kT) + \cdots \quad (7\text{-}3\text{-}1)$$

图 7-3-2 $\tau=0$ 的采样过程   或

$$e^*(t) = e(t) \sum_{k=0}^{\infty} \delta(t-kT) \tag{7-3-2}$$

其中，k 是采样的步数，即采样开关 K 闭合的次数。

Terms and Concepts

Sampled data：Data obtained for the system variables only at discrete intervals. Data obtained once every sampling period.

Sampled-data system：A system where part of the system acts on sampled data (sampled variables).

Sampling period：The period when all the numbers leave or enter the computer. The period for which the sampled variable is held constant.

Stability of a sampled-data system：The stable condition exists when all the poles of the closed-loop transfer function are within the unit circle on the z-plane.

7.3.2 采样定理

连续信号 $e(t)$ 在其有定义的时域内任何时刻都是有确切值的。而 $e(t)$ 经过采样后，只能给出采样时刻的数值 $e(0), e(T), e(2T), \cdots$。从时域上看，在采样间隔内连续信号的信息丢失了。直觉上，如果采样周期 T 越大（对应采样频率 ω_s 越低），连续信号 $e(t)$ 变化越快（对应其最大频率 ω_{max} 越高），则采样后信息的丢失越严重，直至无法从采样信号 $e^*(t)$ 中完全复现出原连续信号 $e(t)$。但是，如果采样频率 ω_s 较高，而连续信号 $e(t)$ 变化缓慢（对应其最大频率 ω_{max} 较低），则可以从采样信号 $e^*(t)$ 中完全复现出原连续信号 $e(t)$。也就是说，要想从采样信号 $e^*(t)$ 中完全复现出采样前的连续信号 $e(t)$，对采样频率 ω_s 应有一定的要求。下面从信号采样前后的信号频谱变化来分析。

设连续信号 $e(t)$ 的频谱 $E(j\omega)$ 为有限带宽，其最大频率为 ω_{max}，如图 7-3-3(a)所示，来看采样后信号 $e^*(t)$ 的频谱。

从式(7-3-1)中可以看出，理想单位脉冲序列 $\delta_T(t)$ 是一个以 T 为周期的周期函数，可以展成傅里叶级数形式：

$$\delta_T(t) = \sum_{k=-\infty}^{\infty} C_k e^{jk\omega_s t} \tag{7-3-3}$$

其中，$\omega_s = \dfrac{2\pi}{T}$ 为采样角频率，C_k 为傅里叶级数，即

$$C_k = \frac{1}{T} \int_{-\frac{T}{2}}^{\frac{T}{2}} \delta_T(t) e^{-jk\omega_s t} dt \tag{7-3-4}$$

由式(7-3-1)和式(7-3-2)可知，在区间 $\left[-\dfrac{T}{2}, \dfrac{T}{2}\right]$ 中，$\delta_T(t)$ 仅在 $t=0$ 处等于 1，其余处都等于零，所以有

$$C_k = \frac{1}{T} \int_{0-}^{0+} \delta_T(t) dt \tag{7-3-5}$$

将式(7-3-5)代入式(7-3-3)得

$$\delta_T(t) = \frac{1}{T}\sum_{k=-\infty}^{\infty} \mathrm{e}^{\mathrm{j}k\omega_s t} \qquad (7\text{-}3\text{-}6)$$

再将式(7-3-5)代入式(7-3-1),可得

$$e^*(t) = \frac{1}{T}\sum_{k=-\infty}^{\infty} e(t)\mathrm{e}^{\mathrm{j}k\omega_s t} \qquad (7\text{-}3\text{-}7)$$

对式(7-3-7)两边取拉普拉斯变化,并由拉普拉斯变换的复数位移定理,可推出

$$E^*(s) = \frac{1}{T}\sum_{k=-\infty}^{\infty} E(s+\mathrm{j}k\omega_s) \qquad (7\text{-}3\text{-}8)$$

式(7-3-8)提供了理想采样器在频域中的特点。如果 $E^*(s)$ 在右半 s 平面没有极点,则由 $s=\mathrm{j}\omega$ 得到采样输出信号 $e^*(t)$ 的傅里叶变换为

$$E^*(\mathrm{j}\omega) = \frac{1}{T}\sum_{k=-\infty}^{\infty} E[\mathrm{j}(\omega+k\omega_s)] \qquad (7\text{-}3\text{-}9)$$

式中,$E(\mathrm{j}\omega)$ 为连续信号 $e(t)$ 的傅里叶变换。

从式(7-3-8)中可以看出,采样器输出信号的频谱量 $E^*(\mathrm{j}\omega)$ 是以采样频率 ω_s 为周期的无穷多个频谱之和,如图 7-3-3(b)所示。其中,$k=0$ 的部分称为采样频谱的主分量,它与连续信号的频谱 $|E(\mathrm{j}\omega)|$ 的形状一致,幅度上改变了 $\frac{1}{T}$ 倍;$k\neq 0$ 的频谱分量都是在采样过程中引进的高频分量,称为采样频谱的补分量。

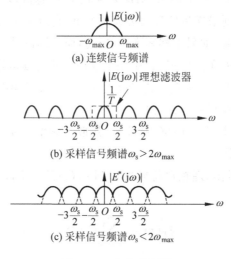

图 7-3-3 信号的频谱

比较图 7-3-3(b)和图 7-3-3(c)可以看出,如果 $\omega_s > 2\omega_{\max}$,则采样器输出信号的频谱 $E^*(\mathrm{j}\omega)$ 的各频谱分量彼此不会重叠,连续信号的频谱 $E(\mathrm{j}\omega)$ 可完整地保存下来。这样,通过一个如图 7-3-3(b)中虚线所示的理想低通滤波器,滤除所有高频分量后,就能复现出原连续信号 $e(t)$;反之,如果 $\omega_s < 2\omega_{\max}$,则采样器输出信号的频谱 $E^*(\mathrm{j}\omega)$ 的各频谱分量彼此重叠在一起,称为频率混叠现象。这样,即使有一个理想滤波器滤去高频部分也不能无失真地恢复原来的连续信号 $e(t)$,如图 7-3-3(c)所示。

由此可知,要想从采样信号 $e^*(t)$ 中完全复现出采样前的连续信号 $e(t)$,采样频率 ω_s 和

连续信号的最高频率 ω_{max} 之间的关系必须满足

$$\omega_s > 2\omega_{max} \tag{7-3-10}$$

这就是采样定理,又称为香农(Shannon)定理,它指明了复现原信号所必需的最低采样频率,是分析和设计采样控制系统的理论依据。香农采样定理的物理意义为:如果选择的采样频率 ω_s 对连续信号的最高频率来说,能够做到在一个周期内采样两次以上,那么经过采样而得到的脉冲序列就能包含连续信号的全部信息。如果采样次数太少,就不可能完整地复现原信号。

7.3.3 采样周期在工程应用中的选择方法

采样定理只是作为数字控制系统确定采样周期的理论指导原则,它给出的是满足不产生混叠效应的最大采样时间间隔 T,即采样频率 ω_s 的下限。如果把采样定理直接用到实际的数字系统中,还存在一些问题,主要是因为系统连续时间函数 $e(t)$ 的最高角频率 ω_{max} 不好确定。所以实际应用中都是根据设计者的经验并由系统实际运行实验最后确定。

在工程应用的实践中,主要根据系统被控对象的物理特性、被控对象的惯性,以及加在该对象中的预期干扰程度和性质来选择采样周期 T。例如,由于温度控制系统的热惯性大、反应慢、调节不宜过于频繁,因此采样周期 T 选得要长一些。而对于一些快速系统,如直流可逆调速系统或随动系统,要求动态响应速度快,抗扰动能力强,因此采样周期要短。总之,根据理论指导原则,结合实际对象,可以得出采样周期 T 选择的实用公式。表 7-3-1 列出了不同被控参数物理量采样周期 T 选择的参考数值。

表 7-3-1 采样周期 T 选择数据参考值

被控物理量	采样周期 T	备 注
流量	1～5s	优先选用 2s
压力	3～10s	优先选用 6s
波面	6～8s	
温度	15～20s	
位置	10～50ms	
电流	1～5ms	优先选用 3.3ms
转速	5～20ms	优先选用 9.9ms

Reading Material

Sampling theorem

Sampling theorem plays a crucial role in signal processing and communications. In the sampling problem, the objective is to reconstruct a signal from its samples. For a bandlimited signal, Shannon sampling theorem provides a full reconstruction by its uniform samples with a sampling rate higher than its Nyquist frequency. For non-bandlimited signals, several sampling criteria have been proposed associated with wavelet transform and Wigner distribution function etc.

In digital signal communication, a continuous signal is usually recovered and processed by using its discrete samples. Sampling theorem is one of the most important mathematical tools used in communication engineering and data processing since it enables engineers to reconstruct signals from some of their sampled data. The fundamental result in interpolation and sampling theorem is the celebrated Whittaker-Kotelnikov-Shannon sampling theorem, which states that if f is a signal(function) band-limited to $[-\sigma, \sigma]$ which means that its Fourier transform is located in compact support $[-\sigma, \sigma]$, then it can be completely reconstructed by its sampled values at the equidistant points $x_k = k\pi/\sigma$, $k = 0, \pm 1, \pm 2, \cdots$ This theorem has several forms. We formulate one of them proved by Kotelnikov.

A/D conversion

In many applications, one needs to convert analog, continuous time signals into quantized discrete time signals. This leads to an important set of questions regarding the best way to represent a signal by a sequence of sampled and quantized values, such that the information loss inherent in the sampling and quantization process is minimized in some sense. In the present work, we are interested in how to quantize a possible non band-limited signal to obtain the lowest possible reconstruction distortion.

We will show that, for a given sampling rate and reconstruction filters, minimization of reconstruction error, in a L^2 sense, can be converted into a discrete time problem. It turns out that if an appropriate pre-filter is used, then all the information required to find the optimal quantized sequence can always be extracted from discrete time samples of its output, even if the continuous time input signal is not band-limited.

7.4 采样信号的复现

为了实现对受控对象的有效控制，必须把采样信号恢复为相应的连续信号，这个过程称为信号的复现。保持器是将采样信号准确地复现为原来连续信号的装置。从数学的角度看，它的任务就是解决两相邻采样时刻间的插值问题。在控制工程中，一般都采用时域外推的原理来复现采样信号，其中零阶保持器采用恒值外推原理，一阶保持器采用线形外推原理。在采样控制系统中结构最简单、应用最广泛的是零阶保持器。

零阶保持器采用恒值外推原理，它把前一采样时刻的采样值 $e(kT)$ 一直保持到下一个采样时刻 $(k+1)T$，使采样信号 $e^*(t)$ 变为阶梯信号 $e_b(t)$，在 $kT \leqslant t \leqslant (k+1)T$ 期间，$e_b(t) = e(kT)$。零阶保持器的输入、输出特性如图 7-4-1 所示。

零阶保持器的输出信号 $e_b(t)$ 是一个阶梯波，它含有高次谐波，故不同于连续信号 $e(t)$。将阶梯信号的各中点连接起来，就可以得到一条比连续信号滞后 $T/2$ 的曲线，这说明零阶保持器的相位具有滞后特性。

设在零阶保持器的输入端加上单位脉冲函数 $\delta(t)$，其输出 $g_b(t)$ 称为零阶保持器的单位脉冲响应，它是一个高度为 1，持续时间为 T 的矩形波，如图 7-4-2(a)所示。这个波形可以分解为两个单位阶跃函数的叠加，如图 7-4-2(b)所示。其表达式为

$$g_b(t) = 1(t) - 1(t-T) \tag{7-4-1}$$

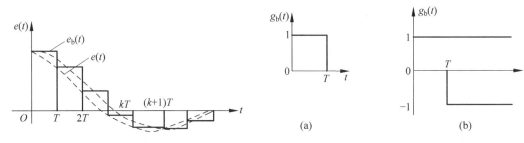

图 7-4-1 零阶保持器的输入、输出特性   图 7-4-2 零阶保持器的单位脉冲响应

对式(7-4-1)求拉普拉斯变换,得零阶保持器的传递函数为

$$G_b(s) = \frac{1}{s} - \frac{e^{-Ts}}{s} = \frac{1 - e^{-Ts}}{s} \tag{7-4-2}$$

零阶保持器的频率特性为

$$\begin{aligned} G_b(j\omega) &= \frac{1 - e^{-j\omega T}}{j\omega} = \frac{-j[1 - \cos(\omega T) + j\sin(\omega T)]}{\omega} \\ &= \frac{\sin(\omega T) - j[1 - \cos(\omega T)]}{\omega} \end{aligned} \tag{7-4-3}$$

幅频特性为

$$|G_b(j\omega)| = \frac{\sqrt{\sin^2(\omega T) + [1 - \cos(\omega T)]^2}}{\omega} = \frac{2}{\omega}\sin(\omega T/2) \tag{7-4-4}$$

相频特性为

$$\angle G_b(j\omega) = \arctan\frac{-[1 - \cos(\omega T)]}{\sin(\omega T)} = -\frac{\omega T}{2} \tag{7-4-5}$$

由以上分析可知,零阶保持器的幅频特性的幅值随频率的增大而衰减,具有低通滤波特性,但它不是理想的滤波器。另外,它的相频特性具有滞后的相位移,因此降低了系统的稳定性。

零阶保持器的传递函数展开为级数形式,即

$$G_b(s) = \frac{1 - e^{-Ts}}{s} = \frac{1}{s}\left(1 - \frac{1}{e^{Ts}}\right) = \frac{1}{s}\left(1 - \frac{1}{1 + Ts + T^2s^2/2 + \cdots}\right) \tag{7-4-6}$$

当 T 很小时,可作近似处理,只取级数的前两项,得

$$G_b(s) = \frac{1}{s}\left(1 - \frac{1}{1+Ts}\right) = \frac{T}{Ts+1} \tag{7-4-7}$$

由式(7-4-7)可见,零阶保持器可近似作为惯性环节。

7.5 离散控制系统的数学模型

要对离散控制系统进行分析研究,首先要建立它的数学模型。线性离散控制系统的数学模型有差分方程、脉冲传递函数或状态变量表达式三种。考虑到其他课程中已讲述差分方程,故在此不再重复叙述。本节主要介绍线性定常离散系统的另一种数学模型——脉冲传递函数。

7.5.1 脉冲传递函数的定义

连续系统中的时域函数和通过拉普拉斯变换所建立起来的传递函数是研究连续系统性能的重要基础。对于采样系统,与连续系统相似,可以通过与传递函数相对应的脉冲传递函数来研究采样系统的性能。

设开环离散系统如图 7-5-1 所示,假设系统的初始条件为零,输入信号为 $x(t)$,采样后 $x^*(t)$ 的 z 变换函数为 $X(z)$,系统连续部分的输出为 $y(t)$,采样后 $y^*(t)$ 的 z 变换函数为 $Y(z)$。脉冲传递函数 $G(z)$ 定义为:在初始条件为零情况下,系统离散输出信号的 z 变换 $Y(z)$ 与输入采样信号的 z 变换 $X(z)$ 之比,即

图 7-5-1 开环离散系统

$$G(z) = \frac{Y(z)}{X(z)} \tag{7-5-1}$$

由式(7-5-1)可求得线性离散系统的输出采样信号为

$$y^*(t) = \mathcal{Z}^{-1}[Y(z)] = \mathcal{Z}^{-1}[G(z)X(z)] \tag{7-5-2}$$

实际上,许多采样系统的输出信号是连续信号 $y(t)$,而不是离散信号 $y^*(t)$,如图 7-5-1 所示。在这种情况下,为了应用脉冲传递函数的概念,可以在系统的输出端虚设一个理想采样开关,如图 7-5-2 中的虚线所示。该虚设采样开关的采样周期与输入端采样开关的采样周期相同。如果系统的实际输出 $y(t)$ 比较平滑,且采样频率较高,则可用 $y^*(t)$ 近似描述 $y(t)$。必须指出,虚设的采样开关是不存在的,它仅表明了脉冲传递函数所能描述的只是输出连续信号 $y(t)$ 在采样时刻上的离散值 $y^*(t)$。

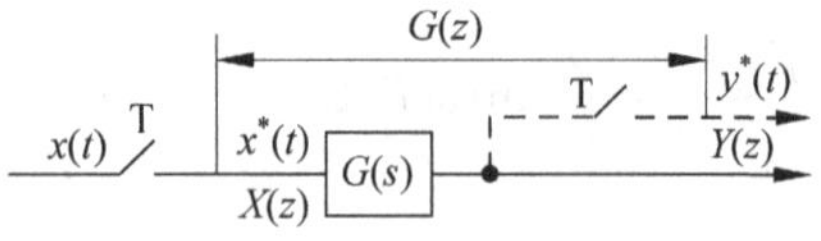

图 7-5-2 实际开环离散系统

连续系统或元件的脉冲传递函数 $G(z)$ 可以通过其传递函数 $G(s)$ 来求取。具体步骤如下。

(1) 对连续传递函数 $G(s)$ 进行拉普拉斯反变换,求得脉冲响应 $g(t)$ 为

$$g(t) = \mathcal{L}^{-1}[G(s)] \tag{7-5-3}$$

(2) 对 $g(t)$ 进行采样,求得离散脉冲响应 $g^*(t)$

$$g^*(t) = \sum_{k=0}^{\infty} g(kT)\delta(t-kT) \tag{7-5-4}$$

(3) 对 $g^*(t)$ 进行 z 变换,即可得到该系统的脉冲传递函数 $G(z)$

$$G(z) = \mathcal{Z}[g^*(t)] = \sum_{k=0}^{\infty} g(kT)z^{-k} \tag{7-5-5}$$

脉冲传递函数也可由给定连续系统的传递函数,经部分分式法通过查表求得。

【例 7-5-1】 设系统的结构图如图 7-5-2 所示,其中连续部分的传递函数为 $G(s) = \dfrac{2}{s(s+2)}$。试求该系统的脉冲传递函数 $G(z)$。

解:(1) 对连续传递函数 $G(s)$ 进行拉普拉斯反变换,求得脉冲响应 $g(t)$ 为

$$g(t) = \mathcal{L}^{-1}\left[\frac{2}{s(s+2)}\right] = \mathcal{L}^{-1}\left[\frac{1}{s} - \frac{1}{s+2}\right] = 1 - e^{-2t}$$

(2) 对 $g(t)$ 进行采样,求得离散脉冲响应 $g^*(t)$

$$g^*(t) = \sum_{k=0}^{\infty} [1(kT) - e^{-2kT}]\delta(t-kT)$$

(3) 对 $g^*(t)$ 进行 z 变换,即可得到该系统的脉冲传递函数 $G(z)$

$$G(z) = \mathcal{Z}[g^*(t)] = \sum_{k=0}^{\infty} [1(kT) - e^{-2kT}]z^{-k}$$

$$= \sum_{k=0}^{\infty} 1(kT) \cdot z^{-k} - \sum_{k=0}^{\infty} e^{-2kT} \cdot z^{-k}$$

$$= \frac{z}{z-1} - \frac{z}{z-e^{-2kT}} = \frac{(1-e^{-2T})z}{z^2-(1+e^{-2T})z+e^{-2T}}$$

本例也可由部分分式法求得,即 $G(s) = \dfrac{1}{s} - \dfrac{1}{s+2}$。查 z 变换表得

$$G(z) = \frac{z}{z-1} - \frac{z}{z-e^{-2T}} = \frac{(1-e^{-2T})z}{z^2-(1+e^{-2T})z+e^{-2T}}$$

7.5.2 开环采样系统的脉冲传递函数

1. 串联环节的脉冲传递函数

在连续系统中,串联环节的传递函数等于各环节传递函数之积。但是对采样系统而言,串联环节的就不一定是这样。根据它们之间有无采样开关,其等效的脉冲传递函数是不相同的。

1) 串联环节之间有采样开关

两个串联环节之间有采样开关分隔的情况,如图 7-5-3 所示。

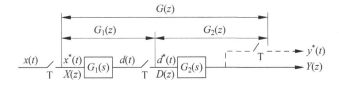

图 7-5-3　串联环节之间有采样开关

在两个串联连续环节 $G_1(s)$ 和 $G_2(s)$ 之间有采样开关时,根据脉冲传递函数的定义,有 $D(z) = G_1(z)X(z),Y(z) = G_2(z)D(z)$。其中,$G_1(z)$ 和 $G_2(z)$ 分别为 $G_1(s)$ 和 $G_2(s)$ 的脉冲传递函数。于是 $Y(z) = G_1(z)G_2(z)X(z)$。所以该系统的脉冲传递函数为

$$G(z) = \frac{Y(z)}{X(z)} = G_1(z)G_2(z) \qquad (7-5-6)$$

式(7-5-6)表明,有采样开关的两个串联环节的脉冲传递函数等于两个环节各自脉冲传递函数的乘积。这一结论可以推广到有采样开关的 n 个串联环节的情形。

2) 串联环节之间没有采样开关

两个串联环节之间无采样开关的情况,如图 7-5-4 所示。

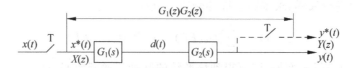

图 7-5-4 串联环节之间无采样开关

由图 7-5-4 可得

$$G(s) = \frac{Y(s)}{X(s)} = G_1(s)G_2(s) \tag{7-5-7}$$

对式(7-5-7)取 z 变换，可得脉冲传递函数

$$G(z) = \frac{Y(z)}{X(z)} = \mathcal{Z}[G_1(s)G_2(s)] = G_1(z)G_2(z) \tag{7-5-8}$$

式(7-5-8)表明，没有采样开关的两个串联环节的脉冲传递函数，等于这两个环节传递函数乘积后的相应 z 变换。同理，此结论适用于没有采样开关的 n 个环节串联的情形。

【例 7-5-2】 试求图 7-5-5(a)和图 7-5-5(b)所示的两个系统脉冲传递函数。

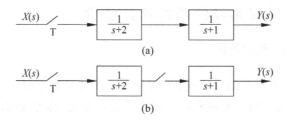

图 7-5-5 例 7-5-2 附图

解：(1) 图 7-5-5(a)中的系统，其脉冲传递函数为

$$G(z) = \mathcal{Z}[G_1(s)G_2(s)] = \mathcal{Z}\left[\frac{1}{(s+2)(s+1)}\right] = \frac{(e^{-T} - e^{-2T})z}{z^2 - (e^{-T} + e^{-2T})z + e^{-3T}}$$

(2) 图 7-5-5(b)中两个环节之间有采样开关，因此其脉冲传递函数为两个串联环节脉冲传递函数的乘积，即

$$G(z) = \mathcal{Z}[G_1(s)] \cdot \mathcal{Z}[G_2(s)] = \mathcal{Z}\left[\frac{1}{s+2}\right] \cdot \mathcal{Z}\left[\frac{1}{s+1}\right] = \frac{z^2}{z^2 - (e^{-T} + e^{-2T}) + e^{-3T}}$$

2. 具有零阶保持器的开环脉冲传递函数

具有零阶保持器的开环离散系统如图 7-5-6 所示。图中零阶保持器的传递函数为 $G_h(s)$，且 $G_h(s) = \dfrac{1 - e^{Ts}}{s}$，$G_p(s)$ 为系统其他连续部分的传递函数，即两个串联环节之间没有同步采样开关。

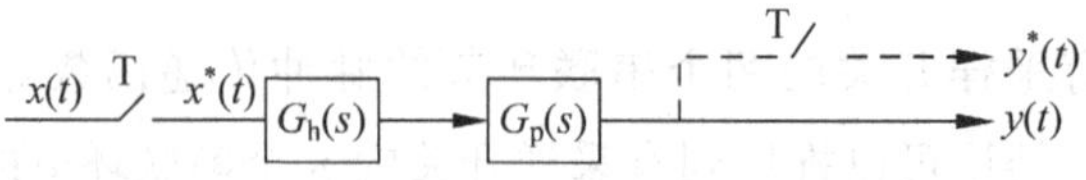

图 7-5-6 具有零阶保持器的开环离散系统

系统的脉冲传递函数为

$$G(s) = G_h(s)G_p(s) = \frac{1-e^{Ts}}{s}G_p(s) = (1-e^{Ts})\frac{G_p(s)}{s} \quad (7\text{-}5\text{-}9)$$

根据 z 变换,得到系统的脉冲传递函数为

$$G(z) = \mathcal{Z}[G_h(s)G_p(s)] = (1-z^{-1})\mathcal{Z}\left[\frac{G_p(s)}{s}\right] \quad (7\text{-}5\text{-}10)$$

【例 7-5-3】 具有零阶保持器的开环离散系统如图 7-5-6 所示,$G_p(s) = \dfrac{2}{s(s+2)}$,试求该系统的脉冲传递函数。

解:因为

$$\frac{G_p(s)}{s} = \frac{2}{s(s+2)}\bigg/s = \frac{2}{s^2(s+2)} = \frac{1}{s^2} - \frac{0.5}{s} + \frac{0.5}{s+2}$$

查 z 变换表,得

$$\mathcal{Z}\left[\frac{G_p(s)}{s}\right] = \mathcal{Z}\left[\frac{1}{s^2} - \frac{0.5}{s} + \frac{0.5}{s+2}\right] = \frac{Tz}{(z-1)^2} - \frac{0.5}{z-1} + \frac{0.5}{z-e^{-2T}}$$

于是得到系统的脉冲传递函数

$$G(z) = (1-z^{-1})\mathcal{Z}\left[\frac{G_p(s)}{s}\right] = \frac{(T-0.5+0.5e^{-2T})z + (0.5-Te^{-2T}-0.5e^{-2T})}{(z-1)(z-e^{-2T})}$$

7.5.3 闭环采样系统的脉冲传递函数

在连续系统中,闭环传递函数与相应的开环传递函数之间存在确定的关系,由于在采样系统中,采样器在闭环系统中可以有多种配置的可能性,因而闭环采样系统的脉冲传递函数还与采样开关的位置有关,只能根据系统具体的实际结构求取。

一种比较常见的误差采样闭环系统框图如图 7-5-7 所示。图中系统的误差为

$$E(s) = X(s) - B(s) \quad (7\text{-}5\text{-}11)$$

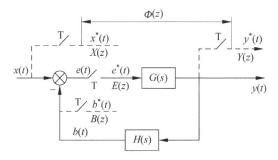

图 7-5-7 闭环采样系统框图

反馈方程为

$$B(s) = H(s)Y(s) \quad (7\text{-}5\text{-}12)$$

输出方程为

$$Y(s) = G(s)E^*(s) \quad (7\text{-}5\text{-}13)$$

将式(7-5-12)和式(7-5-13)代入式(7-5-11),得

$$E(s) = X(s) - H(s)G(s)E^*(s) \tag{7-5-14}$$

于是误差采样信号 $e^*(t)$ 的拉普拉斯变换为

$$E^*(s) = X^*(s) - G(s)H^*(s)E^*(s) \tag{7-5-15}$$

整理得

$$E^*(s) = \frac{X^*(s)}{1+G(s)H^*(s)} \tag{7-5-16}$$

由于

$$Y^*(s) = G^*(s)E^*(s) = \frac{G^*(s)}{1+HG^*(s)}R^*(s) \tag{7-5-17}$$

对式(7-5-16)和式(7-5-17)取 z 变换,可得

$$E(z) = \frac{X(z)}{1+HG(z)} \tag{7-5-18}$$

$$Y(z) = \frac{G(z)}{1+HG(z)}X(z) \tag{7-5-19}$$

根据式(7-5-18),得到闭环采样系统对于输入量的误差脉冲传递函数

$$\Phi_e(z) = \frac{E(z)}{X(z)} = \frac{1}{1+HG(z)} \tag{7-5-20}$$

根据式(7-5-19),得到闭环采样系统对输入量的脉冲传递函数

$$\Phi(z) = \frac{Y(z)}{X(z)} = \frac{G(z)}{1+HG(z)} \tag{7-5-21}$$

式(7-5-20)和式(7-5-21)是研究闭环采样系统时经常用到的两个闭环脉冲传递函数。表 7-5-1 列出了一些典型闭环采样系统的框图及其输出量的 z 变换。

表 7-5-1 典型闭环采样系统及其输出量的 z 变换函数

	系 统 框 图	$Y(z)$
1	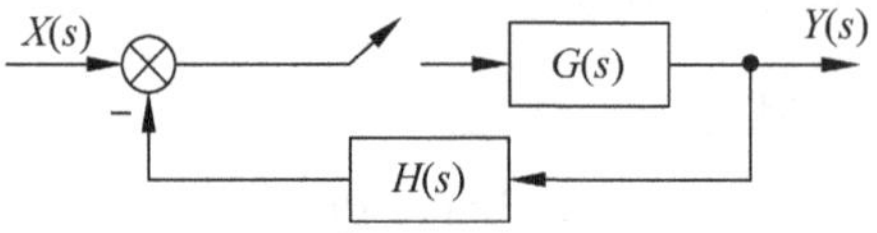	$Y(z) = \dfrac{G(z)X(z)}{GH(z)}$
2	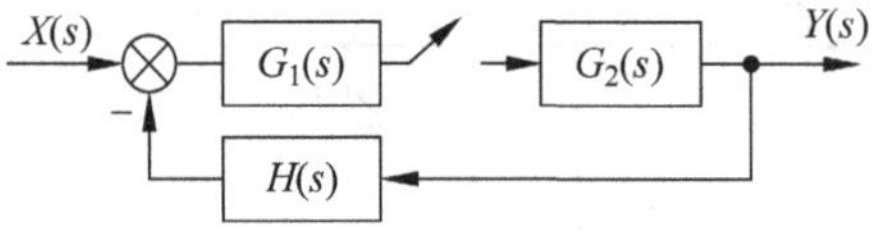	$Y(z) = \dfrac{XG_1(z)G_2(z)}{1+G_2HG_1(z)}$
3	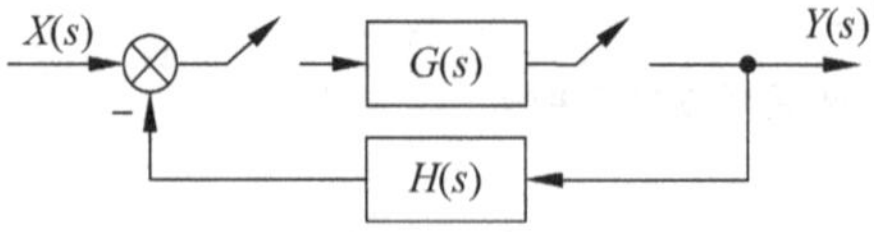	$Y(z) = \dfrac{X(z)G(z)}{1+G(z)H(z)}$
4	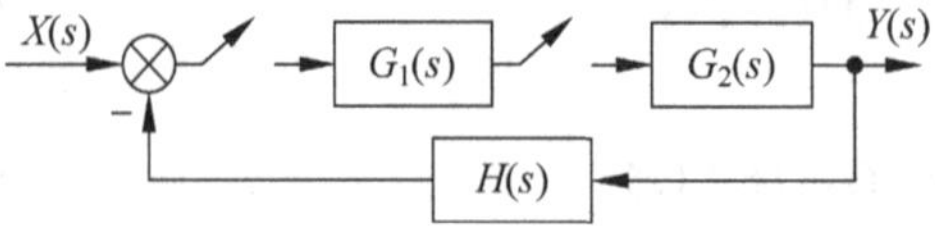	$Y(z) = \dfrac{X(z)G_1(z)G_2(z)}{1+G_1(z)G_2(z)H(z)}$

续表

	系统框图	$Y(z)$
5	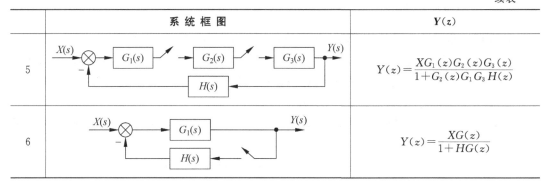	$Y(z) = \dfrac{XG_1(z)G_2(z)G_3(z)}{1+G_2(z)G_1G_3H(z)}$
6		$Y(z) = \dfrac{XG(z)}{1+HG(z)}$

7.6 采样控制系统的稳定性分析

一旦建立了采样控制系统的数字模型——脉冲传递函数，就可以用它来分析系统的稳定性。

在连续系统中，曾经介绍了利用传递函数特征方程的根在 s 域中的分布，来判断系统的稳定性，并提出了各种稳定判据。由于 z 变换与拉普拉斯变换的密切关系，使得我们能够利用已有的稳定判据在 z 域中判别采样系统的稳定性。为此先介绍 s 域和 z 域的关系，并提出稳定条件，然后再判别采样系统的稳定性。

1. $[s]$ 平面与 $[z]$ 平面的关系

曾在定义 z 变换时，已假定

$$z = e^{sT} \tag{7-6-1}$$

其中 s 是复变量，即

$$s = \sigma + j\omega \tag{7-6-2}$$

将式(7-6-1)代入式(7-6-2)，得

$$z = e^{(\sigma+j\omega)T} = e^{\sigma T} \cdot e^{j\omega T} = |z| e^{j\theta} \tag{7-6-3}$$

式中，$|z| = e^{\sigma T}$，$\theta = \omega T$。注意式(7-6-2)与式(7-6-3)，它们分别描写了 $[s]$ 平面与 $[z]$ 平面，其对应关系如表 7-6-1 和图 7-6-1 所示。

表 7-6-1 $[s]$ 平面与 $[z]$ 平面的对应关系

$[s]$ 平面		$[z]$ 平面	
$\sigma=0$	虚轴	$|z|=1$	单位圆周
$\sigma<0$	左半平面	$|z|<1$	单位圆内
$\sigma>0$	右半平面	$|z|>1$	单位圆外

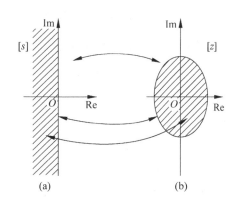

图 7-6-1 $[s]$ 平面与 $[z]$ 的对应关系

2. z 域的稳定条件

连续系统稳定的充要条件是所有闭环极点必须处于[s]平面的左半平面。由于[s]平面与[z]平面的上述关系,可得出采样控制系统稳定的充要条件为:闭环脉冲传递函数的所有极点,必须处在[z]平面的单位圆内。只要有一个极点在单位圆外,系统就不稳定;若有一个极点在单位圆周上,则为临界稳定(实属不稳定)。

【例 7-6-1】 设采样控制系统结构图如图 7-6-2 所示,已知采样周期 $T=1$s,试分析系统稳定性。

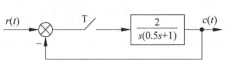

图 7-6-2 例 7-6-1 附图

解:由图 7-6-2 可知

$$G(s)=\frac{2}{s(0.5s+1)}=\frac{4}{s(s+2)}=\frac{2}{s}-\frac{2}{s+2}$$

可求得开环脉冲传递函数

$$G(z)=\mathcal{Z}[G(s)]|_{T=1}=\mathcal{Z}\left[\frac{2}{s}-\frac{2}{s+2}\right]\Big|_{T=1}$$
$$=\left[\frac{2z}{z-1}-\frac{2z}{2-e^{-2T}}\right]\Big|_{T=1}=\frac{2z}{z-1}-\frac{2z}{z-e^{-2}}$$
$$=\frac{1.73z}{(z-1)(z-0.135)}$$

故闭环脉冲传递函数

$$\Phi(z)=\frac{C(z)}{R(z)}=\frac{G(z)}{1+G(z)}=\frac{1.73z}{(z-1)(z-0.135)+1.73z}$$
$$=\frac{1.73z}{z^2+0.595z+0.135}$$

则闭环特征方程

$$D(z)=z^2+0.595z+0.135=0$$

解得 $z_{1,2}=-0.3\pm j0.216=0.36e^{j125.75°}$。

显然,由于 $|z|=0.36<1$,两个极点(特征根)都落在单位圆内,所以系统稳定。

应当指出,这种基于稳定条件判别稳定性的方法,只适应于二阶以下的采样系统。对于特征方程阶次较高的闭环采样系统,由于求解其根较难而不适用。

事实上,对于高阶采样控制系统稳定性的判别,既有自己独立的稳定判据,也可利用连续系统的时域与频域判据,本书仅介绍利用连续系统时域判据的方法。

3. 采样控制系统的时域判据

为了利用连续系统的时域判据,必须将 z 域以 1 单位圆为界的平面转换成以虚轴为界的平面。虽然[s]平面与[z]平面具有这种对应关系,但因 $z=e^{sT}$,会使特征方程 $D(s)=0$ 出现超越函数,求解其根则是不利的。因此,必须要找到新的对应平面,这就是[ω]平面。令

$$z=\frac{\omega+1}{\omega-1} \tag{7-6-4}$$

就可以找到[ω]平面与[z]平面的对应关系。

这样,通过式(7-6-4)的变换,本来以 z 为变量的特征方程 $D(z)=0$ 就转换成以 ω 为变量的特征方程 $D(\omega)=0$ 了,于是可利用连续系统的劳斯判据判别采样系统的稳定性。

【例 7-6-2】 试用劳斯判据判别例 7-6-1 系统的稳定性。

解：由例 7-6-1 所得到闭环采样系统的特征方程
$$z^3 + 0.595z + 0.135 = 0$$
将式(7-6-4)代入上式得
$$\left(\frac{\omega+1}{\omega-1}\right)^2 + 0.595\left(\frac{\omega+1}{\omega-1}\right) + 0.135 = 0$$
整理得到以 ω 为变量的特征方程为
$$1.73\omega^2 + 1.73\omega + 0.54 = 0$$
由于所有系数都大于零,满足稳定的必要条件,列出劳斯阵列表为

s^2	1.73	1.73
s^2	1.73	0
s^0	0.54	

显然因第一列元素没有改变符号,所以闭环采样系统稳定,结论与例 7-6-1 相同。

【例 7-6-3】 已知闭环采样系统脉冲传递函数
$$\Phi(z) = \frac{C(z)}{R(z)} = \frac{z^2 + 4z + 3}{z^3 + 2z^2 - 0.5z - 1}$$
试判断其稳定性。

解：其特征方程
$$z^3 + 2z^2 - 0.5z - 1 = 0$$
将式(7-6-4)代入上式并整理为
$$3\omega^3 + 17\omega^2 - 3\omega - 1 = 0$$
由于系数有负值,故本系统不稳定。

【例 7-6-4】 设采样控制系统结构图如图 7-6-3 所示,试分析采样周期 $T = 0.5s$ 和 $T = 1s$ 时增益 K 的临界值。

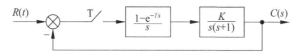

图 7-6-3 例 7-6-4 附图

解：系统的闭环脉冲传递函数为
$$\Phi(z) = \frac{C(z)}{R(z)} = \frac{K[(T-1+e^{-T})z - (1-e^{-T}-Te^{-T})]}{z^2 + [K(T-1+e^{-T}) - (1+e^{-T})]z + [K(1-e^{-T}-Te^{-T}) + e^{-T}]}$$
特征方程为
$$D(z) = z^2 + [K(T-1+e^{-T}) - (1+e^{-T})]z + [K(1-e^{-T}-Te^{-T}) + e^{-T}] = 0$$
(1) 当采样周期为 $T = 0.5s$ 时,特征方程为
$$D(z) = z^2 + (0.107K - 1.607)z + (0.09K + 0.607) = 0$$
经过 ω 变换可得到以 ω 为变量的特征方程
$$D(\omega) = 0.197K\omega^2 + (0.786 - 0.18K)\omega + (3.214 - 0.017K) = 0$$
列出劳斯阵列表为

ω^2	$0.197K$	$3.214-0.017K$
ω	$0.786-0.18K$	
ω^0	$3.214-0.017K$	

由此可得当 $T=0.5\mathrm{s}$ 时,欲使系统稳定,K 的取值范围是
$$0<K<4.37$$
则当 $T=0.5\mathrm{s}$ 时,K 的临界值为 $K_c=4.37$。

(2) 当采样周期为 $T=1\mathrm{s}$ 时,特征方程为
$$D(z)=z^2+(0.368K-1.368)z+(0.264K+0.368)=0$$
经过 ω 变换可得到以 ω 为变量的特征方程
$$D(\omega)=0.632K\omega^2+(1.264-0.528K)\omega+(2.763-0.104K)=0$$
列出劳斯阵列表为

ω^2	$0.632K$	$2.763-0.104K$
ω	$1.264-0.528K$	
ω^0	$2.763-0.104K$	

由此可得当 $T=1\mathrm{s}$ 时,欲使系统稳定,K 的取值范围是
$$0<K<2.39$$
则当 $T=1\mathrm{s}$ 时,K 的临界值为 $K_c=2.39$。

应该指出的是,在例 7-6-4 系统中,如果没有采样开关和零阶保持器,该系统就是一个一阶线性连续系统,无论开环增益 K 取何值,系统始终是稳定的,而二阶线性连续系统却不一定是稳定的,它与系统的参数有关。当开环增益比较小时系统可能稳定,当开环增益比较大,超过临界值时,系统就会不稳定。

另一个值得注意的问题是,采样周期 T 是离散系统的一个重要参数。采样周期变化时,系统的开环脉冲函数、闭环脉冲函数和特征方程都要变化,因此系统的稳定性也要发生变化。一般情况下,缩短采样周期可使线性离散系统的稳定性得到改善,增大采样周期对稳定性不利。这是因为缩短采样周期将导致采样频率的提高,从而增加离散控制系统获取的信息量,使其在特性上更加接近相应的连续系统。

Reading Material

In the present course, we have limited our attention to feedback control systems that are made up of elements and sub-systems(actuators, transducers, control circuitry etc.) having linear and continuous time dynamics only. It has been assumed that the controller function is implemented with analogue hardware.

However, we are now well and truly into the digital era, an era of digital music, digital cameras and digital TVs. The main reason is that signal processing, transmission and storage are most reliably done in the digital domain.

Indeed the impact of digital computers on control systems has been as great as on any branch of technology. It is now common practice to use a small digital computer as the controller even for a single loop. They are widely used in large industrial plants, and a modern control system may involve many computers interlinked in a network.

The problem inherent in the use of the digital processor for the purposes of feedback control and the development of the theoretical and practical backgrounds required to design and implement such digital control systems will be covered elsewhere, but for the time being it is sufficient to acquire an overview of digital control systems, which is the objective of the following activity.

7.7 其他控制系统简介

7.7.1 过程控制系统简介

过程控制通常是指石油、化工、冶金、轻工、纺织、制药、建材等工业生产过程中的自动控制，它是自动化技术的一个极其重要的方面，它的发展与生产过程自身的发展紧密相关，经历了一个由简单到复杂、从低级到高级，并正向纵深发展的过程。过程控制的发展是与控制理论、仪表、计算机以及有关学科的发展紧密相关的，其发展大体上可以分为如表 7-7-1 所示的三个阶段。

表 7-7-1 过程控制发展的三个阶段

阶　　段	20 世纪 70 年代以前	20 世纪 70—80 年代	20 世纪 90 年代
控制理论	经典控制理论	现代控制理论	控制论、信息论、系统论、人工智能等学科交叉
控制工具	常规仪表（气动、液动、电动）	分布式控制计算机（DCS）	计算机网络
控制要求	安全、平稳	优质、高产、低消耗	市场预测、快速响应、柔性生产、创新管理
控制系统	简单控制系统	先进控制系统	综合自动化（CIPS）

第一阶段是初级阶段，包括人工控制，以经典控制理论为主要基础，采用常规气动、液动和电动仪表，对生产过程中的温度、流量、压力和液位进行控制，在诸多控制系统中，以单回路结构、PID 策略为主，同时针对不同的对象与要求，创造了一些专门的控制系统。例如，使物料按比例配制的比值控制，克服大滞后的 Smith 预估器，克服干扰的前馈控制和串级控制等。这个阶段的主要任务是稳定系统，实现定值控制。这与当时生产水平是相适应的。

第二阶段是发展阶段，以现代控制理论为主要基础，以微型计算机和高档仪表为工具，对较复杂的工业过程进行控制。这个阶段的建模理论、在线辨识和实时控制已突破前期的形式，涌现了大量的先进控制系统和高级控制策略。例如，克服对象特性时变和环境干扰等不确定因素影响的自适应控制，消除因模型失配而产生不良影响的预测控制等。这个阶段的主要任务是克服干扰和模型变化，满足复杂的工艺要求，提高控制质量。

1975 年，世界上第一台分散控制系统在美国 Honeywell 公司问世，从而揭开了过程控制崭新的一页。分散控制系统也称做集散控制系统，它综合了计算机技术、控制技术、通信技术和显示技术，采用多层分级的结构形式，按总体分散、管理集中的原则，完成对工业过程的操作、监视和控制。由于采用了分散的结构和冗余等技术,系统的可靠性极高,再加上硬件方面的开放式框架和软件方面的模块化形式,使得其组态、扩展极为方便,还有众多的控制算法(几十至上百种)、较好的人机界面和故障检测报告功能。经过二十多年的发展,它已日臻完善,在众多的控制系统中,显示出出类拔萃的风范。因此,可以毫不夸张地说,分散控制系统是过程控制发展史上的一个里程碑。

第三阶段是高级阶段,自动过程控制(APC)起源于过程工业。在过程工业中许多因素很难通过简单调整达到控制过程质量的目的,如周围的温度、气压等。这些难以控制的因素只能通过反馈和前馈的方法以补偿的方式控制和调整生产过程,这就是所谓的自动过程控制 APC。其原理是通过控制方程自动补偿输出偏差,设为 e_t,t 表示 t 时刻。当输出质量特性与设计目标值的偏差较大时,控制方程就自动补偿输入,使其偏差变小。由于这是通过控制方程自动监测调整,因而称为自动过程控制(APC),其工作原理如图 7-7-1 所示,其中,e_t 为 t 时刻输出质量特性偏差,y_t 为过程输入的补偿量。

一旦找出控制方程,就可得到相应的工作方程。

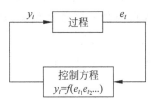

图 7-7-1　APC 工作原理图

当前过程控制正处于第三个发展阶段,并以前所未有的速度和规模飞速前进,纵观这一时期,可以归纳为如下三个主要特点。

第一,简单控制向先进控制发展。早期的控制受经典控制理论和常规仪表的限制,难以处理工业过程中存在的复杂性、耦合性、非线性等,只能按某种原则将复杂系统分解成若干相对独立的单变量系统。这种简单控制是一种分散自治控制。随着企业提出的高效益、高柔性的要求,上述控制方式已不能适应,先进控制便应运而生。先进过程控制(Advanced Process Control)是指一类在动态环境中,基于数学模型,借助充分的计算能力,为工厂获得最大利润而实施的运行和技术策略。

第二,封闭的分布式计算机控制系统转向具有国际统一标准的开放式系统。1975 年 Honeywell 公司推出第一台分布式计算机控制系统(DCS),实现了控制分散、监视集中的功能,提高了系统的可靠性和灵活性,为连续工业自动化建立了丰功伟绩。但 DCS 的一个致命弱点就是封闭性。随着综合自动化的潮流和计算机科学与技术的发展,Fisher-Rosemount、Honeywell 等欧美十余家公司经过激烈的竞争,最后终于联手,将共同推出一种国际标准的现场总线(Field Bus)控制系统。

第三,单一控制系统向综合自动化系统发展。从本质上讲,工业企业自动化在 20 世纪 90 年代以前仍是自动化孤岛模式,进入 20 世纪 90 年代,国内外企业界在国际市场剧烈竞争的刺激下,已把注意力转移到节能降耗、少投入多产出的高效生产模式上,企业开始把提高综合自动化水平作为快速挖潜增效、提高竞争能力的重要途径,集常规控制、先进控制、在线优化、生产调度、企业管理、经营决策等功能于一体的综合自动化成了当前自动化发展的趋势。

Reading Material

Process control deals with the control of equipment and plant in the process industries such as petrochemicals, food, steel, glass, paper and energy. Process control engineer will need to understand how changes in operating conditions affect product quality and how variables within the plant interact. It is in the process industries that early development in the use of digital computers for control occurred.

The idea of using digital computers for process control emerged in the mid 1950s when a group of engineers developed a computer-controlled system for the polymerization unit based on RW-300 computer. The system controlled 26 flows, 72 temperatures, 3 pressures, and 3 compositions.

Early development of the disciplines of process control and servo systems occurred in parallel with very little cross-fertilization. As a result the language and terminology used by process control engineers differed from that used by the servo designer. However, both disciplines were for broadly similar objectives.

In servo systems, the emphasis is on how well the control system can follow changes in the reference signal. In a typical process control system the reference value will not change frequently. The emphasis in process control is on the performance of the loop as a regulator, i.e. disturbance rejection.

For example, the required temperature of a particular product or stream may remain constant for days. During reference value changes, the product is often recycled or dumped until the plant has stabilized. Typically, process control systems have very slow dynamics compared to electrical and electromechanical systems.

7.7.2 机电一体化系统简介

机电一体化的外文名词是 Mechatronics，起源于日本，是日本人创造的英文名词，是取英语 Mechanics 的前半部和 Electronics 的后半部拼合而成的，表示机械学与电子学两种学科的综合。目前，国内外对机电一体化的含义有各种各样的认识，其各自的出发点和着眼点不尽相同，再加上机电一体化本身的含义还在随着生产和科学技术的发展不断被赋予新的内容。到目前为止，较为人们所接受的含义是日本机械振兴协会经济研究所于 1981 年 3 月提出的解释："机电一体化是机械的主功能、动力功能、信息功能和控制功能上引进了微电子技术，并将机械装置与电子装置用相关软件有机融合而构成的系统的总称。"随着微电子技术、传感器技术、精密机械技术、自动控制技术、微型计算机技术、人工智能技术等新技术的发展，以机械为主体的工业产品和民用产品不断采用诸学科的新技术，在机械化的基础上，正向自动化和智能化方向发展。

机电一体化系统(产品)是由若干具有特定功能的机械和电子要素组成的有机整体，具有满足人的使用要求的最佳功能。机电一体化系统(产品)，主要是指机械系统(或部件)与微电子系统(或部件)相互置换和有机结合，从而赋予新的功能和性能的新一代产品，有良好的人机协作关系。一个机电一体化的系统主要是由以下几个基本结构要素构成的。

(1) 机械本体部分。机械本体就像人体的身躯骨架，它是系统所有功能元素的机械支持结构，包括机身、框架、机械连接等。

(2) 动力部分。动力部分像人体内脏产生能量去维持生命运动一样，为系统提供能量和动力功能，驱动执行机构，使系统按照控制要求正常运行。

(3) 传感部分。传感部分就像人的眼、鼻、耳、口等感觉器官,将系统运行中所需要的本身和外界环境的各种参数及状态进行检测,变成可识别的信号,传输到信息处理单元,经过分析、处理后产生相应的控制信息。其功能一般由专门的传感器和仪器仪表完成。

(4) 驱动部分。驱动部分就像人体的肌、肋、腱接受大脑指挥驱动四肢运动一样,在控制信息作用下,驱动执行机构完成各种动作和功能。

(5) 执行部分。执行部分如同人的四肢由大脑支配完成各项工作任务一样,根据控制信息和指令,完成各种动作和功能。执行机构是运动部件,一般采用机械、电磁、电液等机构。

(6) 控制及信息处理部分。控制及信息处理部分犹如人的大脑指挥和控制全身运动并能记忆、思考和判断问题一样,将来自各传感器的检测信息集中、存储、分析、加工,并根据信息处理的结果,按照一定的程序和节奏发出相应的指令控制整个系统有目的地运行。它一般由计算机、可编程序控制器、数控装置、逻辑电路、A/D 与 D/A 转换、I/O(输入输出)接口和计算机外部设备等组成。

机电一体化是集机械、电子、光学、控制、计算机、信息等多学科的交叉综合,它的发展和进步依赖并促进相关技术的发展和进步。因此,机电一体化的主要发展方向如下。

(1) 光机电一体化。一般的机电一体化系统是由传感系统、能源系统、信息处理系统、机械结构等部件组成的,因此,引进光学技术,能有效地改进机电一体化系统的传感系统性能、能源(动力)系统和信息处理系统性能,光机电一体化是机电产品发展的重要趋势。

(2) 全系统化——智能化。今后的机电一体化产品"全息"特征越来越明显,智能化水平越来越高。模拟人类智能,使它具有判断推理、逻辑思维、自主决策等能力,以求得到更高的控制目标。这主要受益于模糊技术、信息技术(尤其是软件及芯片技术)的发展,机器人与数控机床的智能化就是其重要应用。

(3) 技术产品——网络化。20 世纪 90 年代,计算机技术等的突出成就是网络技术。网络技术的兴起和飞速发展给科学技术、工业生产、政治、军事、教育等都带来了巨大的变革。各种网络将全球经济、生产连成一片,企业间的竞争也将全球化。机电一体化新产品一旦研制出来,只要其功能独到,质量可靠,很快就会畅销全球。由于网络的普及,基于网络的各种远程控制和监视技术方兴未艾。而远程控制的终端设备本身就是机电一体化产品。因此,机电一体化产品无疑朝着网络化方向发展。

(4) 微型机电化——微型化。泛指几何尺寸不超过 $1cm^3$ 的机电一体化产品,正向微米、纳米级发展。目前,利用半导体器件制造过程中的蚀刻技术,在实验室中已制造出亚微米级的机械元件,当这一成果用于实际产品时,就没有必要区分机械部分和控制器了。届时机械和电子完全可以"融合",机体、执行机构、传感器、CPU 等可集成在一起,体积很小,并组成一种自律元件。这种微型机械学是机电一体化的重要发展方向。

(5) 自律分配系统化——柔性化。未来的机电一体化产品,控制和执行系统有足够的"冗余度",有较强的"柔性",能较好地应付突发事件,被设计成"自律分配系统"。在自律分配系统中,各个子系统是相互独立工作的,子系统为总系统服务,同时具有本身的"自律性",可根据不同的环境条件做出不同反应。其特点是子系统可产生本身的信息并附加所给信息,在总的前提下,具体"行动"是可以改变的。这样,既明显地增加了系统的适应能力(柔性),又不因某一子系统的故障而影响整个系统。

(6) 设计产品——绿色化。工业的发展给人们生活带来了巨大的变化。一方面，物质丰富，生活舒适；另一方面，资源减少，生态环境受到严重污染。于是，人们呼吁保护环境资源，回归自然。绿色产品概念在这种呼声下应运而生，绿色化是时代的趋势。绿色产品在其设计、制造、使用和销毁的生命过程中，符合特定的环境保护和人类健康的要求，对生态环境无害或危害极少，资源利用率极高。设计绿色的机电一体化产品，具有远大的发展前途。

Reading Material

To practice engineering today, we must understand new ways to process information and be able to utilize semiconductor electronics with our products.

The term mechatronics is used to denote the rapidly developing, interdisciplinary field of engineering dealing with the design of products whose function relies on the integration of mechanical, electrical and electronic components coordinated by a control architecture.

The concept of mechatronics has been continuously evolving. There are numerous definitions by many people. It is therefore useful to look back at the history of mechatronics ever since its birth in 1969 to the present.

7.8 MATLAB 在本章中的应用

The purpose of this digital control tutorial is to show you how to work with discrete functions either in transfer function or state-space form to design digital control systems.

In the above schematic of the digital control system, we see that the digital control system contains both discrete and the continuous portions. When designing a digital control system, we need to find the discrete equivalent of the continuous portion so that we only need to deal with discrete functions.

For this technique, we will consider the following portion of the digital control system and rearrange as follows. And we will give the definition of zero-order hold equivalence and see how it works in MATLAB.

The clock connected to the D/A and A/D converters supplies a pulse every T seconds and each D/A and A/D sends a signal only when the pulse arrives. The purpose of having this pulse is to require that zero-order hold equivalence Hzoh(z) have only samples $u(k)$ to work on and produce only samples of output $y(k)$; thus, Hzoh(z) can be realized as a discrete function.

The philosophy of the design is the following. We want to find a discrete function Hzoh(z) so that for a piecewise constant input to the continuous system $H(s)$, the sampled output of the continuous system equals the discrete output. Suppose the signal $u(k)$ represents a sample of the input signal. There are techniques for taking this sample $u(k)$ and holding it to produce a continuous signal uhat(t). The sketch below shows that the uhat(t) is held constant at $u(k)$ over the interval kT to $(k+1)T$. This operation of holding uhat(t) constant over the sampling time is called zero-order hold.

The zero-order held signal uhat(t) goes through $H(s)$ and A/D to produce the output $y(k)$ that will be the piecewise same signal as if the continuous $u(t)$ goes through $H(s)$ to produce the continuous output $y(t)$.

Now we will redraw the schematic, placing Hzoh(z) in place of the continuous portion. By placing Hzoh(z), we can design digital control systems dealing with only discrete functions.

1. Stability and Transient Response

For continuous systems, we know that certain behaviors results from different pole locations in the s-plane. For instance, a system is unstable when any pole is located to the right of the imaginary axis. For discrete systems, we can analyze the system behaviors from different pole locations in the z-plane.

If you noticed in the z-plane, the stability boundary is no longer imaginary axis, but is the unit circle $|z|=1$. The system is stable when all poles are located inside the unit circle and unstable when any pole is located outside.

2. Continuous to Discrete Conversion

The first step in designing a discrete control system is to convert the continuous transfer function to a discrete transfer function. MATLAB command c2dm will do this for you. The c2dm command requires the following four arguments: the numerator polynomial(num), the denominator polynomial(den), the sampling time(Ts) and the type of hold circuit. In this example, the hold we will use is the zero-order hold('zoh').

From the design requirement, let the sampling time, Ts equal to 0.12 seconds, which is 1/10 the time constant of a system with a settling time of 2 seconds. Let's create a new m-file and enter the following commands:

```
R = 1;L = 0.5;
Kt = 0.01;J = 0.01;
b = 0.1;num = Kt;
den = [(J * L)(J * R) + (L * b)(R * b) + (Kt^2)];

Ts = 0.12;
[numz,denz] = c2dm(num,den,Ts,'zoh')
Running this m-file should return the following:
numz =
         0    0.0092    0.0057
denz =
    1.0000   -1.0877    0.2369
```

From these matrices, the discrete transfer function can be written now.

First, we would like to see what the closed-loop response of the system looks like without any control. If you see the numz matrices shown above, it has one extra zero in the front, we have to get rid of it before closing the loop with the MATLAB cloop command. Add the following code into the end of your m-file.

```
numz = [numz(2) numz(3)];
[numz_cl,denz_cl] = cloop(numz,denz);
```

After you have done this, let's see how the closed-loop step response looks like. The dstep command will generate the vector of discrete output signals and stairs command will connect these signals. Add the following MATLAB code at the end of previous m-file and rerun it.

```
[x1] = dstep(numz_cl,denz_cl,101);
t = 0:0.12:12;stairs(t,x1)
xlabel('Time(seconds)')
ylabel('Velocity(rad/s)')
title('Stairstep Response:Original')
```

3. PID Controller

Recall that the continuous-time transfer function for a PID controller.

We can derive the discrete PID controller with bilinear transformation mapping. Equivalently, the c2dm command in MATLAB will help you to convert the continuous-time PID compensator to discrete-time PID compensator by using the "tustin" method in this case. The "tustin" method will use bilinear approximation to convert to discrete time of the derivative. According to the PID Design Method for the DC Motor, $K_p = 100$, $K_i = 200$ and $K_d = 10$ are satisfied the design requirement. We will use all of these gains in this example. Now add the following MATLAB commands to your previous m-file and rerun it in MATLAB window.

Now let's design a PID controller and add it into the system. First create a new m-file and type in the following commands.

```
J = 0.01;b = 0.1;
K = 0.01;R = 1;
L = 0.5;num = K;
den = [(J*L)((J*R)+(L*b))((b*R)+K^2)];
% (1)The traditional PID control
Kp = 100;Ki = 1;
Kd = 1;
numc = [Kd, Kp, Ki];
denc = [1 0];
numa = conv(num,numc);
dena = conv(den,denc);
[numac,denac] = cloop(numa,dena);
step(numac,denac)
title('PID Control with small Ki and Kd')
% (2)Discrete PID controller with bilinear approximation
Kp = 100;Ki = 200;
Kd = 10;
[dencz,numcz] = c2dm([1 0],[Kd Kp Ki],Ts,'tustin');
```

Note that the numerator and denominator in c2dm were reversed above. The reason is that the PID transfer function is not proper. MATLAB will not allow this. By switching the numerator and denominator the c2dm command can be fooled into giving the right answer. Let's see if the performance of the closed-loop response with the PID compensator satisfies the design requirements. Now add the following code to the end of your m-file and rerun it. You should get the following close-loop stairstep response.

```
numaz = conv(numz,numcz);
denaz = conv(denz,dencz);
[numaz_cl,denaz_cl] = cloop(numaz,denaz);

[x2] = dstep(numaz_cl,denaz_cl,101);
t = 0:0.12:12;
stairs(t,x2)
xlabel('Time(seconds)')
ylabel('Velocity(rad/s)')
title('Stairstep Response:with PID controller')
```

As you can see from the plot, the closed-loop response of the system is unstable. Therefore there must be something wrong with compensated system. So we should take a look at root locus of the compensated system. Let's add the following MATLAB command into the end of your m-file and rerun it.

```
rlocus(numaz,denaz)
title('Root Locus of Compensated System')
```

From this root-locus plot, we see that the denominator of the PID controller has a pole at -1 in the z-plane. We know that if a pole of a system is outside the unit circle, the system will be unstable. This compensated system will always be unstable for any positive gain because there are an even number of poles and zeroes to the right of the pole at -1. Therefore that pole will always move to the left and outside the unit circle. The pole at -1 comes from the compensator, and we can change its location by changing the compensator design. We choose it to cancel the zero at -0.62. This will make the system stable for at least some gains. Furthermore we can choose an appropriate gain from the root locus plot to satisfy the design requirements using rlocfind. Enter the following MATLAB code to your m-file.

```
dencz = conv([1 -1],[1.6  1])
numaz = conv(numz,numcz);
denaz = conv(denz,dencz);
rlocus(numaz,denaz)
title('Root Locus of Compensated System');
[K,poles] = rlocfind(numaz,denaz)
[numaz_cl,denaz_cl] = cloop(K*numaz,denaz);
[x3] = dstep(numaz_cl,denaz_cl,101);
t = 0:0.12:12;
stairs(t,x3)
```

```
xlabel('Time(seconds)')
ylabel('Velocity(rad/s)')
title('Stairstep Response:with PID controller')
```

The new dencz will have a pole at -0.625 instead of -1, which almost cancels the zero of uncompensated system. In the MATLAB window, you should see the command asking you to select the point on the root-locus plot. You should click on the plot as the following.

Then MATLAB will return the appropriate gain and the corresponding compensated poles, and it will plot the closed-loop compensated response.

The plot will show that the settling time is less than 2 seconds and the percent overshoot is around 3%. In addition, the steady state error is zero. Also, the gain, K, from root locus is 0.2425 which is reasonable. Therefore this response satisfies all of the design requirements. You can refer to Reference [24] for more details about this methods.

本章小结

(1) 计算机控制系统具有精度高、抗干扰性能好、通用性强和控制灵活等优点,因而在控制工程中得到了日益广泛的应用。

(2) 在计算机控制系统中,通过采样开关,将连续信号$e(t)$转换成离散信号$e^*(t)$的过程称为采样。$e^*(t)$是一个脉冲序列,可以通过保持器使其复现为原来的连续信号,条件是采样角频率ω_s与采样信号中最高次谐波角频率ω_{max}之间应符合

$$\omega_s > 2\omega_{max}$$

这就是采样定理。

采样定理只是给出了能不失真地复现连续信号的最低要求。在实际应用中,只有当采样周期T很小时,才能使采样信号$e^*(t)$与$e(t)$基本一致,因而采样角频率ω_s一般比ω_{max}要高得多。

(3) 计算机控制系统中设置保持器的目的是使其离散信号复现为相应的连续信号,以控制受控对象。实际应用中一般采用零阶保持器。

(4) 在零初始条件下,系统(或元件)输出、输入信号采样值的z变换之比定义为脉冲传递函数,它也等于系统(或元件)的脉冲响应函数(或传递函数)的z变换。虽然差分方程和脉冲传递函数都是采样系统的数学模型,但在系统分析中,一般采用脉冲传递函数,其原因是后者避免了求解高阶差分方程的麻烦。

(5) 对一个系统来说,根据有无采样开关、采样开关位置的不同,以及有无保持器等,可以得到不同的脉冲传递函数以及不同的闭环输出的z变换表达式。

(6) 与连续控制系统一样,计算机控制系统瞬态响应的模式和系统的稳定性也是由其闭环极点所决定的。采样系统稳定的充要条件是,闭环脉冲传递函数的极点位于[z]平面上以原点为圆心的单位圆内,即$|z_i| \leqslant 1$。通过双线性变换,把z变量变换为ω变量后,就可应用劳斯判据来判别采样系统的稳定性。

(7) 采样控制系统的稳态误差不仅与系统的结构和参数有关,还与输入信号的大小、形

式以及采样周期 T 有关。

Summary and Outcome Checklist

Controls systems technology is constantly evolving and the digital computer has had a very significant role in this evolution. Although control systems in the various industries evolved in their unique ways, it is now possible to build an integrated, but distributed system of control for an entire manufacturing facility.

The emphasis of this course has been on the design of feedback control system. At the outset, we noted that feedback control is a very effective way of achieving the desired control of dynamic systems. We then learnt to configure, model, and synthesis feedback control systems that would meet a given set of performance specifications.

What we have learnt is a systematic procedure to fine-tune the feedback control loop so that the control system will have the desired transient and steady state responses. This is akin to the task of tuning the engine of an automobile so that you can use the vehicle in the most efficient manner. When driving the car, you are of course required (as the controller of the car), to carry out several higher level tasks in real time.

Actuator sizing, selection of the feedback transducer, hardware implementation of the controller transfer function, building of the electronic interfaces, and generation of the command signal in real time are other key technical issues that need to be addressed by the control systems engineer.

Control system technology has many facets depending on what is controlled, how the control is accomplished, and the industry they are used in. The following part introduces you to the various types of control systems that are encountered in practice, and the language associated with each type.

The formal boundaries of traditional engineering disciplines have become fuzzy following the advent of integrated circuits and computers. Mechatronics is defined as the field of study involving the analysis, design, synthesis, and selection of systems that combine electrical, electronic and mechanical components with modern controls and microprocessors.

In process control it is very uncommon to design a fixed controller based on a mathematic model of the process. This is because of the inherent uncertainty in the dynamics of process plants. Changes in the operating conditions significantly affect the process dynamics. Usually a general purpose PID controller is used. The PID controller is placed in the closed loop and its parameters tuned in situ by carrying out some simple experiments on the actual closed loop system.

The problems inherent in the use of the digital processor for the purpose of feedback control and the development of the theoretical and practical backgrounds required to design and implement such digital control systems will be covered elsewhere, but for the time being it is sufficient to acquire an overview of digital control systems, which is the object of the following activity.

Tick the box for each statement with which you agree.

☐ I am able to explain the difference between analogue signals and digital signals.

☐ I am able to explain the advantages of digital signal processing over analogue signal processing, and discuss why digital computers are now the preferred option for implementing the controller function.

☐ I am able to discuss sequential control and differentiate between event-sequenced processes and time-sequenced processes.

☐ I am able to differentiate between centralised control and distributed control and give advantages and disadvantages of each.

☐ I am able to describe process control and explain the operation of a typical process controller.

☐ I am able to discuss the key features of numerical control and robotics.

☐ I can define mechatronics and discuss its relevance to contemporary engineering design.

☐ I am able to give practical examples of mechatronics systems and identify the primary elements in each case.

☐ I can distinguish between the traditional control systems and modern mechatronics systems.

习题

7-1 求下列函数的 z 变换。

(1) $e(t)=a^k$

(2) $e(t)=e^t\sin\omega t$

(3) $G(s)=\dfrac{s+3}{(s+1)(s+2)}$

(4) $G(s)=\dfrac{1-e^{-s}}{s^2(s+1)}$

7-2 求下列函数的 z 反变换。

(1) $F(z)=\dfrac{z}{z-1}$

(2) $F(z)=\dfrac{10z}{(z-1)(z-2)}$

(3) $F(z)=\dfrac{z}{(z+1)^2(z-2)^2}$

(4) $F(z)=\dfrac{(1-e^{-aT})z}{(z-1)(z-e^{-aT})}$

7-3 设系统的结构图如题 7-3 图所示,其中连续部分的传递函数为 $G(s)=\dfrac{2}{s(s+2)}$。试求该系统的脉冲传递函数 $G(z)$。

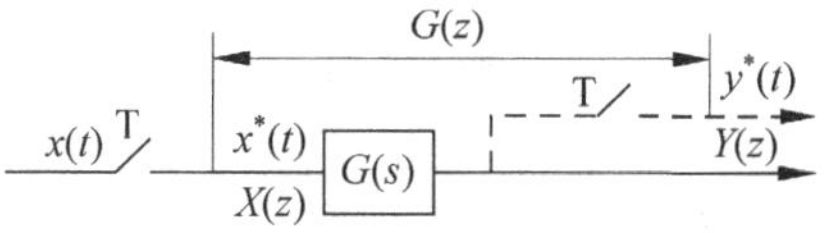

题 7-3 图

7-4 试求如题 7-4 图(a)和题 7-4 图(b)所示的两个开环离散系统的脉冲传递函数 $G(z)$。

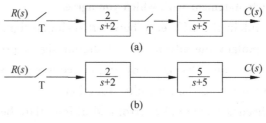

题 7-4 图

7-5 试求如题 7-5 图所示系统的脉冲传递函数 $\Phi(z)$ 或输出 z 变换 $C(z)$。

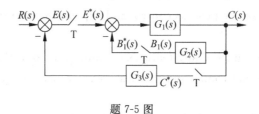

题 7-5 图

7-6 已知采样系统结构如题 7-6 图所示,其中 $T=1\text{s}$,试分析确定适用闭环系统稳定的 K 值范围。

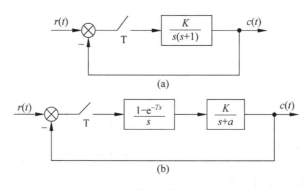

题 7-6 图

7-7 已知采样系统结构如题 7-7 图所示,采样周期 $T=1\text{s}$,试确定使系统稳定的 K 值范围,及 $K=1, r(t)=t$ 时系统的稳态误差。

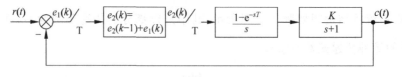

题 7-7 图

7-8 已知采样系统结构如题 7-8 图所示,其中 $T=1\text{s}, K=10$,输入信号 $r(t)=1(t)+t$,试求其稳定误差。

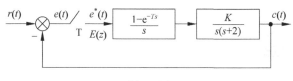

题 7-8 图

7-9 设有单位反馈误差采样系统,连续部分传递函数为
$$G(s)=\frac{1}{s^2(s+5)}$$
输入 $r(t)=1(t)$,采样周期 $T=0.25\text{s}$。

试求:(1) 输出 z 变换 $C(z)$;

(2) 输出响应的终值 $c(\infty)$。

7-10 已知采样系统结构图如题 7-10 图所示,其中 $G_1(s)=\dfrac{1-\mathrm{e}^{Ts}}{s}$,$G_2(s)=\dfrac{K}{(s+1)}$,$H(s)=\dfrac{s+1}{K}$,采样周期 $T=0.1\text{s}$,试确定闭环系统稳定的 K 值范围。

7-11 (哈尔滨工业大学硕士研究生入学考试试题)控制系统如题 7-11 图所示,其中 $G_c(s)$ 为校正环节。

(1) 若用计算机实现校正环节 $G_c(s)$,画出采样系统的框图。

(2) 如采样周期 $T=1\text{s}$,求使采样系统稳定的 K 的取值范围。

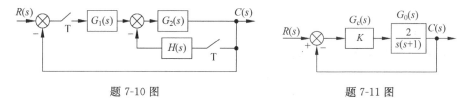

题 7-10 图 题 7-11 图

7-12 (西北工业大学硕士研究生入学考试试题)已知离散系统结构图如题 7-12 图所示,T 为采样周期。

(1) 要求系统在 $r(t)=t$ 作用下的稳态误差 $e_{ss}=0.1T$,试确定相应的开环增益 K。

(2) 当 $K=10$ 时,确定使系统稳定的采样周期 T 的取值范围。

7-13 (南京航空航天大学硕士研究生入学考试试题)闭环采样系统结构图如题 7-13 图所示。

其中:采样周期 $T=1\text{s}$,$K>0$,试求:

(1) 使系统稳定的 K 值范围。

(2) 当 $K=1$ 时,求系统在单位阶跃输入下的输出响应和稳态输出值。

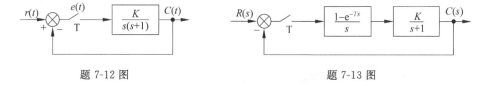

题 7-12 图 题 7-13 图

第 8 章 现代控制理论基础

8.1 线性系统的状态空间描述

经典控制理论中常常采用系统输入和输出之间的关系来描述控制机理,这种方法一般称为控制系统的输入-输出描述。然而这种描述有着很大的局限性,因为它只能从系统的外部概括输入-输出之间的关系。经典控制理论中分析和设计控制系统所采用的方法主要有时域法、根轨迹法、频域法。这些方法用来分析简单的线性定常单变量系统是非常有效的,但对于非线性系统、时变系统、多变量系统等复杂系统,经典控制理论就显得无能为力了。状态空间描述是现代控制理论的基础。它不仅可以描述系统的输入-输出关系,还可以反映系统的内部特性,既适用于多输入-多输出系统,又适合于时变、非线性系统和随机控制系统。因此,状态空间描述是对系统的一种完全描述,在控制系统分析与设计中有着广泛的应用。

本章的主要内容:
- □ 线性系统的状态空间描述
- □ 线性定常系统状态方程的解
- □ 线性定常系统的可控性和可观测性
- □ 线性定常系统的状态反馈和状态观测器
- □ MATLAB 在本章的应用

8.1.1 状态空间描述的基本概念

下面介绍在系统状态空间描述方法中涉及的一些概念。

1. 状态

状态是指系统的运动状态,可以理解为系统在时间域中的行为或运动状况的信息集合。这个时间域包括系统的过去、现在和将来。状态也可以理解为系统记忆,$t=t_0$ 时刻的初始状态能记忆系统在 t_0 时的全部输入信息。例如,质点的机械运动状态要由质点的位置和动量来确定;由一定质量气体组成的系统的热力学状态由系统的温度、压强和体积来确定。在外界作用下,物质系统的状态将随时间而变化。

2. 状态变量

状态变量是指足以完全描述系统运动状态的最小个数的一组变量。完全描述是指如果给定了 $t=t_0$ 时刻的一组变量值 $x_1(t_0), x_2(t_0), \cdots, x_n(t_0)$,以及 $t \geqslant t_0$ 时刻输入的时间函数

$u(t)$，那么系统在 $t \geq t_0$ 的任意时刻的状态就可以完全确定了。最小个数则意味着这组变量是相互独立的。一个用 n 阶微分方程描述的含有 n 个独立变量的系统，当求得 n 个独立变量随时间变化的规律时，系统状态就可完全确定。若变量个数大于 n 时，多余的变量就不是独立变量；若少于 n，又不足以描述系统状态。

状态变量的选取并不是唯一的，可选某一组或其他组变量。状态变量不一定要像系统输出量那样，在物理上是可测量或可观察的量，但在实用上毕竟还是选择容易测量的一些量，以便满足实现状态反馈，改善系统性能的需要。例如，机械系统中常选取线（角）位移和线（角）速度作为变量，RLC 网络中则常选取流经电感的电流和电容的端电压作为状态变量。状态变量常用符号 $x_1(t), x_2(t), \cdots, x_n(t)$ 表示。

3. 状态向量

若以几个状态变量 $x_1(t), x_2(t), \cdots, x_n(t)$ 作为向量 $\boldsymbol{x}(t)$ 的分量时，称为状态向量。即

$$\boldsymbol{x}(t) = [x_1(t), x_2(t), \cdots, x_n(t)]^{\mathrm{T}} \tag{8-1-1}$$

则称向量 $\boldsymbol{x}(t)$ 为 n 维状态向量，给定 $t=t_0$ 时刻的初始状态向量及 $t \geq t_0$ 时刻的输入向量 $u(t)$，则 $t \geq t_0$ 时刻的状态由状态向量 $\boldsymbol{x}(t)$ 唯一确定。

4. 状态空间

以状态变量 $x_1(t), x_2(t), \cdots, x_n(t)$ 为坐标轴构成的 n 维空间，称为状态空间。在某一时刻 t，状态向量 (t) 是状态空间的一个点，以 $\boldsymbol{x}(t) = \boldsymbol{x}(t_0)$ 为起点，随着时间的推移，在状态空间中能描绘出一条轨迹。这条轨迹也称为状态轨迹或状态轨线。

5. 状态方程

状态向量的一阶微分方程与状态变量、输入变量关系的数学表达式称为状态方程，它不含输入的积分项。一般形式为：

$$\boldsymbol{x}(t) = f[\boldsymbol{x}(t), u(t), t] \tag{8-1-2}$$

一般情况下，状态方程既是非线性的，又是时变的。

6. 输出方程

在指定输出变量的情况下，该输出变量与状态变量和输入变量之间的函数关系称为输出方程。一般形式为

$$\boldsymbol{y}(t) = g[\boldsymbol{x}(t), u(t), t] \tag{8-1-3}$$

它反映了系统中输出变量与状态变量和输入变量之间的因果关系。

7. 状态空间表达式

状态方程与输出方程的组合称为状态空间表达式。下面给出状态空间表达式的标准描述：

$$\begin{cases} \dot{\boldsymbol{x}}(t) = f[\boldsymbol{x}(t), u(t), t] \\ \boldsymbol{y}(t) = g[\boldsymbol{x}(t), u(t), t] \end{cases} \tag{8-1-4a}$$

或离散形式：

$$\begin{cases} \boldsymbol{x}(t_{k+1}) = f[\boldsymbol{x}(t_k), u(t_k), t_k] \\ \boldsymbol{y}(t_k) = g[\boldsymbol{x}(t_k), u(t_k), t_k] \end{cases} \tag{8-1-4b}$$

则线性连续时间系统状态空间表达式的一般形式为：

$$\begin{cases} \dot{\boldsymbol{x}}(t) = \boldsymbol{A}(t)x(t) + \boldsymbol{B}(t)u(t) \\ \boldsymbol{y}(t) = \boldsymbol{C}(t)x(t) + \boldsymbol{D}(t)u(t) \end{cases} \tag{8-1-5a}$$

式中，设状态 x、输入 u、输出 y 的维数分别为 n,p,q。矩阵 $\boldsymbol{A}(t)$ 称为系统矩阵或状态矩阵，$\boldsymbol{B}(t)$ 称为控制矩阵或输入矩阵，$\boldsymbol{C}(t)$ 称为输出矩阵或观测矩阵，$\boldsymbol{D}(t)$ 称为前馈矩阵或输入矩阵。

对于离散时间系统，由于在实践中常取 $t_k = kT$（T 为采样周期），其状态空间表达式的一般形式可写为：

$$\begin{cases} x(k+1) = \boldsymbol{G}(k)x(k) + \boldsymbol{H}(k)u(k) \\ y(k) = \boldsymbol{C}(k)x(k) + \boldsymbol{D}(k)u(k) \end{cases} \tag{8-1-5b}$$

若系统矩阵 $\boldsymbol{A}(t),\boldsymbol{B}(t),\boldsymbol{C}(t),\boldsymbol{D}(t)$ 或 $\boldsymbol{G}(k),\boldsymbol{H}(k),\boldsymbol{C}(k),\boldsymbol{D}(k)$ 的各元素都是常数，则称该系统为线性定常系统，线性定常系统状态空间表达式的一般形式为：

$$\begin{cases} \dot{\boldsymbol{x}}(t) = \boldsymbol{A}x(t) + \boldsymbol{B}u(t) \\ \boldsymbol{y}(t) = \boldsymbol{C}x(t) + \boldsymbol{D}u(t) \end{cases} \tag{8-1-6a}$$

或

$$\begin{cases} x(k+1) = \boldsymbol{G}x(k) + \boldsymbol{H}u(k) \\ y(k) = \boldsymbol{C}x(k) + \boldsymbol{D}u(k) \end{cases} \tag{8-1-6b}$$

线性系统的状态空间表达式常用结构图表示。线性连续时间系统的结构图如图 8-1-1 所示，线性离散时间系统的结构图如图 8-1-2 所示。图中 $\boldsymbol{I}$ 为 $(n \times n)$ 单位矩阵，s 为拉普拉斯算子，z 为单位延时算子，s 和 z 均为标量。在后续章节中，为表述简便，经常在状态空间表达式中省略域 k。

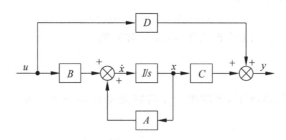

图 8-1-1 线性连续时间系统结构图

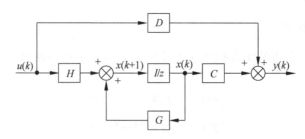

图 8-1-2 线性离散时间系统结构图

Terms and Concepts

The State of a system is a set of variables, whose values together with the input signals and the equations describing the dynamics, will provide the future state and output of the system.

State variable feedback: The control signal for the process is a direct function of all the state variables.

State variable: The set of variables that describe the system. The state variables describe the present configuration of a system and can be used to determine the future response, given the excitation inputs and the equations describing the dynamics.

State vector: The vector containing all n state variables, $x_1, x_2, \cdots, x_n$.

State-space representation: A time-domain model comprised of the state differential equation, $\dot{\boldsymbol{x}} = \boldsymbol{A}x + \boldsymbol{B}u$, and the output equation, $\boldsymbol{y} = \boldsymbol{C}x + \boldsymbol{D}u$.

State differential equation: The differential equation for the state vector: $\dot{\boldsymbol{x}} = \boldsymbol{A}x + \boldsymbol{B}u$.

状态空间描述的结构图绘制步骤如下。

(1) 画出所有积分器。积分器的个数等于状态变量数,每个积分器的输出表示相应的某个状态变量。

(2) 根据状态方程和输出方程,画出相应的加法器和比例器。

(3) 用箭头将这些元件连接起来。

【例 8-1-1】 画出一阶微分方程的状态结构图。

若微分方程为:
$$\dot{x}(t) = ax(t) + bu(t)$$
则状态结构图如图 8-1-3 所示。

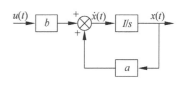

图 8-1-3 状态结构图

【例 8-1-2】 画出下列系统状态空间表达式的系统结构图。

$$\begin{cases} \dot{x}(t) = \begin{bmatrix} 0 & 1 & 0 \\ 0 & 0 & 1 \\ -6 & -3 & -2 \end{bmatrix} x(t) + \begin{bmatrix} 0 \\ 0 \\ 1 \end{bmatrix} u(t) \\ y(t) = \begin{bmatrix} 1 & 1 & 0 \end{bmatrix} x(t) \end{cases}$$

解:由其状态空间表达式可得出其状态方程和输出方程分别为:

$$\begin{cases} \dot{x}_1(t) = x_2(t) \\ \dot{x}_2(t) = x_3(t) \\ \dot{x}_3(t) = -6x_1(t) - 3x_2(t) - 2x_3(t) + u(t) \end{cases}$$

$$y(t) = x_1(t) + x_2(t)$$

则系统结构图如图 8-1-4 所示。

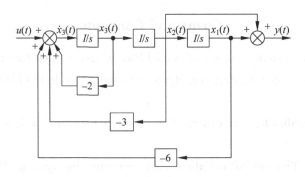

图 8-1-4 系统结构图

8.1.2 线性定常连续系统状态空间表达式的建立

建立状态空间描述有以下三个途径。
(1) 由系统框图建立。
(2) 由系统物理或化学机理进行推导。
(3) 由微分方程或传递函数演化而得。

1. 由系统框图建立状态空间描述

关键：将积分部分单独表示出来，对结构图进行等效变换。

【**例 8-1-3**】 系统框图如图 8-1-5 所示。

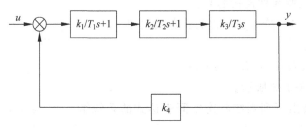

图 8-1-5 系统框图

解：将图 8-1-5 进行等效变换如图 8-1-6 所示。

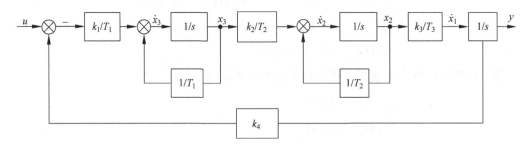

图 8-1-6 等效系统框图

由图可知，图中有三个积分环节，则取三个状态变量如图 8-1-6 所示(选择积分环节后面的变量为状态变量)，则有：

$$\begin{cases} \dot{x}_1 = \dfrac{k_3}{T_1} x_2 \\ \dot{x}_2 = -\dfrac{1}{T_2} x_2 + \dfrac{k_2}{T_2} x_3 \\ \dot{x}_3 = -k_4 \cdot \dfrac{k_1}{T_1} x_1 - \dfrac{1}{T_1} x_3 + \dfrac{k_1}{T_1} u \end{cases}$$

$$y = x_1$$

再将其写成矩阵形式：

$$\dot{\boldsymbol{x}} = \begin{bmatrix} 0 & \dfrac{k_3}{T_3} & 0 \\ 0 & -\dfrac{1}{T_2} & \dfrac{k_2}{T_2} \\ -\dfrac{k_1 k_4}{T_1} & 0 & -\dfrac{1}{T_1} \end{bmatrix} \begin{bmatrix} x_1 \\ x_2 \\ x_3 \end{bmatrix} + \begin{bmatrix} 0 \\ 0 \\ \dfrac{k_1}{T_1} \end{bmatrix} u$$

$$\boldsymbol{y} = \begin{bmatrix} 1 & 0 & 0 \end{bmatrix} \begin{bmatrix} x_1 \\ x_2 \\ x_3 \end{bmatrix}$$

2．由系统机理建立状态空间描述

下面通过例题来介绍根据系统机理建立线性定常连续系统状态空间表达式的方法。

【**例 8-1-4**】 试列写如图 8-1-7 所示 RLC 网络的电路方程，选择几组状态变量并建立相应的状态空间表达式。

解：由电路定律可列写如下方程：

$$Ri + L\dfrac{\mathrm{d}i}{\mathrm{d}t} + \dfrac{1}{C}\int i\mathrm{d}t = e$$

电路输出量为：

$$y = e_c = \dfrac{1}{C}\int i\mathrm{d}t$$

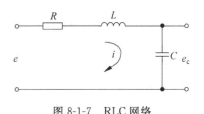

图 8-1-7 RLC 网络

（1）设状态变量 $x_1 = i, x_2 = \dfrac{1}{C}\int i\mathrm{d}t$，则状态方程为

$$\begin{cases} \dot{x}_1 = -\dfrac{R}{L} x_1 - \dfrac{1}{L} x_2 + \dfrac{1}{L} e \\ \dot{x}_2 = \dfrac{1}{C} x_1 \end{cases}$$

输出方程为： $y = x_2$

则向量-矩阵形式为

$$\begin{bmatrix} \dot{x}_1 \\ \dot{x}_2 \end{bmatrix} = \begin{bmatrix} -\dfrac{R}{L} & -\dfrac{1}{L} \\ \dfrac{1}{C} & 0 \end{bmatrix} \begin{bmatrix} x_1 \\ x_2 \end{bmatrix} + \begin{bmatrix} \dfrac{1}{L} \\ 0 \end{bmatrix} e$$

$$\boldsymbol{y} = \begin{bmatrix} 0 & 1 \end{bmatrix} \begin{bmatrix} x_1 \\ x_2 \end{bmatrix}$$

简记为
$$\dot{x} = Ax + Be$$
$$y = Cx$$

式中：
$$\dot{x} = \begin{bmatrix} \dot{x}_1 \\ \dot{x}_2 \end{bmatrix}, \quad x = \begin{bmatrix} x_1 \\ x_2 \end{bmatrix}, \quad A = \begin{bmatrix} -\dfrac{R}{L} & -\dfrac{1}{L} \\ \dfrac{1}{C} & 0 \end{bmatrix}, \quad B = \begin{bmatrix} \dfrac{1}{L} \\ 0 \end{bmatrix}$$

$$C = \begin{bmatrix} 0 & 1 \end{bmatrix}$$

(2) 设状态变量 $x_1 = i, x_2 = \int i \mathrm{d}t$，则有
$$\begin{bmatrix} \dot{x}_1 \\ \dot{x}_2 \end{bmatrix} = \begin{bmatrix} -\dfrac{R}{L} & -\dfrac{1}{LC} \\ 1 & 0 \end{bmatrix} \begin{bmatrix} x_1 \\ x_2 \end{bmatrix} + \begin{bmatrix} \dfrac{1}{L} \\ 0 \end{bmatrix} e, \quad y = \begin{bmatrix} 0 & 1 \end{bmatrix} \begin{bmatrix} x_1 \\ x_2 \end{bmatrix}$$

(3) 设状态变量 $x_1 = \dfrac{1}{C}\int i \mathrm{d}t + Ri, x_2 = \dfrac{1}{C}\int i \mathrm{d}t$，则
$$\begin{cases} \dot{x}_1 = \dot{x}_2 + R\dfrac{\mathrm{d}i}{\mathrm{d}t} = \dfrac{1}{RC}(x_1 - x_2) + \dfrac{R}{L}(-x_1 + e) \\ \dot{x}_2 = \dfrac{1}{C}i = \dfrac{1}{RC}(x_1 - x_2) \end{cases}$$
$$y = x_2$$

其向量-矩阵形式为
$$\begin{bmatrix} \dot{x}_1 \\ \dot{x}_2 \end{bmatrix} = \begin{bmatrix} \dfrac{1}{RC} - \dfrac{R}{L} & -\dfrac{1}{RC} \\ \dfrac{1}{RC} & -\dfrac{1}{RC} \end{bmatrix} \begin{bmatrix} x_1 \\ x_2 \end{bmatrix} + \begin{bmatrix} \dfrac{R}{L} \\ 0 \end{bmatrix} e$$
$$y = \begin{bmatrix} 0 & 1 \end{bmatrix} \begin{bmatrix} x_1 \\ x_2 \end{bmatrix}$$

由此可见，系统的状态空间表达式不具有唯一性，选取不同的状态变量，便会有不同的状态空间表达式，但是它们都描述了同一个系统。

【例 8-1-5】 试列出在外力作用下，以质量 M_1, M_2 的位移 y_1, y_2 为输出的状态空间描述。

解：由图 8-1-8 可知，该系统有 4 个独立的储能元件，取状态变量如下：
$$x_1 = y_1, \quad x_2 = y_2, \quad x_3 = \dot{y}_1 = v_1, \quad x_4 = \dot{y}_2 = v_2$$

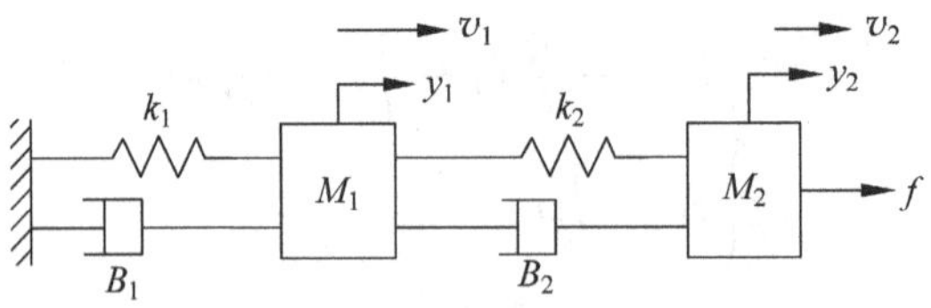

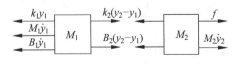

图 8-1-8 机械位移系统

则质量块受力如图 8-1-9 所示。

图 8-1-9 受力图

则有：
$$M_1 \ddot{y}_1 + B_1 \dot{y}_1 + k_1 y_1 = k_2(y_2 - y_1) + B_2(\dot{y}_2 - \dot{y}_1)$$
及
$$M_2 \ddot{y}_2 + B_2(\dot{y}_2 - \dot{y}_1) + k_2(y_2 - y_1) = f$$

将所选的状态变量代入上式并整理得：

状态方程：
$$\begin{cases} \dot{x}_1 = x_3 \\ \dot{x}_2 = x_4 \\ \dot{x}_3 = -\dfrac{k_1 + k_2}{M_1} x_1 + \dfrac{k_2}{M_1} x_2 - \dfrac{B_1 + B_2}{M_1} x_3 + \dfrac{B_2}{M_1} x_4 \\ \dot{x}_4 = \dfrac{k_2}{M_2} x_1 - \dfrac{k_2}{M_2} x_2 + \dfrac{B_2}{M_2} x_3 - \dfrac{B_2}{M_2} x_4 + \dfrac{1}{M_2} f \end{cases}$$

输出方程：
$$\begin{cases} y_1 = x_1 \\ y_2 = x_2 \end{cases}$$

写成矩阵形式则有：
$$\dot{\boldsymbol{x}} = \begin{bmatrix} 0 & 0 & 1 & 0 \\ 0 & 0 & 0 & 1 \\ -\dfrac{k_1+k_2}{M_1} & \dfrac{k_2}{M_1} & -\dfrac{B_1+B_2}{M_1} & \dfrac{B_2}{M_1} \\ \dfrac{k_2}{M_1} & -\dfrac{k_2}{M_2} & \dfrac{B_2}{M_2} & -\dfrac{B_2}{M_2} \end{bmatrix} \begin{bmatrix} x_1 \\ x_2 \\ x_3 \\ x_4 \end{bmatrix} + \begin{bmatrix} 0 \\ 0 \\ 0 \\ \dfrac{1}{M_2} \end{bmatrix} f$$

$$\boldsymbol{y} = \begin{bmatrix} 1 & 0 & 0 & 0 \\ 0 & 1 & 0 & 0 \end{bmatrix} \begin{bmatrix} x_1 \\ x_2 \\ x_3 \\ x_4 \end{bmatrix}$$

3. 由系统微分方程建立状态空间描述

1) 微分方程中不含输入量的导数项（传递函数中没有零点）

在这种情况下，微分方程的一般形式为：
$$y^n + a_{n-1} y^{(n-1)} + a_{n-2} y^{(n-2)} + \cdots + a_1 \dot{y} + a_0 y = b_0 u \tag{8-1-7}$$

系统的传递函数为：
$$W(s) = \frac{b_0}{s^n + a_{n-1} s^{n-1} + \cdots + a_1 s + a_0} \tag{8-1-8}$$

选取几个状态变量为 $x_1 = y, x_2 = \dot{y}, \cdots, x_n = y^{(n-1)}$，则有

$$\begin{cases} \dot{x}_1 = x_2 \\ \dot{x}_2 = x_3 \\ \vdots \\ \dot{x}_{n-1} = x_n \\ \dot{x}_n = -a_0 x_1 - a_1 x_2 - \cdots - a_{n-1} x_n + b_0 u \\ y = x_1 \end{cases} \tag{8-1-9}$$

则其向量矩阵形式为

$$\begin{cases} \dot{\boldsymbol{x}} = \boldsymbol{A}\boldsymbol{x} + \boldsymbol{b}u \\ \boldsymbol{y} = \boldsymbol{c}\boldsymbol{x} \end{cases} \tag{8-1-10}$$

其中

$$\boldsymbol{x} = \begin{bmatrix} x_1 \\ x_2 \\ \vdots \\ x_{n-1} \\ x_n \end{bmatrix} \quad \boldsymbol{A} = \begin{bmatrix} 0 & 1 & 0 & \cdots & 0 \\ 0 & 0 & 1 & \cdots & 0 \\ \vdots & \vdots & \vdots & \ddots & \vdots \\ 0 & 0 & 0 & \cdots & 1 \\ -a_0 & -a_1 & -a_2 & \cdots & -a_{n-1} \end{bmatrix}$$

$$\boldsymbol{b} = \begin{bmatrix} 0 \\ 0 \\ \vdots \\ 0 \\ b_0 \end{bmatrix}, \quad \boldsymbol{c} = \begin{bmatrix} 1 & 0 & \cdots & 0 \end{bmatrix}$$

则对应的状态空间表达式的结构图如图 8-1-10 所示。

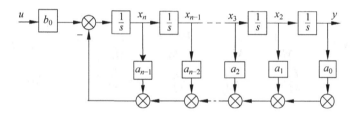

图 8-1-10　系统的状态变量图

2) 微分方程中含有输入量的导数项(传递函数中有零点)

这种情况下,微分方程的一般形式为:

$$y^{(n)} + a_{n-1}y^{(n-1)} + \cdots + a_1\dot{y} + a_0 y = b_n u^{(n-1)} + \cdots + b_1\dot{u} + b_0 u \tag{8-1-11}$$

此时可选取以下几个变量作为一组状态变量。

即

$$\begin{cases} x_1 = y - h_0 u \\ x_2 = \dot{x}_1 - h_1 u = \dot{y} - h_0 \dot{u} - h_1 u \\ \vdots \\ x_n = \dot{x}_{n-1} - h_{n-1} u = y^{n-1} - h_0 u^{n-1} - h_1 u^{n-2} - \cdots - h_{n-1} u \end{cases} \tag{8-1-12}$$

式(8-1-12)可改写成:

$$\begin{cases} \dot{x}_1 = x_2 + h_1 u \\ \dot{x}_2 = x_3 + h_2 u \\ \vdots \\ \dot{x}_{n-1} = x_n + h_{n-1} u \end{cases} \tag{8-1-13}$$

则输出方程为:

$$y = x_1 + h_0 u \tag{8-1-14}$$

对式(8-1-12)中最后一个方程求导并考虑式(8-1-11),有

$\dot{x}_n = -a_{n-1}y^{(n-1)} - \cdots - a_1\dot{y} - a_0 y + b_n u^{(n)} + \cdots + b_0 u - h_0 u^{(n)} - h_1 u^{(n-1)} - \cdots - h_{n-1}\dot{u}$

$= -a_0 x_1 - \cdots - a_{n-1}x_n + (b_n - h_0)u^{(n)} + (b_{n-1} - h_1 - a_{n-1}h_0)u^{(n-1)}$

$+ \cdots + (b_1 - h_{n-1} - a_{n-1}h_{n-2} - \cdots - a_1 h_0)\dot{u} + (b_0 - a_{n-1}h_{n-1} - \cdots - a_1 h_1 - a_0 h_0)u$

令上式中 u 的各阶导数的系数为零,可确定 h 的值,即

$$\begin{cases} h_0 = b_n \\ h_1 = b_{n-1} - a_{n-1}h_0 \\ \vdots \\ h_n = b_0 - a_{n-1}h_{n-1} - \cdots - a_1 h_1 - a_0 h_0 \end{cases} \tag{8-1-15}$$

故 $\dot{x}_n = -a_0 x_1 - \cdots - a_{n-1}x_n + h_n u$

则系统的动态方程为

$$\begin{cases} \dot{\boldsymbol{x}} = \boldsymbol{A}\boldsymbol{x} + \boldsymbol{B}u \\ y = \boldsymbol{C}\boldsymbol{x} + \boldsymbol{D}u \end{cases} \tag{8-1-16}$$

式(8-1-16)中

$$\boldsymbol{A} = \begin{bmatrix} 0 & 1 & 0 & \cdots & 0 \\ 0 & 0 & 1 & \cdots & 0 \\ \vdots & \vdots & \vdots & \ddots & \vdots \\ 0 & 0 & 0 & \cdots & 1 \\ -a_0 & -a_1 & -a_2 & \cdots & -a_{n-1} \end{bmatrix}, \quad \boldsymbol{B} = \begin{bmatrix} h_1 \\ h_2 \\ \vdots \\ h_{n-1} \\ h_n \end{bmatrix}$$

$$\boldsymbol{C} = \begin{bmatrix} 1 & 0 & 0 & \cdots & 0 \end{bmatrix}, \quad \boldsymbol{D} = [h_0]$$

【例 8-1-6】 已知系统的微分方程 $\dddot{y} + 2\ddot{y} + 3\dot{y} + 5y = 0.5\ddot{u} + \dot{u} + 1.5u$,求系统的状态空间表达式。

解:由式(8-1-11)可得 $a_0 = 5, a_1 = 3, a_2 = 2, b_0 = 1.5, b_1 = 1, b_2 = 0.5, b_3 = 0$。

由式(8-1-15),计算可得 $h_1 = 0.5, h_2 = 1, h_3 = 1.5$。

根据式(8-1-16),可得

$$\begin{bmatrix} \dot{x}_1 \\ \dot{x}_2 \\ \dot{x}_3 \end{bmatrix} = \begin{bmatrix} 0 & 1 & 0 \\ 0 & 0 & 1 \\ -5 & -3 & -2 \end{bmatrix} \begin{bmatrix} x_1 \\ x_2 \\ x_3 \end{bmatrix} + \begin{bmatrix} 0.5 \\ 1 \\ 1.5 \end{bmatrix} u, \quad y = \begin{bmatrix} 1 & 0 & 0 \end{bmatrix} \begin{bmatrix} x_1 \\ x_2 \\ x_3 \end{bmatrix}$$

为了更好地求解系统的状态变量模型,下面将通过英文材料的形式,给出如何求解系统状态输出的一般方法。

Reading Material

The state variables describe the present configuration of a system and can be used to determine the future response, given the excitation inputs and the equations describing the dynamics.

The general form of a dynamic system is shown in Figure 8-1-11. A simple example of a state variable is the state of an on-off light switch. The switch can be in either the on or the off position, and thus the state of the switch can assume one of the two possible values. Thus, if we know the present state (position) of the switch at t_0 and if an input is applied, we are able to determine the future value of the state of the element.

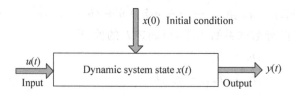

Figure 8-1-11　Dynamic system

The concept of a set of state variables that represent a dynamic system can be illustrated in terms of the spring-mass-damper system shown in Figure 8-1-12. The number of state variables chosen to represent this system should be as small as possible in order to avoid redundant state variables. A set of state variables sufficient to describe this system includes the position and the velocity of the mass.

Therefore, we will define a set of state variables as (x_1, x_2), where

$$x_1(t) = y(t) \quad \text{and} \quad x_2(t) = \frac{\mathrm{d}y(t)}{\mathrm{d}t}$$

The differential equation describes the behavior of the system and is usually written as

$$M\frac{\mathrm{d}^2 y}{\mathrm{d}t^2} + b\frac{\mathrm{d}y}{\mathrm{d}t} + ky = u(t) \quad (8\text{-}1\text{-}17)$$

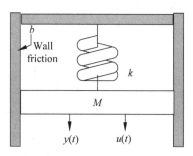

Figure 8-1-12　spring-mass-damper system

To write Equation (8-1-17) in terms of the state variables, we substitute the state variables as already defined and obtain

$$M\frac{\mathrm{d}x_2}{\mathrm{d}t} + bx_2 + kx_1 = u(t) \quad (8\text{-}1\text{-}18)$$

Therefore, we can write the equations that describe the behavior of the spring-mass-damper system as the set of two first-order differential equations

$$\frac{\mathrm{d}x_1}{\mathrm{d}t} = x_2 \quad (8\text{-}1\text{-}19)$$

and

$$M\frac{\mathrm{d}x_2}{\mathrm{d}t} = \frac{-b}{M}x_2 - \frac{k}{M}x_1 + \frac{1}{M}u \quad (8\text{-}1\text{-}20)$$

This set of differential equations describes the behavior of the state of the system in terms of the rate of change of each state variable.

4. 由传递函数建立状态空间描述

微分方程式(8-1-11)对应的系统传递函数为

$$G(s) = \frac{Y(s)}{U(s)} = \frac{b_n s^n + b_{n-1} s^{n-1} + \cdots + b_1 s + b_0}{s^n + a_{n-1} s^{n-1} + \cdots + a_1 s + a_0}$$

$$= b_n + \frac{\beta_{n-1} s^{n-1} + \cdots + \beta_1 s + \beta_0}{s^n + a_{n-1} s^{n-1} + \cdots + a_1 s + a_0}$$

$$= b_n + \frac{N(s)}{D(s)} \tag{8-1-21}$$

式(8-1-21)中，当分母阶数大于分子阶数时，$b_n = 0$，$\dfrac{N(s)}{D(s)}$ 为严格的有理真分式。分子各次项系数分别为

$$\begin{cases} \beta_0 = b_0 - a_0 b_n \\ \beta_1 = b_1 - a_1 b_n \\ \vdots \\ \beta_{n-1} = b_{n-1} - a_{n-1} b_n \end{cases} \tag{8-1-22}$$

1) $\dfrac{N(s)}{D(s)}$ 串联分解的情况

此时取中间变量 Z，有

$$Z^{(n)} + a_{n-1} Z^{(n-1)} + \cdots + a_1 \dot{Z} + a_0 Z = u$$

$$y = \beta_{n-1} Z^{(n-1)} + \cdots + \beta_1 \dot{Z} + \beta_0 Z$$

如图 8-1-13 所示。

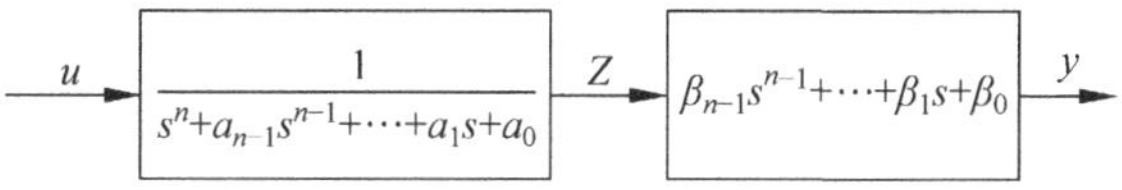

图 8-1-13　串联分解

此时选状态变量

$$x_1 = Z, x_2 = \dot{Z}, \cdots, x_n = Z^{(n-1)}$$

则状态方程为

$$\begin{cases} \dot{x}_1 = x_2 \\ \dot{x}_2 = x_3 \\ \vdots \\ \dot{x}_n = -a_0 x_1 - a_1 x_2 - \cdots - a_{n-1} x_n + u \end{cases} \tag{8-1-23}$$

输出方程为

$$y = \beta_0 x_1 + \beta_1 x_2 + \cdots + \beta_{n-1} x_n \tag{8-1-24}$$

其中

$$\boldsymbol{A} = \begin{bmatrix} 0 & 1 & 0 & \cdots & 0 \\ 0 & 0 & 1 & \cdots & 0 \\ \vdots & \vdots & \vdots & \ddots & \vdots \\ 0 & 0 & 0 & \cdots & 1 \\ -a_0 & -a_1 & -a_2 & \cdots & -a_{n-1} \end{bmatrix}$$

$$B = \begin{bmatrix} 0 \\ 0 \\ \vdots \\ 0 \\ 1 \end{bmatrix}$$

$$C = \begin{bmatrix} \beta_0 & \beta_1 & \cdots & \beta_{n-1} \end{bmatrix}$$

则此时相应的状态方程称为能控标准型。

若取 $A_0 = A^T, B_0 = C^T, C_0 = B^T$，则可构造出新的状态方程：

$$\begin{cases} \dot{x} = A_0 x + B_0 u \\ y = C_0 x \end{cases} \quad (8\text{-}1\text{-}25)$$

式(8-1-25)中

$$A_0 = \begin{bmatrix} 0 & 0 & \cdots & 0 & -a_0 \\ 1 & 0 & \cdots & 0 & -a_1 \\ 0 & 1 & \cdots & 0 & -a_2 \\ \vdots & \vdots & \ddots & \vdots & \vdots \\ 0 & 0 & \cdots & 1 & -a_{n-1} \end{bmatrix}$$

$$B_0 = \begin{bmatrix} \beta_0 \\ \beta_1 \\ \vdots \\ \beta_{n-1} \end{bmatrix}$$

$$C_0 = \begin{bmatrix} 0 & \cdots & 0 & 1 \end{bmatrix}$$

则其所对应的方程称为能观测标准型。

2) $\dfrac{N(s)}{D(s)}$ 只含单实极点的情况

当 $\dfrac{N(s)}{D(s)}$ 只含单实极点时，设 $D(s) = (s-\lambda_1)(s-\lambda_2)\cdots(s-\lambda_i)$。则

$$G(s) = \frac{N(s)}{D(s)} = \sum_{i=1}^{n} \frac{c_i}{s-\lambda_i} = \frac{Y(s)}{U(s)}$$

且有

$$Y(s) = \frac{c_i}{s-\lambda_i} U(s)$$

若令状态变量

$$x_i(s) = \frac{1}{s-\lambda_i} U(s) \quad (i=1,2,\cdots,n)$$

则其反变换结果为

$$\begin{cases} \dot{x}_i(t) = \lambda_i x_i(t) + u(t) \\ y(t) = \sum_{i=1}^{n} c_i x_i(t) \end{cases} \quad (8\text{-}1\text{-}26)$$

再将其展开,可得

$$\begin{cases} \dot{x}_1 = \lambda_1 x_1 + u \\ \dot{x}_2 = \lambda_2 x_2 + u \\ \vdots \\ \dot{x}_n = \lambda_n x_n + u \\ y = c_1 x_1 + c_2 x_2 + \cdots + c_n x_n \end{cases} \tag{8-1-27}$$

则矩阵-向量形式为

$$\begin{bmatrix} \dot{x}_1 \\ \dot{x}_2 \\ \vdots \\ \dot{x}_n \end{bmatrix} = \begin{bmatrix} \lambda_1 & & & 0 \\ & \lambda_2 & & \\ & & \ddots & \\ 0 & & & \lambda_n \end{bmatrix} \begin{bmatrix} x_1 \\ x_2 \\ \vdots \\ x_n \end{bmatrix} + \begin{bmatrix} 1 \\ 1 \\ \vdots \\ 1 \end{bmatrix} u$$

$$\boldsymbol{y} = \begin{bmatrix} c_1 & c_2 & \cdots & c_n \end{bmatrix} \begin{bmatrix} x_1 \\ x_2 \\ \vdots \\ x_n \end{bmatrix} \tag{8-1-28}$$

3) $\dfrac{N(s)}{D(s)}$ 含重实极点的情况

此时 $D(s)=(s-\lambda_1)^r(s-\lambda_{r+1})\cdots(s-\lambda_n)$,其中系统矩阵是一个含约当块的矩阵。

则

$$\frac{Y(s)}{U(s)} = \frac{N(s)}{D(s)} = \sum_{i=1}^{r} \frac{c_i}{(s-\lambda_1)^r} + \sum_{j=r+1}^{n} \frac{c_j}{s-\lambda_i}$$

同理,可得矩阵-向量形式的动态方程为:

$$\dot{\boldsymbol{x}} = \begin{bmatrix} \lambda_1 & 1 & 0 & & & & \\ 0 & \ddots & 1 & & 0 & & \\ 0 & & \lambda_1 & & & & \\ & & & \lambda_{r+1} & 0 & 0 & \\ & 0 & & & \ddots & 0 & \\ & & & 0 & 0 & \lambda_n \end{bmatrix} x + \begin{bmatrix} 0 \\ \vdots \\ 1 \\ 1 \\ \vdots \\ 1 \end{bmatrix} u$$

$$\boldsymbol{y} = \begin{bmatrix} c_{1r} & \cdots & c_{11} & c_{r+1} & \cdots & c_n \end{bmatrix} x \tag{8-1-29}$$

【例 8-1-7】 已知系统的传递函数为 $G(s)=\dfrac{4s^2+17s+16}{(s+2)^2(s+3)}$,求系统的动态方程,使系统矩阵为约当标准型。

解:将传递函数 $G(s)$ 展开成部分分式为 $G(s)=\dfrac{c_{11}}{(s+2)^2}+\dfrac{c_{12}}{(s+2)}+\dfrac{c_{13}}{s+3}$

$$c_{11} = \lim_{s \to -2}(s+2)^2 G(s) = -2$$

$$c_{12} = \lim_{s \to -2} \frac{\mathrm{d}}{\mathrm{d}s}[(s+2)^2 G(s)] = 3$$

$$c_{13} = \lim_{s \to -3}(s+3) G(s) = 1$$

则相应的动态方程为

$$\begin{bmatrix} \dot{x}_1 \\ \dot{x}_2 \\ \dot{x}_3 \end{bmatrix} = \begin{bmatrix} -2 & 1 & 0 \\ 0 & -2 & 0 \\ 0 & 0 & -3 \end{bmatrix} \begin{bmatrix} x_1 \\ x_2 \\ x_3 \end{bmatrix} + \begin{bmatrix} 0 \\ 1 \\ 1 \end{bmatrix} u$$

$$y = \begin{bmatrix} -2 & 3 & 1 \end{bmatrix} \begin{bmatrix} x_1 \\ x_2 \\ x_3 \end{bmatrix}$$

5. 由差分方程或脉冲函数建立状态空间描述

离散系统差分方程的一般形式为：

$$y(k+n) + a_{n-1}y(k+n-1) + \cdots + a_1 y(k+1) + a_0 y(k)$$
$$= b_n u(k+n) + b_{n-1} u(k+n-1) + \cdots + b_1 u(k+1) + b_0 u(k) \tag{8-1-30}$$

式中，k 表示 kT 时刻，T 为采样周期；$y(k)$ 为 kT 时刻的输出量，$u(k)$ 为 kT 时刻的输入量；a_i, b_i 是与系统特性有关的常系数。

初始条件为0，离散函数的子变换关系为：

$$z[y(k)] = y(z), \quad z[y(k+i)] = z^i y(z)$$

因此对式(8-1-30)两端取子变换，并整理，得：

$$G(z) = \frac{Y(z)}{U(z)} = \frac{b_n z^n + b_{n-1} z^{n-1} + \cdots + b_1 z + b_0}{z^n + a_{n-1} z^{n-1} + \cdots + a_1 z + a_0}$$
$$= b_n + \frac{\beta_{n-1} z^{n-1} + \cdots + \beta_1 z + \beta_0}{z^n + a_{n-1} z^{n-1} + \cdots + a_1 z + a_0} \tag{8-1-31}$$

连续系统的动态方程的建立方法与离散系统是同样适用的，因而可得动态方程为

$$\begin{bmatrix} x_1(k+1) \\ x_2(k+1) \\ \vdots \\ x_{n-1}(k+1) \\ x_n(k+1) \end{bmatrix} = \begin{bmatrix} 0 & 1 & 0 & 0 \\ 0 & 0 & 1 & 0 \\ \vdots & \vdots & \vdots & \vdots \\ 0 & 0 & 0 & 1 \\ -a_0 & -a_1 & -a_2 & -a_{n-1} \end{bmatrix} \begin{bmatrix} x_1(k) \\ x_2(k) \\ \vdots \\ x_{n-1}(k) \\ x_n(k) \end{bmatrix} + \begin{bmatrix} 0 \\ 0 \\ \vdots \\ 0 \\ 1 \end{bmatrix} u(k)$$

$$y(k) = \begin{bmatrix} \beta_0 & \beta_1 & \cdots & \beta_{n-1} \end{bmatrix} x(k) + b_n u(k) \tag{8-1-32}$$

简记为

$$\begin{cases} x(k+1) = Gx(k) + Hu(k) \\ y(k) = Cx(k) + Du(k) \end{cases} \tag{8-1-33}$$

式(8-1-33)中，G 为友矩阵，H, G 是可控标准型，由式(8-1-33)可看出，离散系统状态方程描述了 $(k+1)T$ 时刻的状态与 kT 时刻的状态及输入量之间的关系，其输出方程描述了 kT 时刻的输出量与 kT 时刻的状态及输入量之间的关系。

【例 8-1-8】 离散系统的差分方程为

$$y(k+3) + 2y(k+2) + 5y(k+1) + 6y(k)$$
$$= 2u(k+3) + 3u(k+2) + 11u(k+1) + 13u(k)$$

试写出该离散系统的一个状态空间描述。

解：由差分方程写出相应的脉冲传递函数：

$$G(z) = \frac{2z^3 + 3z^2 + 11z + 13}{z^3 + 2z^2 + 5z + 6} = 2 + \frac{-z^2 + z + 1}{z^3 + 2z^2 + 5z + 6}$$

因此就可以直接写出它的一个状态空间描述为：

$$x(k) = \begin{bmatrix} 0 & 1 & 1 \\ 0 & 0 & 1 \\ -6 & -5 & -2 \end{bmatrix} x(k) + \begin{bmatrix} 0 \\ 0 \\ 1 \end{bmatrix} u(k) = Gx(k) + Hu(k)$$

$$y(k) = \begin{bmatrix} 1 & 1 & -1 \end{bmatrix} x(k) + 2u(k) = Cx(k) + Du(k)$$

其中

$$G = \begin{bmatrix} 0 & 1 & 1 \\ 0 & 0 & 1 \\ -6 & -5 & -2 \end{bmatrix}, \quad H = \begin{bmatrix} 0 \\ 0 \\ 1 \end{bmatrix}, \quad C = \begin{bmatrix} 1 & 1 & -1 \end{bmatrix}, \quad D = \begin{bmatrix} 2 \end{bmatrix}$$

6．状态变量组的非唯一性

由前面几节的讨论可知，对于同一系统，选取不同的状态变量便有不同形式的动态方程。但对同一系统，无论如何选取，状态变量的个数总是相同的。由于它们描述的都是同一个系统，因此它们之间必然存在着某种联系。下面就来研究这二者之间的确定关系。

设系统动态方程为 $\dot{x} = Ax + Bu, y = Cx$。令 $x = T \cdot \bar{x}$，得两组状态变量的关系为

$$\begin{cases} \dot{x} = Ax + Bu \\ y = Cx \end{cases} \xrightarrow{x = T\bar{x}} \begin{cases} \dot{\bar{x}} = \bar{A}\bar{x} + \bar{B}u \\ y = \bar{C}\bar{x} \end{cases} \quad (8\text{-}1\text{-}34)$$

其中，$\bar{A} = T^{-1}AT, \bar{B} = T^{-1}B, \bar{C} = CT$。

由于系统状态变量的选取不是唯一的，非奇异线性变换矩阵 T 也不是唯一的。但对于同一系统，由不同的动态方程转换成的传递函数或传递函数矩阵却是相同的。这称为传递函数的不变性。

8.2 线性定常系统状态方程的解

当系统的状态空间模型建立起来以后，在一定的初始条件和某种输入信号作用下，就可以求解它的状态响应和输出响应。

8.2.1 线性定常齐次状态方程的解

所谓齐次状态方程，即为下列不考虑输入的方程，即 $\dot{x} = Ax$。

齐次方程满足初始状态 $x(t)|_{t=t_0} = x(t_0)$ 的解，也就是由初始时刻 t_0 的初始状态 $x(t_0)$ 所引起的无输入强迫项（无外力）时的自由运动。

常用的求解方法有级数展开法和拉普拉斯变换法两种。

1．级数展开法

设齐次方程的解是时间 t 的向量幂级数，即

$$x(t) = b_0 + b_1 t + b_2 t^2 + \cdots + b_k t^k + \cdots \text{ 且 } x(0) = b_0$$

对 t 求导可得

$$\begin{aligned}\dot{x}(t) &= b_1 + 2b_2 + \cdots + k b_k t^{k-1} + \cdots \\ &= A(b_0 + b_1 t + b_2 t^2 + \cdots + b_k t^k + \cdots)\end{aligned} \quad (8\text{-}2\text{-}1)$$

由左右两边对应项系数相等可知

$$Ab_0 = b_1, \quad 2b_2 = Ab_1, \cdots, k b_k = Ab_{k-1}$$

所以 $b_1 = Ab_0, b_2 = \dfrac{A^2}{2} b_0, \cdots, b_k = \dfrac{A^k}{k!} b_0$

所以 $x(t) = b_0 + Ab_0 t + \cdots + \dfrac{A^k}{k!} b_0 t^k + \cdots = x(0)\left(1 + At + \dfrac{A^2}{2} t^2 + \cdots + \dfrac{A^k}{k!} t^k\right) \quad (8\text{-}2\text{-}2)$

定义

$$e^{At} = 1 + At + \dfrac{A^2}{2} t^2 + \cdots + \dfrac{A^k}{k!} t^k$$

则

$$x(t) = e^{At} \cdot x(0) \quad (8\text{-}2\text{-}3)$$

记 $\boldsymbol{\Phi}(t) = e^{At}$。$e^{At}$ 又称为状态转移矩阵。

2. 拉式变换法

对齐次状态方程 $\dot{x}(t) = Ax(t)$，设初始时刻 $t_0 = 0$，且初始状态 $x(t) = x(0)$，对方程两边取拉普拉斯变换，可得 $sx(s) - x(0) = Ax(s)$，则有 $x(s) = (sI - A)^{-1} x(0)$。

进行拉普拉斯反变换，则有

$$x(t) = L^{-1}[(sI - A)^{-1}] x(0) \quad (8\text{-}2\text{-}4)$$

由式(8-2-3)与式(8-2-4)可得

$$\boldsymbol{\Phi}(t) = e^{At} = L^{-1}[(sI - A)^{-1}] \quad (8\text{-}2\text{-}5)$$

【例 8-2-1】 试求如下状态方程在初始状态 x_0 下的解：

$$\begin{bmatrix} \dot{x}_1(t) \\ \dot{x}_2(t) \end{bmatrix} = \begin{bmatrix} 0 & 1 \\ -2 & -3 \end{bmatrix} \begin{bmatrix} x_1(t) \\ x_2(t) \end{bmatrix} \quad x_0(t) = \begin{bmatrix} 1 \\ 2 \end{bmatrix}$$

解：

$$|sI - A| = s^2 + 3s + 2 = (s+2)(s+1)$$

$$(sI - A)^{-1} = \dfrac{\operatorname{adj}(sI - A)}{|sI - A|} = \dfrac{1}{(s+1)(s+2)} \begin{bmatrix} s+3 & 1 \\ -2 & s \end{bmatrix}$$

$$= \begin{bmatrix} \dfrac{2}{s+1} - \dfrac{1}{s+2} & \dfrac{1}{s+1} - \dfrac{1}{s+2} \\ -\dfrac{2}{s+1} + \dfrac{1}{s+2} & -\dfrac{1}{s+1} + \dfrac{2}{s+2} \end{bmatrix}$$

则

$$e^{At} = L^{-1}[(sI - A)^{-1}] = \begin{bmatrix} 2e^{-t} - e^{-2t} & e^{-t} - e^{-2t} \\ -2e^{-t} + 2e^{-2t} & -e^{-t} + 2e^{-2t} \end{bmatrix}$$

则状态方程的解为

$$x(t) = e^{At} x_0 = \begin{bmatrix} 4e^{-t} - 3e^{-2t} \\ -4e^{-t} + 6e^{-2t} \end{bmatrix}$$

求解齐次状态方程的问题，就是计算状态转移矩阵$\boldsymbol{\Phi}(t)$的问题。它包含由齐次状态方程描述的系统运动的全部信息，因此有必要对此状态转移矩阵的各种性质做深入研究。

Reading Material

In mathematics, the Laplace transform is an integral transform named after its discoverer Pierre-Simon Laplace. It takes a function of a real variable t (often time) to a function of a complex variable s (frequency).

The Laplace transform is very similar to the Fourier transform. While the Fourier transform of a function is a complex function of a real variable (frequency), the Laplace transform of a function is a complex function of a complex variable. Laplace transforms are usually restricted to functions of t with $t>0$. A consequence of this restriction is that the Laplace transform of a function is a holomorphic function of the variable s. Unlike the Fourier transform, the Laplace transform of a distribution is generally a well-behaved function. Also techniques of complex variables can be used directly to study Laplace transforms. As a holomorphic function, the Laplace transform has a power series representation. This power series expresses a function as a linear superposition of moments of the function. This perspective has applications in probability theory.

The Laplace transform is invertible on a large class of functions. The inverse Laplace transform takes a function of a complex variable s (often frequency) and yields a function of a real variable t (time). Given a simple mathematical or functional description of an input or output to a system, the Laplace transform provides an alternative functional description that often simplifies the process of analyzing the behavior of the system, or in synthesizing a new system based on a set of specifications. So, for example, Laplace transformation from the time domain to the frequency domain transforms differential equations into algebraic equations and convolution into multiplication. It has many applications in the sciences and technology.

8.2.2 线性定常系统的状态转移矩阵

【定义】 对于线性定常连续系统$\dot{\boldsymbol{x}}(t)=\boldsymbol{A}\boldsymbol{x}(t)$，当初始时刻$t_0=0$时，满足如下矩阵微分方程和初始条件：$\dot{\boldsymbol{\Phi}}(t)=\boldsymbol{A}\boldsymbol{\Phi}(t)$，$\boldsymbol{\Phi}(t)|_{t=0}=\boldsymbol{I}$的解$\boldsymbol{\Phi}(t)$为线性定常连续系统的状态转移矩阵。

【性质】

(1) $\boldsymbol{\Phi}(0)=\mathrm{e}^{\boldsymbol{A}0}=\boldsymbol{I}$

(2) $\mathrm{e}^{\boldsymbol{A}(t+s)}=\mathrm{e}^{\boldsymbol{A}t}\mathrm{e}^{\boldsymbol{A}s}$，$\boldsymbol{\Phi}(t+s)=\boldsymbol{\Phi}(t)\cdot\boldsymbol{\Phi}(s)$

【证明】 由指数矩阵函数的展开式，有

$$\begin{aligned}\mathrm{e}^{\boldsymbol{A}t}\cdot\mathrm{e}^{\boldsymbol{A}s}&=\left(1+\boldsymbol{A}t+\frac{\boldsymbol{A}^2t^2}{2!}+\cdots+\frac{\boldsymbol{A}^kt^k}{k!}+\cdots\right)\left(1+\boldsymbol{A}s+\frac{\boldsymbol{A}^2s^2}{2!}+\cdots+\frac{\boldsymbol{A}^ks^k}{k!}+\cdots\right)\\&=1+\boldsymbol{A}(t+s)+\frac{\boldsymbol{A}^2}{2!}(t^2+2ts+s^2)+\cdots+\frac{\boldsymbol{A}^k}{k!}(t+s)^k+\cdots\\&=\mathrm{e}^{\boldsymbol{A}(t+s)}\end{aligned}$$

(3) $[\boldsymbol{\Phi}(t_2-t_1)]^{-1} = \boldsymbol{\Phi}(t_1-t_2)$

【证明】 $[e^{A(t_2-t_1)}]^{-1} = e^{-A(t_2-t_1)} = e^{A(t_1-t_2)} = \boldsymbol{\Phi}(t_1-t_2)$

(4) 对于 $n \times n$ 阶方阵 $\boldsymbol{A}$ 和 $\boldsymbol{B}$，当 $\boldsymbol{AB} = \boldsymbol{BA}$ 成立时才有：
$$e^{(A+B)t} = e^{At} e^{Bt}$$

(5) $\dfrac{\mathrm{d}}{\mathrm{d}t} e^{At} = \boldsymbol{A} e^{At} = e^{At} \boldsymbol{A}, \dot{\boldsymbol{\Phi}}(t) = \boldsymbol{A}\boldsymbol{\Phi}(t) = \boldsymbol{\Phi}(t)\boldsymbol{A}$

(6) $[\boldsymbol{\Phi}(t)]^n = \boldsymbol{\Phi}(nt), (e^{At})^n = e^{nAt}, n$ 为正整数

(7) $\boldsymbol{\Phi}(t_2-t_1)\boldsymbol{\Phi}(t_1-t_0) = \boldsymbol{\Phi}(t_2-t_0)$

【例 8-2-2】 求如下系统的状态转移矩阵的逆矩阵

$$\begin{bmatrix} \dot{x}_1(t) \\ \dot{x}_2(t) \end{bmatrix} = \begin{bmatrix} 0 & 1 \\ -2 & -3 \end{bmatrix} \begin{bmatrix} x_1(t) \\ x_2(t) \end{bmatrix}$$

解：在例 8-2-1 中，已经求得状态转移矩阵为

$$\boldsymbol{\Phi}(t) = e^{At} = \begin{bmatrix} 2e^{-t} - e^{-2t} & e^{-t} - e^{-2t} \\ -2e^{-t} + 2e^{-2t} & -e^{-t} + 2e^{-2t} \end{bmatrix}$$

由于 $\boldsymbol{\Phi}^{-1}(-t) = \boldsymbol{\Phi}(t)$，所以可以求得状态转移矩阵的逆矩阵为

$$\boldsymbol{\Phi}^{-1}(t) = \boldsymbol{\Phi}(-t) = \begin{bmatrix} 2e^{-t} - e^{-2t} & e^{-t} - e^{-2t} \\ -2e^{-t} + 2e^{-2t} & -e^{-t} + 2e^{-2t} \end{bmatrix}$$

8.2.3 非齐次状态方程的解

当线性定常连续系统具有外加输入作用时，其状态方程为如下非齐次状态方程：
$$\dot{\boldsymbol{x}}(t) = \boldsymbol{A}\boldsymbol{x}(t) + \boldsymbol{B}\boldsymbol{u}(t)$$

该状态方程在初始状态 $x(t)|_{t=t_0} = x(t_0)$ 下的解，也就是由初始状态 $x(t_0)$ 和输入作用 $u(t)$ 所引起的系统状态的运动轨迹。

1. 直接求解法

将状态方程 $\dot{\boldsymbol{x}}(t) = \boldsymbol{A}\boldsymbol{x}(t) + \boldsymbol{B}\boldsymbol{u}(t)$ 移项可得 $\dot{\boldsymbol{x}}(t) - \boldsymbol{A}\boldsymbol{x}(t) = \boldsymbol{B}\boldsymbol{u}(t)$。

将上式两边同时乘以 e^{At}，可得 $e^{At}[\dot{\boldsymbol{x}}(t) - \boldsymbol{A}\boldsymbol{x}(t)] = e^{At} \cdot \boldsymbol{B}\boldsymbol{u}(t)$

即
$$\mathrm{d}[e^{-At}x(t)]/\mathrm{d}t = e^{-At}\boldsymbol{B}\boldsymbol{u}(t)$$

对其在区间 $[t_0, t]$ 内积分，则有 $\displaystyle\int_{t_0}^{t} \dfrac{\mathrm{d}}{\mathrm{d}t}[e^{-A\tau}x(\tau)]\mathrm{d}\tau = \int_{t_0}^{t} e^{-A\tau}Bu(\tau)\mathrm{d}\tau$

即
$$e^{-At}x(t) - e^{-At_0}x(t_0) = \int_{t_0}^{t} e^{-A\tau}\boldsymbol{B}u(\tau)\mathrm{d}\tau$$

因此
$$x(t) = e^{A(t-t_0)}x(t_0) + \int_{t_0}^{t} e^{A(t-\tau)}\boldsymbol{B}u(\tau)\mathrm{d}\tau \tag{8-2-6}$$

当 $t_0 = 0$ 时，解
$$x(t) = e^{At}x(0) + \int_{0}^{t} e^{A(t-\tau)}\boldsymbol{B}u(\tau)\mathrm{d}\tau$$

$$= \boldsymbol{\Phi}(t)x(0) + \int_0^t \boldsymbol{\Phi}(t-\tau)\boldsymbol{B}u(\tau)\mathrm{d}\tau \tag{8-2-7}$$

2. 拉普拉斯变换法

将非齐次状态方程两边取拉普拉斯变换,可得
$$sX(s) - x(0) = \boldsymbol{A}X(s) + \boldsymbol{B}u(s)$$

即
$$X(s) = (s\boldsymbol{I} - \boldsymbol{A})^{-1}x(0) + (s\boldsymbol{I} - \boldsymbol{A})^{-1}\boldsymbol{B}u(s) \tag{8-2-8}$$

再对式(8-2-8)两边取拉普拉斯变换,有
$$x(t) = L^{-1}[(s\boldsymbol{I} - \boldsymbol{A})^{-1}x(0)] + L^{-1}[(s\boldsymbol{I} - \boldsymbol{A})^{-1}\boldsymbol{B}u(s)]$$
$$= \mathrm{e}^{\boldsymbol{A}t}x(0) + \int_0^t \mathrm{e}^{\boldsymbol{A}(t-\tau)}\boldsymbol{B}u(\tau)\mathrm{d}\tau \tag{8-2-9}$$

有非齐次状态方程的解 $x(t)$,可得输出方程 $y = \boldsymbol{C}x(t) + \boldsymbol{D}u(t)$。

其输出响应为:
$$y(t) = \boldsymbol{C}\mathrm{e}^{\boldsymbol{A}(t-t_0)}x(t_0) + \int_{t_0}^t \boldsymbol{C}\mathrm{e}^{\boldsymbol{A}(t-\tau)}\boldsymbol{B}u(\tau)\mathrm{d}\tau + \boldsymbol{D}u(t) \tag{8-2-10}$$

或
$$y(t) = \boldsymbol{C}\mathrm{e}^{\boldsymbol{A}t}x(0) + \int_0^t \boldsymbol{C}\mathrm{e}^{\boldsymbol{A}(t-\tau)}\boldsymbol{B}u(\tau)\mathrm{d}\tau + \boldsymbol{D}u(t) \tag{8-2-11}$$

或
$$y(t) = \boldsymbol{C}\boldsymbol{\Phi}(t)x(0) + \int_0^t \boldsymbol{C}\boldsymbol{\Phi}(t-\tau)\boldsymbol{B}u(\tau)\mathrm{d}\tau + \boldsymbol{D}u(t) \tag{8-2-12}$$

【例 8-2-3】 已知线性定常系统为
$$\begin{bmatrix} \dot{x}_1(t) \\ \dot{x}_2(t) \end{bmatrix} = \begin{bmatrix} 0 & 1 \\ -2 & -3 \end{bmatrix} \begin{bmatrix} x_1(t) \\ x_2(t) \end{bmatrix} + \begin{bmatrix} 0 \\ 1 \end{bmatrix} u \quad x(0) = \begin{bmatrix} x_1(0) & x_2(0) \end{bmatrix}^{\mathrm{T}} = \begin{bmatrix} 1 \\ 2 \end{bmatrix}$$

试求系统在单位阶跃响应输入作用下状态方程的解。

解:在例 8-2-1 中已求出状态转移矩阵为
$$\boldsymbol{\Phi}(t) = \begin{bmatrix} 2\mathrm{e}^{-t} - \mathrm{e}^{-2t} & \mathrm{e}^{-t} - \mathrm{e}^{-2t} \\ -2\mathrm{e}^{-t} + 2\mathrm{e}^{-2t} & -\mathrm{e}^{-t} + 2\mathrm{e}^{-2t} \end{bmatrix}$$

于是系统在阶跃 $u(t) = 1(t)$ 作用下的解为
$$x(t) = \boldsymbol{\Phi}(t)x(0) + \int_0^t \boldsymbol{\Phi}(t-\tau)\boldsymbol{B}u(\tau)\mathrm{d}\tau$$
$$= \begin{bmatrix} 2\mathrm{e}^{-t} - \mathrm{e}^{-2t} & \mathrm{e}^{-t} - \mathrm{e}^{-2t} \\ -2\mathrm{e}^{-t} + 2\mathrm{e}^{-2t} & -\mathrm{e}^{-t} + 2\mathrm{e}^{-2t} \end{bmatrix} \begin{bmatrix} 1 \\ 2 \end{bmatrix}$$
$$+ \int_0^t \begin{bmatrix} 2\mathrm{e}^{-(t-\tau)} - \mathrm{e}^{-2(t-\tau)} & \mathrm{e}^{-(t-\tau)} - \mathrm{e}^{-2(t-\tau)} \\ -2\mathrm{e}^{-(t-\tau)} + 2\mathrm{e}^{-2(t-\tau)} & -\mathrm{e}^{-(t-\tau)} + 2\mathrm{e}^{-2(t-\tau)} \end{bmatrix} \begin{bmatrix} 0 \\ 1 \end{bmatrix} \mathrm{d}\tau$$
$$= \begin{bmatrix} 4\mathrm{e}^{-t} - 3\mathrm{e}^{-2t} \\ -4\mathrm{e}^{-t} + 6\mathrm{e}^{-2t} \end{bmatrix} + \int_0^t \begin{bmatrix} \mathrm{e}^{-(t-\tau)} - \mathrm{e}^{-2(t-\tau)} \\ -2\mathrm{e}^{-(t-\tau)} + 2\mathrm{e}^{-2(t-\tau)} \end{bmatrix} \begin{bmatrix} 0 \\ 1 \end{bmatrix} \mathrm{d}\tau$$
$$= \begin{bmatrix} \dfrac{1}{2} + 3\mathrm{e}^{-t} - \dfrac{5}{2}\mathrm{e}^{-2t} \\ -3\mathrm{e}^{-t} + 5\mathrm{e}^{-2t} \end{bmatrix}$$

8.2.4 线性离散系统的解

离散系统的动态方程如下：

$$\begin{cases} x(k+1) = Gx(k) + Hu(k) \\ y(k) = Cx(k) + Du(k) \end{cases}$$

令状态方程中 $k=0,1,\cdots,k-1$，可得

$k = 0$，$x(1) = \boldsymbol{\Phi}(T)x(0) + G(T)u(0)$

$k = 1$，$x(2) = G(T)x(1) + H(T)u(1)$
$\qquad\qquad = G^2(T)x(0) + G(T)H(T)u(0) + H(T)u(1)$

$k = 2$，$x(3) = G(T)x(2) + H(T)u(2)$
$\qquad\qquad = G^3(T)x(0) + G^2(T)H(T)u(0) + G(T)H(T)u(0) + H(T)u(2)$

$k = k-1$，$x(k) = G(T)x(k-1) + H(T)u(k-1)$
$$\approx G^k(T)x(0) + \sum_{i=0}^{k-1} G^{k-1-i}(T)H(T)u(i)$$

$y(k) = Cx(k) + Dx(k)$
$$= CG^k(T)x(0) + C\sum_{i=0}^{k-1} G^{k-1-i}(T)H(T)u(i) + Du(k)$$

于是，系统的解为 $x(k) = G^k(T)x(0) + \sum_{i=0}^{k-1} G^{k-1-i}(T)H(T)u(i)$

$$y(k) = CG^k(T)x(0) + C\sum_{i=0}^{k-1} G^{k-1-i}(T)H(T)u(i) + Du(k) \tag{8-2-13}$$

Terms and Concepts

Matrix exponential function：An important matrix function. Defined as $e^{At} = I + At + (At)^2/2! + \cdots + (At)^k/k! + \cdots$ that plays a role in the solution of linear constant coefficient differential equations.

State transition matrix, $\boldsymbol{\Phi}(t)$：The matrix exponential function that describes the unforced response of the system.

8.3 线性定常系统的可控性和可观测性

可控性与可观测性深刻地揭示了系统的内部结构关系。这些概念是由 R. E. Kalman 于 20 世纪 60 年代初首先提出来的。它们是用状态空间描述系统引申出来的两个非常重要的新概念。粗略地说，所谓系统的可控性问题是指：对于一个系统，控制作用能否对系统所有状态产生影响，从而能对系统的状态实现控制。所谓系统可观测性是指：一个系统，能否在有限的时间内通过观测输出量，识别出系统的所有状态。

经典控制理论中用传递函数描述系统的输入输出特性，被控量就是输出量。只要系统是因果系统且是稳定的，输出量便可以受控，且输出量总是可以被测量的，因而不需要提出

可控性和可观测性的概念。现代控制理论中用状态方程和输出方程描述系统,输入与输出构成系统的外部变量,而状态为系统的内部变量,这就存在着系统内的所有状态是否可受输入影响和是否可由输出反映的问题,这就是可控性和可观测性问题。

在现代控制理论的研究与实践中,可控性和可观测性具有极其重要的意义。事实上,可控性和可观测性通常决定了最优控制问题解决的存在性。例如,在极点配置问题中,状态反馈的存在性将由系统的可控性决定;在观测器设计和最优估计中,将涉及系统的可观测性条件。同时可知,可控性和可观测性不仅是研究线性系统控制问题必不可少的重要概念,而且对于许多最优控制、最优估计和自适应控制问题,也是经常用到的概念之一。

8.3.1 可控性和可观测性的定义

可控性和可观测性是在动态方程的基础上建立起来的两个非常重要的基本概念。对于线性定常系统

$$\begin{cases} \dot{x}(t) = Ax(t) + Bu(t) \\ y(t) = Cx(t) \end{cases} \quad (8\text{-}3\text{-}1)$$

可控性决定输入 $u(t)$ 是否对状态 $x(t)$ 有控制能力;可观测性回答了状态 $x(t)$ 是否能从输出 $y(t)$ 的测量值重新构造,它们都是由系统结构决定的系统的内在性质。为了使系统能达到良好的性能指标,一般都采用闭环控制。为了使系统的闭环极点能在 S 面上任意配置从而获得理想的动态性能,简单的输出反馈是不够的,必须采用状态反馈,而状态反馈的先决条件是系统的每一个状态变量都可控,这就是研究系统可控性和可观测性的意义。

1. 可控性

对于式(8-3-1)所示的单输入线性连续定常系统,状态可控性定义为:在有限时间间隔 $t \in [t_0, t_f]$ 内,如果存在无约束的分段连续控制函数 $u(t)$,能使系统从任意初态 $x(t_0)$ 转移到任意终态 $x(t_f)$,则称该系统是状态完全可控的,简称是可控的。只要有一个状态变量不可控,则称系统状态不完全可控,简称系统不可控。

【例 8-3-1】 系统状态方程为

$$\begin{bmatrix} \dot{x}_1(t) \\ \dot{x}_2(t) \end{bmatrix} = \begin{bmatrix} -3 & 1 \\ 0 & -2 \end{bmatrix} \begin{bmatrix} x_1(t) \\ x_2(t) \end{bmatrix} + \begin{bmatrix} 1 \\ 0 \end{bmatrix} u(t)$$

判断系统是否可控。

解:从状态方程可以看出,输入 $u(t)$ 对状态变量 $x_1(t)$ 有控制作用;而状态变量 $x_2(t)$ 既不直接受 $u(t)$ 影响,又与 $x_1(t)$ 没有关系,因而不受 $u(t)$ 控制,所以系统不可控。

【例 8-3-2】 系统状态方程为

$$\begin{bmatrix} \dot{x}_1(t) \\ \dot{x}_2(t) \end{bmatrix} = \begin{bmatrix} -2 & 1 \\ 0 & -2 \end{bmatrix} \begin{bmatrix} x_1(t) \\ x_2(t) \end{bmatrix} + \begin{bmatrix} 1 \\ 1 \end{bmatrix} u(t)$$

判断系统是否可控。

解:系统的状态转移矩阵为

$$\boldsymbol{\Phi}(t) = e^{At} = \frac{1}{2} \begin{bmatrix} e^{-t} + e^{-3t} & e^{-t} - e^{-3t} \\ e^{-t} - e^{-3t} & e^{-t} + e^{-3t} \end{bmatrix}$$

设该系统的初始状态为 $x(0) = 0$

$$\begin{aligned}
\boldsymbol{x}(t) &= \mathrm{e}^{At}x(0) + \int_0^t \mathrm{e}^{A(t-\tau)}\boldsymbol{B}u(\tau)\mathrm{d}\tau \\
&= \boldsymbol{\Phi}(t)x(0) \int_0^t \boldsymbol{\Phi}(t-\tau)\boldsymbol{B}u(\tau)\mathrm{d}\tau \\
&= \frac{1}{2}\int_0^t \begin{bmatrix} \mathrm{e}^{-(t-\tau)} + \mathrm{e}^{-3(t-\tau)} & \mathrm{e}^{-(t-\tau)} - \mathrm{e}^{-3(t-\tau)} \\ \mathrm{e}^{-(t-\tau)} - \mathrm{e}^{-3(t-\tau)} & \mathrm{e}^{-(t-\tau)} + \mathrm{e}^{-3(t-\tau)} \end{bmatrix} \begin{bmatrix} 1 \\ 1 \end{bmatrix} u(\tau)\mathrm{d}\tau \\
&= \frac{1}{2}\begin{bmatrix} 1 \\ 1 \end{bmatrix} \int_0^t \mathrm{e}^{-(t-\tau)} u(\tau)\mathrm{d}\tau
\end{aligned}$$

从 $x(t)$ 的表达式可以看出,无论给系统施加何种控制作用 $u(t)$,系统的两个状态变量总是相等的,$x_1(t) = x_2(t)$,即状态不可控转移到状态空间中 $x_1(t) \neq x_2(t)$ 任意点上。因此,系统是不可控制的。

2. 可观测性

给定系统的动态方程(8-3-1),对于任意初始时刻 t_0,若能在有限的时间 $t > t_0$ 之内,根据从 t_0 到 t 系统的输出 $y(t)$ 测量值,能唯一地确定在初始时刻 t_0 的状态 $x(t_0)$,则称系统状态完全可观测,简称系统可观测。只要有一个状态变量在初始时刻 t_0 的值不能由输出唯一地确定,则称系统状态不完全可观测,简称系统不可观测。

【例 8-3-3】 已知系统的状态方程为

$$\begin{bmatrix} \dot{x}_1(t) \\ \dot{x}_2(t) \end{bmatrix} = \begin{bmatrix} -3 & 1 \\ 0 & -2 \end{bmatrix} \begin{bmatrix} x_1(t) \\ x_2(t) \end{bmatrix} + \begin{bmatrix} 3 \\ 1 \end{bmatrix} u(t)$$

$$\boldsymbol{y}(t) = \begin{bmatrix} 1 & 0 \end{bmatrix} \begin{bmatrix} x_1(t) \\ x_2(t) \end{bmatrix}$$

判断系统是否可观测。

解: 从输出方程可知,输出量 $y(t)$ 等于 $x_1(t)$;从状态方程可知,$x_2(t)$ 与 $x_1(t)$ 没有关系,$y(t)$ 中不包含任何关于 $x_2(t)$ 的信息,即由输出 $y(t)$ 不能确定状态变量 $x_2(t)$。因此,本例的系统不可观测。

【例 8-3-4】 已知系统的状态方程为

$$\begin{bmatrix} \dot{x}_1(t) \\ \dot{x}_2(t) \end{bmatrix} = \begin{bmatrix} -2 & 1 \\ 0 & -2 \end{bmatrix} \begin{bmatrix} x_1(t) \\ x_2(t) \end{bmatrix} + \begin{bmatrix} 1 \\ 0 \end{bmatrix} u(t)$$

$$\boldsymbol{y}(t) = \begin{bmatrix} 1 & -1 \end{bmatrix} \begin{bmatrix} x_1(t) \\ x_2(t) \end{bmatrix}$$

判断系统是否可观测。

解: 系统的状态转移矩阵为:

$$\boldsymbol{\Phi}(t) = \mathrm{e}^{At} = \frac{1}{2}\begin{bmatrix} \mathrm{e}^{-t} + \mathrm{e}^{-3t} & \mathrm{e}^{-t} - \mathrm{e}^{-3t} \\ \mathrm{e}^{-t} - \mathrm{e}^{-3t} & \mathrm{e}^{-t} + \mathrm{e}^{-3t} \end{bmatrix}$$

假设 $t_0 = 0$ 时,系统的初始状态 $x(0) = x_0 \neq 0$,并令 $u(t) = 0$,则

$$y(t) = Cx(T) = C\Phi(t)x(0)$$
$$= \frac{1}{2}\begin{bmatrix} 1 & -1 \end{bmatrix}\begin{bmatrix} e^{-t}+e^{-3t} & e^{-t}-e^{-3t} \\ e^{-t}-e^{-3t} & e^{-t}+e^{-3t} \end{bmatrix}\begin{bmatrix} x_{10} \\ x_{20} \end{bmatrix}$$
$$= (x_{10}-x_{20})e^{-3t}$$

式中,$x_{10}=x_1(0)$,$x_{20}=x_2(0)$。

从 $y(t)$ 的表达式可知,由输出 $y(t)$ 只能确定差值 $(x_{10}-x_{20})$,而无法唯一确定 x_{10} 和 x_{20} 数值,因此,系统不可观测。

Reading Material

Controllability is an important property of a control system, and the controllability property plays a crucial role in many control problems, such as stabilization of unstable systems by feedback, or optimal control.

Controllability and observability are dual aspects of the same problem. Roughly, the concept of controllability denotes the ability to move a system around in its entire configuration space using only certain admissible manipulations. The exact definition varies slightly within the framework or the type of models applied.

Formally, a system is said to be observable, if for any possible sequence of state and control vectors, the current state can be determined in finite time using only the outputs. (This definition is slanted towards the state space representation.) Less formally, this means that one can determine the behavior of the entire system from the system's outputs. If a system is not observable, this means that the current values of some of its states cannot be determined through output sensors. This implies that their value is unknown to the controller (although they can be estimated by various means).

8.3.2 线性定常连续系统的可控性判别

1. 秩判据

(1) 考虑 n 阶线性定常连续系统的状态方程为
$$\dot{x}(t) = Ax(t) + Bu(t) \tag{8-3-2}$$
可根据 A 和 B 给出系统可控性的秩判据。

线性定常连续系统式(8-3-2)完全可控的充分必要条件是
$$\text{rank}\begin{bmatrix} B & AB & \cdots & A^{n-1}B \end{bmatrix} = n \tag{8-3-3}$$
其中,n 为矩阵 A 的维数,$S_c = \begin{bmatrix} B & AB & \cdots & A^{n-1}B \end{bmatrix}$ 称为系统的可控性判别阵。

(2) 考虑 n 阶定常离散系统的状态方程为
$$x(k+1) = Gx(k) + Hu(k) \tag{8-3-4}$$
可根据 G 和 H 给出系统可控性的秩判据。

线性定常离散系统式(8-3-4)完全可控的充分必要条件是
$$\text{rank}\begin{bmatrix} H & GH & \cdots & G^{n-1}H \end{bmatrix} = n \tag{8-3-5}$$
其中,n 为矩阵 G 的维数,$Q_k = \begin{bmatrix} H & GH & \cdots & G^{n-1}H \end{bmatrix}$ 称为系统的可控性判别阵。

【例 8-3-5】 判断下列状态方程的可控性：

$$\begin{bmatrix} \dot{x}_1(t) \\ \dot{x}_2(t) \\ \dot{x}_3(t) \end{bmatrix} = \begin{bmatrix} 1 & 3 & 2 \\ 0 & 2 & 0 \\ 0 & 1 & 3 \end{bmatrix} \begin{bmatrix} x_1 \\ x_2 \\ x_3 \end{bmatrix} + \begin{bmatrix} 2 & 1 \\ 1 & 1 \\ -1 & -1 \end{bmatrix} \begin{bmatrix} u_1 \\ u_2 \end{bmatrix}$$

解：

$$S_C = \begin{bmatrix} B & AB & A^2B \end{bmatrix} = \begin{bmatrix} 2 & 1 & 3 & 2 & 5 & 4 \\ 1 & 1 & 2 & 2 & 4 & 4 \\ -1 & -1 & -2 & -2 & -4 & -4 \end{bmatrix}$$

显然 S_C 矩阵的第二行和第三行元素的绝对值相同，rank $S_C = 2 < 3$，所以系统不可控。

2. A 矩阵为对角阵或约当阵的可控性判据

(1) 设线性定常连续系统 $\dot{x}(t) = Ax(t) + Bu(t)$ 具有互异的特征值 $\lambda_1, \lambda_2, \cdots, \lambda_n$，则系统状态完全可控的充分必要条件是系统经非奇异变换后的对角规范形式

$$\dot{\tilde{x}}(t) = \begin{bmatrix} \lambda_1 & & & 0 \\ & \lambda_2 & & \\ & & \ddots & \\ 0 & & & \lambda_n \end{bmatrix} \tilde{x}(t) + \widetilde{B} u(t) \tag{8-3-6}$$

式中，$\widetilde{B}$ 不包含元素全为零的行。

(2) 设线性定常连续系统 $\dot{x}(t) = Ax(t) + Bu(t)$ 具有多重特征值 $\lambda_1(m_1 重), \lambda_2(m_2 重), \cdots,$ $\lambda_k(m_k 重)$，$\sum_{i=1}^{k} m_i = n, \lambda_i \neq \lambda_j (i \neq j)$，则系统状态完全可控的充分必要条件是系统经非奇异变换后的对角规范形式：

$$\dot{\tilde{x}}(t) = \begin{bmatrix} \widetilde{A}_1 & & & 0 \\ & \widetilde{A}_2 & & \\ & & \ddots & \\ 0 & & & \widetilde{A}_n \end{bmatrix} \tilde{x}(t) + \widetilde{B} u(t) \tag{8-3-7}$$

式中，$\widetilde{B}$ 与每一个约当块 $\widetilde{A}_i (i=1,2,\cdots,k)$ 的最后一行相应的那些行的所有元素不全为零。即当 A 矩阵为约当阵且相同特征值分布在一个约当块时，只需要根据输入矩阵中与约当块最后一行所对应的行不是全零行，即可判断系统可控，与输入矩阵中其他行是否为零行是无关的。

【例 8-3-6】 试分析下列系统的可控性。

(1) $\begin{bmatrix} \dot{x}_1 \\ \dot{x}_2 \\ \dot{x}_3 \end{bmatrix} = \begin{bmatrix} -2 & 0 & 0 \\ 0 & -3 & 0 \\ 0 & 0 & -1 \end{bmatrix} \begin{bmatrix} x_1 \\ x_2 \\ x_3 \end{bmatrix} + \begin{bmatrix} 1 \\ 2 \\ 7 \end{bmatrix} u$

$y = [1, 2, 3] x$

(2) $\begin{bmatrix} \dot{x}_1 \\ \dot{x}_2 \\ \dot{x}_3 \end{bmatrix} = \begin{bmatrix} -7 & 0 & 0 \\ 0 & -5 & 0 \\ 0 & 0 & -1 \end{bmatrix} \begin{bmatrix} x_1 \\ x_2 \\ x_3 \end{bmatrix} + \begin{bmatrix} 2 & 1 \\ 0 & 0 \\ 3 & -2 \end{bmatrix} \begin{bmatrix} u_1 \\ u_2 \end{bmatrix}$

$y = \begin{bmatrix} 0 & 1 & 2 \\ 0 & 2 & 3 \end{bmatrix} x$

(3) $\begin{bmatrix} \dot{x}_1 \\ \dot{x}_2 \\ \dot{x}_3 \end{bmatrix} = \begin{bmatrix} -3 & 0 & 0 \\ 0 & -3 & 0 \\ 0 & 0 & -1 \end{bmatrix} \begin{bmatrix} x_1 \\ x_2 \\ x_3 \end{bmatrix} + \begin{bmatrix} 0 \\ 5 \\ 7 \end{bmatrix} u$

$y = \begin{bmatrix} 2 & 3 & 1 \end{bmatrix} x$

(4) $\begin{bmatrix} \dot{x}_1 \\ \dot{x}_2 \\ \dot{x}_3 \end{bmatrix} = \begin{bmatrix} -3 & 1 & 0 \\ 0 & -3 & 0 \\ 0 & 0 & 1 \end{bmatrix} \begin{bmatrix} x_1 \\ x_2 \\ x_3 \end{bmatrix} + \begin{bmatrix} 2 & -1 \\ 0 & 0 \\ 3 & 2 \end{bmatrix} \begin{bmatrix} u_1 \\ u_2 \end{bmatrix}$

$y = \begin{bmatrix} 1 & 0 & 3 \end{bmatrix} x$

解：(1)、(3)状态完全可控；(2)状态不完全可控，且状态 x_2 不可控；(4)状态不完全可控，且状态 x_2 不可控。

8.3.3 线性定常系统可观测性判别

1. 秩判据

（1）考虑 n 阶线性定常连续系统的状态方程为

$$\begin{cases} \dot{x}(t) = Ax(t) + Bu(t) \\ y(t) = Cx(t) \end{cases} \tag{8-3-8}$$

可根据 A 和 B 给出系统可观测的秩判据。

线性定常连续系统式(8-3-7)完全可观测的充分必要条件是

$$\mathrm{rank} \begin{bmatrix} C \\ CA \\ CA^2 \\ \vdots \\ CA^{n-1} \end{bmatrix} = n \tag{8-3-9}$$

或

$$\mathrm{rank} \begin{bmatrix} C^{\mathrm{T}} & A^{\mathrm{T}}C^{\mathrm{T}} & (A^{\mathrm{T}})^2 C^{\mathrm{T}} & \cdots & (A^{\mathrm{T}})^{n-1} C^{\mathrm{T}} \end{bmatrix} = n \tag{8-3-10}$$

其中，n 为矩阵 A 的维数，$S_0 = \begin{bmatrix} C^{\mathrm{T}} & A^{\mathrm{T}}C^{\mathrm{T}} & (A^{\mathrm{T}})^2 C^{\mathrm{T}} & \cdots & (A^{\mathrm{T}})^{n-1} C^{\mathrm{T}} \end{bmatrix}$ 称为系统的可观测性判别阵。

（2）考虑 n 阶线性离散系统的状态方程为

$$\begin{cases} x(k+1) = Gx(k) + Hu(k) \\ y(k) = Cx(k) \end{cases} \tag{8-3-11}$$

可根据 G 和 H 给出系统可观测性的秩判据。

线性离散系统式(8-3-11)完全可观测的充分必要条件是

$$\text{rank}\begin{bmatrix} C \\ CG \\ \vdots \\ CG^{n-1} \end{bmatrix} = n \qquad (8\text{-}3\text{-}12)$$

或

$$\text{rank}[C^T \quad G^T C^T \quad (G^T)^2 C^T \quad \cdots \quad (G^T)^{n-1} C^T] = n \qquad (8\text{-}3\text{-}13)$$

其中,n 为矩阵 A 的维数,$Q_g = [C^T \quad G^T C^T \quad (G^T)^2 C^T \quad \cdots \quad (G^T)^{n-1} C^T]$ 称为系统的可观测性判别阵。

2. A 为对角阵或约当阵时的可观测性判据

当系统矩阵 A 已化成对角阵或约当阵时,由可观测性矩阵能导出更简洁直观的可观测性判据。

(1) 设线性定常连续系统 $\dot{x}(t) = Ax(t) + Bu(t)$,$y(t) = Cx(t)$ 具有互异的特征值 λ_1,λ_2,$\cdots$,λ_n,则系统状态完全可观测的充分必要条件是系统经过非奇异变换后的对角规范形式为

$$\begin{cases} \dot{\tilde{x}}(t) = \begin{bmatrix} \lambda_1 & & & 0 \\ & \lambda_2 & & \\ & & \ddots & \\ 0 & & & \lambda_n \end{bmatrix} \tilde{x}(t) + \tilde{B}u(t) \\ y(t) = \tilde{C}x(t) \end{cases} \qquad (8\text{-}3\text{-}14)$$

式中的输出矩阵 $\tilde{C}$ 中不包括元素全为零的列。

(2) 设线性定常连续系统 $\dot{x}(t) = Ax(t) + Bu(t)$,$y(t) = Cx(t)$ 具有多重特征值 $\lambda_1(m_1$ 重),$\lambda_2(m_2$ 重),$\cdots$,$\lambda_k(m_k$ 重),$\sum_{i=1}^{k} m_i = n$,$\lambda_i \neq \lambda_j (i \neq j)$,则系统完全可观测的充分必要条件是系统经非奇异变换后的对角规范形式为

$$\begin{cases} \dot{\tilde{x}}(t) = \begin{bmatrix} \tilde{A}_1 & & & 0 \\ & \tilde{A}_2 & & \\ & & \ddots & \\ 0 & & & \tilde{A}_n \end{bmatrix} \tilde{x}(t) + \tilde{B}u(t) \\ y(t) = \tilde{C}x(t) \end{cases} \qquad (8\text{-}3\text{-}15)$$

式中,$\tilde{C}$ 与每一个约当块 $\tilde{A}_i (i=1,2,\cdots,k)$ 的首列相应的那些列的所有元素不全为零。

【例 8-3-7】 试分析例 8-3-6 所给系统的可观测性。

解:系统(1),(3),(4)是系统完全可观测的,系统(2)是状态不完全可观测的且状态 x_1 不可观测。

Terms and Concepts

A system is completely controllable if there exists an unconstrained control $u(t)$ that can transfer any initial state $x(t_0)$ to any other desired location $x(t)$ in a time, $t_0 \leqslant t \leqslant T$.

A system is completely observable if and only if there exists a finite time T such that the initial state $x(0)$ can be determined from the observation history $y(t)$ given the control, $0 \leqslant t \leqslant T$.

Controllability matrix: A linear system is (completely) controllable if and only if the controllability matrix $\boldsymbol{P}_C = [\boldsymbol{B} \quad \boldsymbol{AB} \quad \boldsymbol{A}^2\boldsymbol{B} \quad \cdots \quad \boldsymbol{A}^{n-1}\boldsymbol{B}]$ has full rank. Where $\boldsymbol{A}$ is an $n \times n$ matrix. For single-input, single-output linear systems, the system is controllable if and only if the determinant of the $n \times n$ controllability matrix $\boldsymbol{P}_C$ is nonzero.

Observability matrix: A linear system is (completely) observable if and only if the observability matrix $\boldsymbol{P}_O = [\boldsymbol{C}^T \quad (\boldsymbol{CA})^T \quad (\boldsymbol{CA}^2)^T \quad \cdots \quad (\boldsymbol{CA}^{n-1})^T]^T$ has full rank. Where $\boldsymbol{A}$ is an $n \times n$ matrix. For single-input, single-output linear systems, the system is observable if and only if the determinant of the $n \times n$ observability matrix $\boldsymbol{P}_C$ is nonzero.

Observable system: A system is observable on the interval $[t_0, t_f]$ if any state $x(t_0)$ is uniquely determined by observing output $y(t)$ on the interval $[t_0, t_f]$.

Observer: A dynamic system used to estimate the state of another dynamic system given knowledge of the system inputs and measurements of the system outputs.

State variable feedback: The use of control signal formed as a direct function of all the state variables.

8.3.4 可控性、可观测性与传递函数矩阵的关系

考虑系统动态方程为

$$\begin{cases} \dot{\boldsymbol{x}}(t) = \boldsymbol{A}x(t) + \boldsymbol{B}u(t) \\ \boldsymbol{y}(t) = \boldsymbol{C}x(t) \end{cases} \quad (8\text{-}3\text{-}16)$$

单输入-单输出系统可控、可观测的充分必要条件是：在传递函数或传递函数矩阵中不出现相约现象。如果发生相约，那么在被约去的模态中，系统不可控或输出不可观测了。所以，当且仅当系统是状态可控或可观测时，其传递函数才没有相约因子。这就意味着，可相约的传递函数不具备表征动态系统的所有信息。只有当系统是可控又是可观测时，传递函数描述与状态空间描述才是等价的。

【例 8-3-8】 已知下列动态方程，试研究其可控性、可观测性与传递函数的关系。

(1) $\dot{\boldsymbol{x}} = \begin{bmatrix} 0 & 1 \\ 2.5 & -1.5 \end{bmatrix} x + \begin{bmatrix} 0 \\ 1 \end{bmatrix} u$

$\boldsymbol{y} = [2.5 \quad 1]x$

(2) $\dot{\boldsymbol{x}} = \begin{bmatrix} 0 & 2.5 \\ 1 & -1.5 \end{bmatrix} x + \begin{bmatrix} 2.5 \\ 1 \end{bmatrix} u$

$\boldsymbol{y} = [0 \quad 1]x$

(3) $\dot{x} = \begin{bmatrix} 1 & 0 \\ 0 & 2.5 \end{bmatrix} x + \begin{bmatrix} 1 \\ 0 \end{bmatrix} u$

$y = \begin{bmatrix} 1 & 0 \end{bmatrix} x$

解：三个系统的传递函数均为 $G(s) = \dfrac{Y(s)}{U(s)} = \dfrac{s+2.5}{(s+2.5)(s-1)}$，存在相约现象。①系统可控不可观测；②系统可观测，不可控；③系统不可控，不可观测。因此以上三个系统的传递函数和状态空间在描述上不是等价的。

8.4 线性定常系统的状态反馈和状态观测器

闭环系统的性能与系统极点(特征值)密切相关。经典控制理论的系统综合中,不管是频率法还是根轨迹法,本质上都可视为极点配置问题。在状态空间方法的分析与综合中,大都采用状态反馈来配置极点,有时也可以利用输出反馈配置极点。状态反馈有较好的抗干扰性或鲁棒性,能提供更多的校正信息,在形成最优控制规律,抑制或消除扰动影响,实现系统解耦控制等方面得到了广泛的应用。为了利用状态进行反馈,状态必须是可用传感器直接测量的。但是,在有的情况下,无法直接测量状态,这就产生了利用状态测量器来估计状态的问题。所以,状态反馈和状态观测器的设计是状态空间方法综合设计的主要内容。

8.4.1 线性定常系统常用反馈结构

反馈是控制系统设计的主要手段。经典控制理论中采用输出作为反馈量,称为输出反馈,而在现代控制理论中,除了输出反馈外,广泛采用状态作为反馈量,称为状态反馈。

1. 状态反馈

给定单输入-单输出线性定常被控系统

$$\begin{cases} \dot{x} = Ax + Bu \\ y = Cx \end{cases} \tag{8-4-1}$$

式中, $x(t) \in \mathbf{R}^n, u(t) \in \mathbf{R}^p, A \in \mathbf{R}^{n \times p}, y \in \mathbf{R}^q, C \in \mathbf{R}^{q \times n}$。选取线性反馈控制律为

$$u = r - Kx \tag{8-4-2}$$

式中, $K \in \mathbf{R}^{p \times n}$ 为状态增益矩阵或线性状态反馈矩阵, r 为与 u 同维的 p 维参考输入向量。加入状态反馈的系统结构图如图 8-4-1 所示。

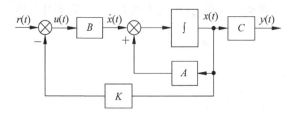

图 8-4-1 状态反馈系统结构图

将式(8-4-2)代入式(8-4-1)可得状态反馈系统动态方程

$$\begin{cases} \dot{x} = (A - BK)x + Br \\ y = Cx \end{cases} \quad (8\text{-}4\text{-}3)$$

其传递函数矩阵为

$$G_K(s) = C(sI - A + BK)^{-1}B \quad (8\text{-}4\text{-}4)$$

2. 输出反馈

输出有两种形式,一种是将输出量反馈到状态微分处,另一种是将输出量仅反馈到参考输入。下面以多输出-单输出系统为例。

输出量反馈到状态微分的系统结构图如图 8-4-2 所示。

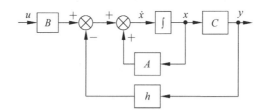

图 8-4-2 输出反馈到状态微分

设被控对象的状态方程为

$$\begin{cases} \dot{x} = Ax + Bu \\ y = Cx \end{cases} \quad (8\text{-}4\text{-}5)$$

输出反馈系统的状态方程为

$$\begin{cases} \dot{x} = Ax + Bu - hy \\ y = Cx \end{cases} \quad (8\text{-}4\text{-}6)$$

故

$$\begin{cases} \dot{x} = (A - hC)x + Bu \\ y = Cx \end{cases} \quad (8\text{-}4\text{-}7)$$

式中,h 为 $n \times p$ 输出反馈矩阵。

输出量反馈到参考输入的系统结构图如图 8-4-3 所示。

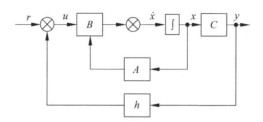

图 8-4-3 输出反馈到参考输入

其中

$$u = r - hy \quad (8\text{-}4\text{-}8)$$

该系统的动态方程为
$$\begin{cases} \dot{x} = (A - Bhc)x + Br \\ y = Cx \end{cases} \quad (8\text{-}4\text{-}9)$$

式中，h 为输出反馈矩阵。若令 $hc = K$，该输出反馈便等价于状态反馈。

8.4.2 状态反馈与极点配置

设单输入系统的动态方程为
$$\begin{cases} \dot{x} = Ax + Bu \\ y = Cx \end{cases} \quad (8\text{-}4\text{-}10)$$

状态向量 x 通过待设计的状态反馈矩阵 K 负反馈到控制输入处，于是
$$u = r - Kx \quad (8\text{-}4\text{-}11)$$

由以上两式可得：
$$\begin{cases} \dot{x} = (A - BK)x + Br \\ y = Cx \end{cases} \quad (8\text{-}4\text{-}12)$$

式中，$(A-BK)$ 为闭环状态矩阵，则 $|\lambda I - (A-BK)|$ 为闭环特征多项式。显然可见，引入状态反馈后，只改变了系统矩阵及其特征值，B、C 矩阵均无改变。

【极点配置定理】 采用状态反馈能够任意配置闭环极点的充要条件是式(8-4-10)所表示的系统 (A, B, C) 可控。

计算状态反馈矩阵的步骤如下。
(1) 检验系统的可控性矩阵；
(2) 计算特征多项式和特征值；
(3) 与具有希望特征值的特征多项式比较，从而确定 K 矩阵。

【例 8-4-1】 考虑如下线性定常系统，
$$\dot{x} = Ax + Bu$$

式中，$A = \begin{bmatrix} 0 & 1 & 0 \\ 0 & 0 & 1 \\ -1 & -5 & -6 \end{bmatrix}$，$B = \begin{bmatrix} 0 \\ 0 \\ 1 \end{bmatrix}$。

利用状态反馈控制 $u = r - Kx$，希望得到该系统的闭环极点为 $s = -2 \pm j4$ 和 $s = -10$。试确定状态反馈增益矩阵 K。

解：首先需检验该系统的可控性矩阵。由于可控性矩阵为
$$S_C = \begin{bmatrix} B & AB & A^2B \end{bmatrix} = \begin{bmatrix} 0 & 0 & 1 \\ 0 & 1 & -6 \\ 1 & -6 & 31 \end{bmatrix}$$

所以得出 $\det S_C = -1$。因此，$\text{rank} S_C = 3$。因此该系统是状态完全可控的，可任意配置极点。

设状态反馈矩阵为
$$K = \begin{bmatrix} k_0 & k_1 & k_2 \end{bmatrix}$$

而状态反馈系统特征方程为

$$|\lambda I - (A - BK)| = \lambda^3 + (6+k_2)\lambda^2 + (5+k_1)\lambda + k_0$$

期望闭环极点对应的系统特征方程为

$$(\lambda + 10)(\lambda + 2 + j4)(\lambda + 2 - j4) = \lambda^3 + 14\lambda^2 + 60\lambda + 199$$

由两个特征方程同幂项系数应相同,可得

$$k_0 = 199, \quad k_1 = 55, \quad k_2 = 8$$

所以,$K = [199\ 55\ 8]$。

8.4.3 状态观测器

在许多实际情况中,不是所有的状态变量都可用于反馈,这时需要估计不可直接物理测量的状态变量。需特别强调,应避免将一个状态变量微分产生另一个状态变量,因为噪声通常比控制信号变化更迅速,所以信号的微分总是减小信噪比。不可物理测量的状态变量的估计通常称为观测。估计或者观测状态变量的装置称为状态观测器。如果状态观测器可观测到系统的所有状态变量,不管其是否能直接测量,这种状态观测器均称为全维状态观测器。有时,只需观测不可观测的状态变量,而不是可直接观测的状态变量。例如,由于输出变量是可观测的,并且它们与状态变量线性相关,所以无须观测所有的状态变量,而只观测 $n-m$ 个状态变量,其中,n 是状态向量的维数,m 为输出向量的维数。因此,估计小于 n 个状态变量的观测器称为降维状态观测器,或简称降维观测器。如果降维观测器的阶数是最小的,则称该观测器为最小维状态观测器或最小维观测器。本节将讨论全维状态观测器。

1. 全维状态观测器

考虑如下线性定常系统

$$\begin{cases} \dot{x} = Ax + Bu \\ y = Cx \end{cases} \tag{8-4-13}$$

从理论上可以构造一个动态方程,其形式与式(8-4-13)相同,用计算机实现的模拟受控系统为

$$\begin{cases} \dot{\tilde{x}} = A\tilde{x} + Bu \\ \tilde{y} = C\tilde{x} \end{cases} \tag{8-4-14}$$

式中,$\tilde{x}, \tilde{y}$ 分别为模拟系统的状态向量估计值和模拟输出向量。当模拟系统与受控对象的初始状态向量相同时,在同一输入向量的作用下,有 $\tilde{x} = x$,可用作状态反馈所需的信息。但是受控对象的初始状态可能不相同,模拟系统中的积分器初始条件的设置只能估计,因而两个系统的初始值总是有差异,即使系统的 A, B, C 矩阵完全一样,也必然存在估计状态与受控对象实际状态的误差 $\tilde{x} - x$,难以实现所需要的状态反馈。这就是说,力图采用开环形式的状态估计器是不适用的。由于估计状态的误差 $\tilde{x} - x$ 必然在输出误差 $\tilde{y} - y$ 中反映出来,根据反馈控制原理,可利用 $\tilde{y} - y$,并将其反馈到 $\dot{\tilde{x}}$ 处,控制 $\tilde{y} - y$ 尽快趋近于零,从而使得 $\tilde{x} - x$ 尽快趋近于零,这时就可利用 $\tilde{x}$ 形成状态反馈了。此时的状态向量 x 由如下动态方程

$$\dot{\tilde{x}} = A\tilde{x} + Bu + K_e(y - C\tilde{x}) \tag{8-4-15}$$

中的状态 $\tilde{x}$ 来近似，该式表示状态观测器。注意到状态观测器的输入为 y 和 u，输出为 $\tilde{x}$。式(8-4-15)的右端最后一项包含实际输出与观测输出 $C\tilde{x}$ 之间差的修正项，修正项能减小动态模型与实际系统之间的差异。矩阵 K_e 起到加权矩阵的作用。如图 8-4-4 所示为系统和全维状态观测器的框图。

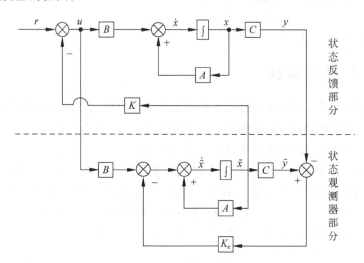

图 8-4-4　全维状态观测器框图

2. 观测器的引入对闭环系统的影响——分离定理

在极点配置的设计过程中，是假设真实状态 $x(t)$ 可用于反馈。然而实际上，真实状态 $x(t)$ 可能无法测量，所以必须设计一个状态观测器，并且将观测到的状态 $\tilde{x}(t)$ 用于反馈，如图 8-4-4 所示。因此，该设计过程分为两个阶段，第一个阶段是确定反馈增益矩阵，从而产生所期望的特征方程；第二个阶段是确定观测器的增益矩阵 K_e，以产生所期望的观测特征方程。

考虑如下线性定常系统

$$\begin{cases} \dot{x} = Ax + Bu \\ y = Cx \end{cases} \tag{8-4-16}$$

且假定该系统状态完全可控和完全可观测。

对基于观测状态 $\tilde{x}$ 的状态反馈控制

$$u = r - K\tilde{x} \tag{8-4-17}$$

利用该控制，状态方程为

$$\dot{x} = Ax - B(r - K\tilde{x}) = (A - Bk)x + Bk(x - \tilde{x}) + Br \tag{8-4-18}$$

定义误差

$$e(t) = x(t) - \tilde{x}(t) \tag{8-4-19}$$

将式(8-4-19)代入式(8-4-18)，得

$$\dot{x} = (A - Bk)x + Bke + Br \tag{8-4-20}$$

同时，用式(8-4-13)减去式(8-4-15)，可得

$$\dot{x} - \dot{\tilde{x}} = Ax - A\tilde{x} - ke(Cx - C\tilde{x}) = (A - keC)(x - \tilde{x}) \tag{8-4-21}$$

所以，结合式(8-4-19)将式(8-4-21)改写为
$$\dot{e} = (A - k_e C)e \tag{8-4-22}$$
将式(8-4-20)和式(8-4-22)合并，可得
$$\begin{bmatrix} \dot{x} \\ \dot{e} \end{bmatrix} = \begin{bmatrix} A - Bk & Bk \\ 0 & A - k_e C \end{bmatrix} \begin{bmatrix} x \\ e \end{bmatrix} + \begin{bmatrix} B \\ 0 \end{bmatrix} r \tag{8-4-23}$$

式(8-4-23)描述了观测-状态反馈控制系统的动态特征。该系统的特征方程为
$$\begin{vmatrix} sI - (A - Bk) & -Bk \\ 0 & sI - (A - k_e C) \end{vmatrix} = 0 \tag{8-4-24}$$
或
$$|sI - (A - BK)||sI - (A - k_e C)| = 0 \tag{8-4-25}$$

注意，观测-状态反馈控制系统的闭环极点包括由极点配置单独设计产生的极点和由观测器单独设计产生的极点。这意味着，极点配置和观测器的设计是相互独立的，它们可分别进行设计，并合并为观测-状态反馈控制系统。

【分离定理】 若受控系统(A, B, C)可控、可观测，用状态观测器估计值形成状态反馈时，其系统的极点配置和观测器设计可分别独立进行。

【例 8-4-2】 考虑如下的线性定常系统
$$\begin{cases} \dot{x} = Ax + Bu \\ y = Cx \end{cases}$$

式中，$A = \begin{bmatrix} 0 & 20.6 \\ 1 & 0 \end{bmatrix}$，$B = \begin{bmatrix} 0 \\ 1 \end{bmatrix}$，$C = \begin{bmatrix} 0 & 1 \end{bmatrix}$。

设系统结构和如图 8-4-4 所示的相同，又设观测器的期望特征值为
$$u_1 = -1.8 + j2.4, \quad u_2 = -1.8 - j2.4$$

设计一个全维状态观测器。

解：由于状态观测器的设计实际上可总结为确定一个合适的观测器增益矩阵K_e，为此先检验可观测性矩阵，即
$$S_0 = [C^T, A^T C^T] = \begin{bmatrix} 0 & 1 \\ 1 & 0 \end{bmatrix}$$

显然可见，$\text{rank} S_0 = 2$。因此该系统是完全可观测的，并且可确定期望的观测器增益矩阵K_e。

定义
$$K_e = \begin{bmatrix} K_{e1} \\ K_{e2} \end{bmatrix}$$

则特征方程为
$$\begin{vmatrix} \begin{bmatrix} s & 0 \\ 0 & s \end{bmatrix} - \left(\begin{bmatrix} 0 & 20.6 \\ 1 & 0 \end{bmatrix} - \begin{bmatrix} K_{e1} \\ K_{e2} \end{bmatrix} \begin{bmatrix} 0 & 1 \end{bmatrix} \right) \end{vmatrix}$$
$$= \begin{vmatrix} s & -20.6 + K_{e1} \\ -1 & s + K_{e2} \end{vmatrix}$$
$$= s^2 + K_{e2} s - 20.6 + K_{e1}$$

所期望的特征方程为

$$(s+1.8-j2.4)(s+1.8+j2.4) = s^2+3.6s+9 = 0$$

比较以上两式,可得

$$K_{e1} = 29.6, \quad K_{e2} = 3.6$$

所以观测器增益矩阵 $K_e = \begin{bmatrix} 29.6 \\ 3.6 \end{bmatrix}$。

由状态反馈(极点配置)选择所产生的期望闭环极点,应使系统能满足性能要求。观测器极点的选取通常使得观测响应比系统响应快得多。一个经验法则是选择观测器的响应至少比系统响应快 2~5 倍。使观测状态迅速收敛到真实状态。观测器的最大响应速度通常只受到控制系统中的噪声和灵敏度的限制。注意在极点配置过程中,观测器极点位于所期望的闭环极点的左边,所以后者在响应中起主导作用。

另外,也可以采用全维状态观测器的设计方法,来设计和实现最小阶观测器。

Reading Material

In control theory, a state observer is a system that provides an estimate of the internal state of a given real system, from measurements of the input and output of the real system. It is typically computer-implemented, and provides the basis of many practical applications.

Knowing the system state is necessary to solve many control theory problems; for example, stabilizing a system using state feedback. In most practical cases, the physical state of the system cannot be determined by direct observation. Instead, indirect effects of the internal state are observed by way of the system outputs. A simple example is that of vehicles in a tunnel: the rates and velocities at which vehicles enter and leave the tunnel can be observed directly, but the exact state inside the tunnel can only be estimated. If a system is observable, it is possible to fully reconstruct the system state from its output measurements using the state observer.

8.5 MATLAB 在本章中的应用

A dynamic system is most commonly described in one of three ways: by a set of state-space equations and the corresponding matrices, by a transfer function with the numerator and denominator polynomials, or by a list of poles and zeros and the associated gain. Here we present a state space tutorial according to the reference attached to this book.

8.5.1 State-space equations

There are several different ways to describe a system of linear differential equations. The state-space representation is given by the equations:

$$\frac{d\boldsymbol{x}}{dt} = \boldsymbol{A}\boldsymbol{x} + \boldsymbol{B}u \quad \text{and} \quad \boldsymbol{y} = \boldsymbol{C}\boldsymbol{x} + \boldsymbol{D}u \qquad (8\text{-}5\text{-}1)$$

where x is an n by 1 vector representing the state (commonly position and velocity variable in mechanical systems), u is a scalar representing the input (commonly a force or torque in mechanical systems), and y is a scalar representing the output. The matrices **A** (n by n), **B** (n by 1), and **C** (1 by n) determine the relationships between the state and input and output variable. Note that there are n first-order differential equations. State space representation can also be used for systems with multiple inputs and outputs (MIMO), but we will only use single-input, single-output (SISO) systems in these tutorials.

To introduce the state space design method, we will use the magnetically suspended ball as an example. As the figure 8-5-1, the current through the coils induces a magnetic force which can balance the force of gravity and cause the ball (which is made of a magnetic material) to be suspended in midair. The modeling of this system has been established in many control text books (including *Automatic Control*

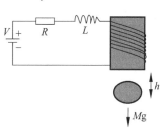

Figure 8-5-1 Control Systems

Systems by B. C. Kuo, the seventh edition). The equations for the system are given by

$$M\frac{d^2 h}{dt^2} = Mg - \frac{Ki^2}{h} \quad \text{and} \quad V = L\frac{di}{dt} + iR \tag{8-5-2}$$

Where h is the vertical position of the ball, i is the current through the electromagnet, V is the applied voltage, M is the mass of the ball, g is gravity, L is the inductance, R is the resistance, and K is a coefficient that determines the magnetic force exerted on the ball. For simplicity, we will choose values $M = 0.05\text{kg}, K = 0.0001, L = 0.01\text{H}, R = 1\Omega$, $g = 9.81\text{m/s}^2$. The system is at equilibrium (the ball is suspended in midair) whenever $h = Ki^2/Mg$ (at which point $dh/dt = 0$). We linearize the equations about the point $h = 0.01\text{m}$ (where the nominal current is about 7 amp) and get the state space equations:

$$\frac{d\boldsymbol{x}}{dt} = \boldsymbol{Ax} + \boldsymbol{Bu} \quad \text{and} \quad \boldsymbol{y} = \boldsymbol{Cx} + \boldsymbol{Du} \tag{8-5-3}$$

Where: $x = [\Delta h \quad \Delta \dot{h} \quad \Delta i]^T$ is the set of state variable for the system (a 3×1 vector), u is the input voltage (delta V), and y (the output), is delta h. Enter the system matrices into a m-file.

```
A = [0 1 0; 980 0 -2.8; 0 0 -100];
B = [0; 0; 100];
C = [1 0 0];
```

One of the first things you want to do with the state equations is find the poles of the system; these are the values of s where $\det(s\boldsymbol{I} - \boldsymbol{A}) = 0$, or the eigenvalues of the **A** matrix:

```
poles = eig(A)
```

You should get the following three poles:

```
poles = 31.3050; -31.3050; -100.0000
```

One of the poles is in the right-half plane, which means that the system is unstable in open-loop.

To check out what happens to this unstable system when there is a nonzero initial condition, add the following lines to your m-file, Figure 8-5-2 shows the open-loop response to non-zero condition linear simulation result.

```
t = 0:0.01:2;
u = 0*t;
x0 = [0.005 0 0];
sys = ss(A,B,C,0);
[y,t,x] = lsim(sys,u,t,x0);
figure
box on
hold on;
plot(t,y)
title('Open-loop response to non-zero condition');
xlabel('Time/s');
ylabel('Ball position');
```

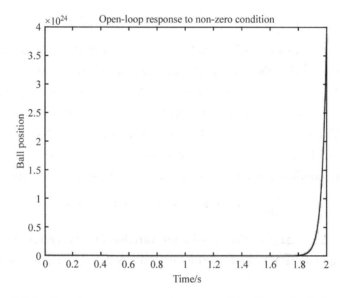

Figure 8-5-2 Open-loop response to non-zero condition linear simulation result

It looks like the distance between the ball and the electromagnet will go to infinity, but probably the ball hits the table or the floor first (and also probably goes out of the range where our linearization is valid).

8.5.2 Control design using pole placement

Let's build a controller for this system. The schematic of a full-state feedback system is the following Figure 8-5-3:

Recall that the characteristic polynomial for this closed-loop system is the determinant of $(s\bm{I}-(\bm{A}-\bm{BK}))$. Since the matrices $\bm{A}$ and $\bm{B}\times\bm{K}$ are both 3 by 3 matrices, there will be 3 poles for the system. By using full-state feedback e can place the poles anywhere we want. We could use the MATLAB function *place* to find the control matrix, $\bm{K}$, which will give the desired poles. Before attempting this method, we have to decide where

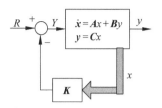

Figure 8-5-3 The schematic of a full-state feedback system

we want the closed-loop poles to be. Suppose the criteria for the controller were settling time <0.5s and overshoot $<5\%$, then we might try to place the two dominant poles at $-10+10$i, $-10-10$i(at zeta $=0.7$ or 45 degrees with sigma $=10>4.6\times2$). The third pole we might place at -50 to start, and we can change it later depending on what the closed-loop behavior is. Remove the lsim command from your m-file and everything after it, then add the following lines to your m-file,

```
A = [0   1   0
     980 0   -2.8
     0   0   -100];
B = [0
     0
     100];
C = [1   0   0];
t = 0:0.01:2;
u = 0 * t;
x0 = [0.005 0 0];
p1 = -10 + 10i;
p2 = -10 - 10i;
p3 = -50;
K = place(A,B,[p1 p2 p3]);
sys_cl = ss(A - B * K,B,C,0);
lsim(sys_cl,u,t,x0);
title('Linear Simulation Result');
xlabel('Time(sec)');
ylabel('Ball position(m)');
```

From Figure 8-5-4, we can see the overshoot is too large (there are also zeros in the transfer function which can increase the overshoot; you do not see the zeros in the state-space formulation). Try placing the poles further to the left to see if the transient response improves (this should also make the response faster).

```
p1 = -20 + 20i;
p2 = -20 - 20i;
p3 = -100;
K = place(A,B,[p1 p2 p3]);
sys_cl = ss(A - B * K,B,C,0);
lsim(sys_cl,u,t,x0);
```

From Figure 8-5-4, the overshoot is smaller. Consult your textbook for further

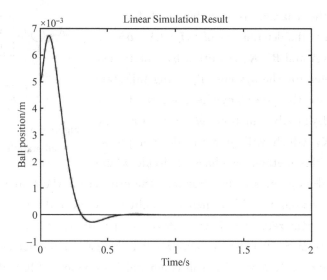

Figure 8-5-4 Closed-loop system linear simulation result

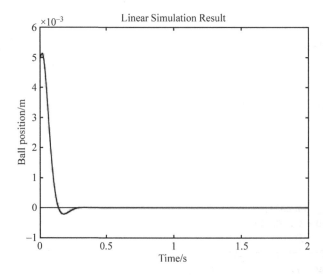

Figure 8-5-5 Control system linear simulation result

suggestions on choosing the desired closed-loop poles.

Compare the control effort required (K) in both cases. In general, the farther you move the poles, the more control effort it takes.

Note: If you want to place two or more poles at the same position, place will not work. You can use a function called acker which works similarly to place:

$$K = \text{acker}(A, B[p_1 \quad p_2 \quad p_3]) \tag{8-5-4}$$

8.5.3 Introducing the reference input

Now, we will take the control system as defined above and apply a step input (we choose a small value for the step, so we remain in the region where our linearization is valid). To show the result in Figure 8-5-6, you need replace t, u and lsim in your m-file with the following,

```
t = 0:0.01:2;
u = 0.001 * ones(size(t));
sys_cl = ss(A - B * K, B, C, 0);
lsim(sys_cl, u, t)
```

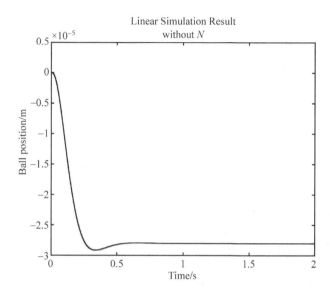

Figure 8-5-6 Control system linear simulation result (without N)

The system does not track the step well at all; not only is the magnitude not one, but it is negative instead of positive!

Recall the schematic above, we don't compare the output to the reference.

Instead we measure all the states, multiply by the gain vector $\boldsymbol{K}$, and then subtract this result from the reference. There is no reason to expect that $\boldsymbol{K} \times x$ will be equal to the desired output. To eliminate this problem, we can scale the reference input to make it equal to $\boldsymbol{K} \times x$ steady state. This scale factor is often called Nbar; it is introduced as shown in the following Figure 8-5-7:

We can get Nbar from MATLAB by using the function *rscale* (place the following line of code after $\boldsymbol{K} = \cdots$).

Nbar = rscale(sys, K); % Nbar = -285.7143
$$(8-5-5)$$

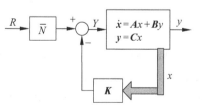

Figure 8-5-7 Control system

Note that this function is not standard in MATLAB. You will need to copy it to a new m-file to use it. Now, if we want to find the response of the system under state feedback with this introduction of the reference, we simply note the fact that the input is multiplied by this new factor, Nbar:

$$\text{lism}(\text{sys\_cl}, \text{Nbar} * u, t) \tag{8-5-6}$$

And now a step can be tracked reasonably well in Figure 8-5-8.

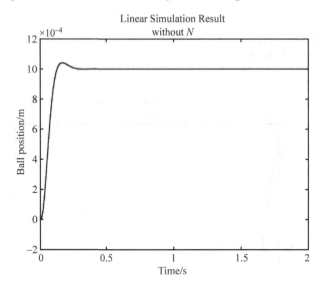

Figure 8-5-8 Control system linear simulation result (without N)

8.5.4 Observer design

When we can't measure all the states x (as is commonly the case), we can build an observer to estimate them, while measuring only the output $y = Cx$. For the magnetic ball example, we will add three new, estimated states to the system. The schematic is as follows Figure 8-5-9:

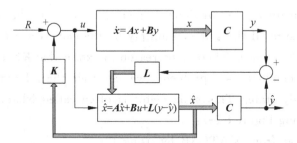

Figure 8-5-9 Build an observer to estimate states to the system

The observer is basically a copy of the plant; it has the same input and almost the same differential equation. An extra term compares the actual measured output y to the estimated output $\hat{y}$; this will cause the estimated states $\hat{x}$ to approach the values of the actual states x. The error dynamics of the observer are given by the poles of $(A - L \times C)$.

First we need to choose the observer gain L. Since we want the dynamics of the observer to be much faster than the system itself, we need to place the poles at least five times farther to the left than the dominant poles of the system. If we want to use *place*, we need to put the three observer poles at different locations.

```
op1 =-100;
op2 =-101;
op3 =-102;
```

Because of the duality between controllability and observability, we can use the same technique used to find the control matrix, but replacing the matrix **B** by the matrix **C** and taking the transposes of each matrix (consult your text book for the derivation):

```
L = place(A',C',[op1 op2 op3])';
```

The equations in the block diagram above are given for $\hat{x}$. It is conventional to write the combined equations for the system plus observer using the original state x plus the error state: $e = x - \hat{x}$. We use as state feedback $u = -\boldsymbol{K}\hat{x}$. After a little bit of algebra (consult your textbook for more details), we arrive at the combined state and error equations with the full-state feedback and an observer:

```
At = [A - B*K         B*K
      zeros(size(A))  A - L*C];
Bt = [B*Nbar
      zeros(size(B))];
Ct = [ C  zeros(size(C))];
```

To see how the response looks to a nonzero initial condition with no reference input, add the following lines into your m-file. We typically assume that the observer begins with zero initial condition, $\hat{x}=0$. This gives us that the initial condition for the error is equal to the initial condition of the state.

```
sys = ss(At,Bt,Ct,0);
lsim(sys,zeros(size(t)),t,[x0 x0]);
```

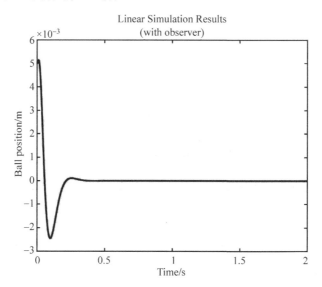

Figure 8-5-10　Control system with linear simulation result (without N)

Responses of all the states are plotted in Figure 8-5-10, Figure 8-5-11 and Figure 8-5-12. Recall that lsim gives us x and e; to get $\hat{x}$ we need to compute $x-e$.

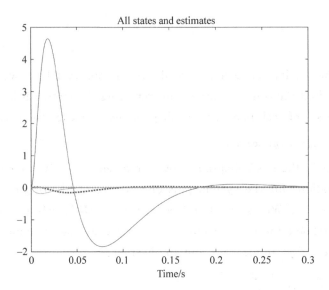

Figure 8-5-11　Control system with all states and estimates

Zoom in to see some detail:

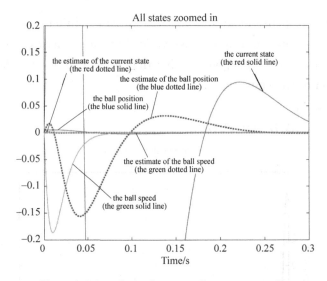

Figure 8-5-12　Control system all states zoomed in

The blue solid line is the response of the ball position Δh, the blue dotted line is the estimated state $\Delta \hat{h}$;

The green solid line is the response of the ball speed $\Delta \dot{h}$, the green dotted line is the estimated state $\Delta \dot{\hat{h}}$;

The red solid line is the response of the current Δi, the red dotted line is the estimated state $\Delta \hat{i}$.

We can see that the observer estimates the states quickly and tracks the states reasonably well in the steady-state.

The plot above can be obtained by using the plot command as follows:

```
[y,t,x] = lsim(sys,zeros(size(t)),t,[x0 x0]);
plot(t,x)
axis([0,.3,-.2,.2])
```

本章小结

(1) 本章介绍了建立在状态空间法基础上的现代控制理论,它是自动控制理论的一个主要组成部分。在现代控制理论中,对控制系统的分析和设计主要是通过对系统的状态变量的描述来进行的,基本的方法是时间域方法。现代控制理论比经典控制理论所能处理的控制问题要广泛得多,包括线性系统和非线性系统,定常系统和时变系统,单变量系统和多变量系统。它所采用的方法和算法也更适合用计算机处理。现代控制理论还为设计和构造具有指定性能指标的最优控制系统提供了可能性。现代控制理论所包含的学科内容十分广泛,主要的方面有:线性系统理论、非线性系统理论、最优控制理论、随机控制理论和适应控制理论。

(2) 本章介绍了状态空间表示式的基本概念,控制系统结构图的绘制,如何由传递函数或微分方程求取状态空间表达式等;引入了状态转移矩阵的概念和性质,讨论了线性定常齐次状态方程的解法和控制系统全状态响应的解法;给出系统能控或能观测的含义与条件,控制系统的能控性或能观测性判定,状态空间表达式的能控标准型和能观测标准型的实现方法,传递函数阵的实现问题解法等;最后介绍了线性系统基本结构特性,极点配置和状态观测器的设计分析。

(3) 本章需要学生掌握采用现代控制理论的时域建模方法(状态空间表达式)以及模型的规范描述形式;掌握系统性能的基本分析方法——求解状态空间表达式、系统的能控性和能观测性以及系统的稳定性分析;掌握在时域进行反馈控制系统的设计,包括系统的极点配置以及观测器的设计等。

Summary and Outcome Checklist

In this chapter, we consider system modeling using time-domain methods. As before, we will consider physical systems described by an nth-order ordinary differential equation. Utilizing a (nonunique) set of variables, known as state variables, we can obtain a set of first-order differential equations. We group these first-order equations using a compact matrix notation in a model known as the state variable model. The time-domain state variable model lends itself readily to computer solution and analysis. The relationship between transfer models and state variable models will be investigated. Several interesting physical systems are presented and analyzed.

Tick the box for each statement with which you agree:

☐ I am able to understand state variable, state differential equations, and output equations.

- ☐ I am able to recognize that state variable models can describe the dynamic behavior of physical systems and can be represented by block diagrams and signal flow graphs.
- ☐ I am able to know how to obtain the transfer function model from a state variable model, and vice versa.
- ☐ I am able to be aware of solution methods for state variable models and the role of the state transition matrix in obtaining the time response.
- ☐ I am able to understand the important role of state variable modeling in control system design.
- ☐ I am able to be familiar with the concepts of controllability and observability.
- ☐ I am able to design full-state feedback controllers and observers.
- ☐ I am able to appreciate pole-placement methods and the application of Ackermann's formula.
- ☐ I am able to understand the separation principle and how to construct state variable compensators.
- ☐ I am able to have a working knowledge of reference inputs, optimal control, and internal model design.

习题

8-1 考虑由下式确定的系统：
$$\frac{Y(s)}{U(s)} = \frac{s+3}{s^2+3s+2}$$
试求其状态空间表达式之能控标准形、能观测标准形和对角线标准形。

8-2 试求如下线性定常系统
$$\begin{bmatrix} \dot{x}_1 \\ \dot{x}_2 \end{bmatrix} = \begin{bmatrix} 0 & 1 \\ -2 & -3 \end{bmatrix} \begin{bmatrix} x_1 \\ x_2 \end{bmatrix}$$
的状态转移矩阵 $\Phi(t)$ 和状态转移矩阵的逆矩阵 $\Phi^{-1}(t)$。

8-3 求下列系统的时间响应：
$$\begin{bmatrix} \dot{x}_1 \\ \dot{x}_2 \end{bmatrix} = \begin{bmatrix} 0 & 1 \\ -2 & -3 \end{bmatrix} \begin{bmatrix} x_1 \\ x_2 \end{bmatrix} + \begin{bmatrix} 0 \\ 1 \end{bmatrix} u(t)$$
式中，$u(t)$ 为 $t=0$ 时作用于系统的单位阶跃函数，即 $u(t)=1(t)$。

8-4 试判断由式
$$\begin{bmatrix} \dot{x}_1 \\ \dot{x}_2 \end{bmatrix} = \begin{bmatrix} 1 & 1 \\ -2 & -1 \end{bmatrix} \begin{bmatrix} x_1 \\ x_2 \end{bmatrix} + \begin{bmatrix} 0 \\ 1 \end{bmatrix} u$$
$$y = \begin{bmatrix} 1 & 0 \end{bmatrix} \begin{bmatrix} x_1 \\ x_2 \end{bmatrix}$$
所描述的系统是否为能控和能观测的。

8-5 试简述现代控制理论与经典控制理论的区别及现代控制理论的主要内容。

拉普拉斯变换及反变换

1. 拉普拉斯变换的基本性质

拉普拉斯变换的基本性质如表 A-1 所示。

表 A-1 拉普拉斯变换的基本性质

序号	定理	性质
1	线性定理	$\mathcal{L}[\alpha_1 f_1(t) \pm \alpha_2 f_2(t)] = \alpha_1 F_1(s) \pm \alpha_2 F_2(s)$
2	微分定理	$\mathcal{L}\left[\dfrac{\mathrm{d}f(t)}{\mathrm{d}t}\right] = sF(s) - f(0)$ $\mathcal{L}\left[\dfrac{\mathrm{d}^2 f(t)}{\mathrm{d}t^2}\right] = s^2 F(s) - sf(0) - f'(0)$ $\vdots$ $\mathcal{L}\left[\dfrac{\mathrm{d}^n f(t)}{\mathrm{d}t^n}\right] = s^n F(s) - \sum\limits_{k=1}^{n} s^{n-k} f^{(k-1)}(0)$
3	积分定理	$\mathcal{L}\left[\int f(t)\mathrm{d}t\right] = \dfrac{F(s)}{s} + \dfrac{\left[\int f(t)\mathrm{d}t\right]_{t=0}}{s}$ $\mathcal{L}\left[\iint f(t)(\mathrm{d}t)^2\right] = \dfrac{F(s)}{s^2} + \dfrac{\left[\int f(t)\mathrm{d}t\right]_{t=0}}{s^2} + \dfrac{\left[\iint f(t)(\mathrm{d}t)^2\right]_{t=0}}{s}$ $\vdots$ $\mathcal{L}\left[\overbrace{\int\cdots\int}^{共n个} f(t)(\mathrm{d}t)^n\right] = \dfrac{F(s)}{s^n} + \sum\limits_{k=1}^{n}\dfrac{1}{s^{n-k+1}}\overbrace{\left[\int\cdots\int f(t)(\mathrm{d}t)^n\right]}^{共k个}_{t=0}$
4	延迟定理(t 域平移定理)	$\mathcal{L}[f(t-T)u(t-T)] = \mathrm{e}^{-Ts}F(s)$
5	衰减定理(s 域平移定理)	$\mathcal{L}[f(t)\mathrm{e}^{-at}] = F(s+a)$
6	终值定理	$\lim\limits_{t\to\infty} f(t) = \lim\limits_{s\to 0} sF(s)$
7	初值定理	$\lim\limits_{t\to 0^*} f(t) = \lim\limits_{s\to\infty} sF(s)$
8	卷积定理	$\mathcal{L}\left[\int_0^t f_1(t-\tau)f_2(\tau)\mathrm{d}\tau\right] = \mathcal{L}\left[\int_0^t f_1(t)f_2(t-\tau)\mathrm{d}\tau\right] = F_1(s)F_2(s)$

2. 用查表法进行拉普拉斯反变换

用查表法进行拉普拉斯反变换的关键在于将变换式进行部分分式展开,然后逐项查表进行反变换。设 $F(s)$ 是 s 的有理真分式,即

$$F(s) = \frac{B(s)}{A(s)} = \frac{b_m s^m + b_{m-1} s^{m-1} + \cdots + b_1 s + b_0}{a_n s^n + a_{n-1} s^{n-1} + \cdots + a_1 s + a_0} \quad n > m$$

式中,系数 $a_0, a_1, \cdots, a_{n-1}, a_n$ 和 $b_0, b_1, \cdots, b_{m-1}, b_m$ 都是实常数;m, n 是正整数。按代数定理可将 $F(s)$ 展开为部分分式。分为以下两种情况讨论。

(1) $A(s) = 0$ 无重根。这时,$F(s)$ 可展开为 n 个简单的部分分式之和的形式,即

$$F(s) = \frac{c_1}{s - s_1} + \frac{c_2}{s - s_2} + \cdots + \frac{c_i}{s - s_i} + \cdots + \frac{c_n}{s - s_n} = \sum_{i=1}^{n} \frac{c_i}{s - s_i} \quad \text{(A-1)}$$

式中,$s_1, s_2, \cdots, s_n$ 是特征方程 $A(s) = 0$ 的根;c_i 为待定常数,称为 $F(s)$ 在 s_i 处的留数,可按下列两式计算。

$$c_i = \lim_{s \to s_i} (s - s_i) F(s) \quad \text{(A-2)}$$

或

$$c_i = \left. \frac{B(s)}{A'(s)} \right|_{s = s_i} \quad \text{(A-3)}$$

式中,$A'(s)$ 为 $A(s)$ 对 s 的一阶导数。根据拉普拉斯变换的性质,从式(A-1)可求得原函数为

$$f(t) = \mathcal{L}^{-1}[F(s)] = \mathcal{L}^{-1}\left[\sum_{i=1}^{n} \frac{c_i}{s - s_i}\right] = \sum_{i=1}^{n} c_i e^{s_i t} \quad \text{(A-4)}$$

(2) $A(s) = 0$ 有重根。设 $A(s) = 0$ 有 r 重根 s_1,$F(s)$ 可写为

$$F(s) = \frac{B(s)}{(s - s_1)^r (s - s_{r+1}) \cdots (s - s_n)}$$

$$= \frac{c_r}{(s - s_1)^r} + \frac{c_{r-1}}{(s - s_1)^{r-1}} + \cdots + \frac{c_1}{(s - s_1)} + \frac{c_{r+1}}{s - s_{r+1}} + \cdots + \frac{c_i}{s - s_i} + \cdots + \frac{c_n}{s - s_n}$$

式中,s_1 为 $F(s)$ 的 r 重根,$s_{r+1}, \cdots, s_n$ 为 $F(s)$ 的 $(n-r)$ 个单根;其中,$c_{r+1}, \cdots, c_n$ 仍按式(A-2)或式(A-3)计算,$c_r, c_{r-1}, \cdots, c_1$ 则按式(A-5)计算。

$$\left. \begin{aligned} c_r &= \lim_{s \to s_1} (s - s_1)^r F(s) \\ c_{r-1} &= \lim_{s \to s_i} \frac{\mathrm{d}}{\mathrm{d}s}[(s - s_1)^r F(s)] \\ &\vdots \\ c_{r-j} &= \frac{1}{j!} \lim_{s \to s_1} \frac{\mathrm{d}^{(j)}}{\mathrm{d}s^{(j)}} (s - s_1)^r F(s) \\ &\vdots \\ c_1 &= \frac{1}{(r-1)!} \lim_{s \to s_1} \frac{\mathrm{d}^{(r-1)}}{\mathrm{d}s^{(r-1)}} (s - s_1)^r F(s) \end{aligned} \right\} \quad \text{(A-5)}$$

原函数 $f(t)$ 为

$$\begin{aligned} f(t) &= \mathcal{L}^{-1}[F(s)] \\ &= \mathcal{L}^{-1}\left[\frac{c_r}{(s - s_1)^r} + \frac{c_{r-1}}{(s - s_1)^{r-1}} + \cdots + \frac{c_1}{(s - s_1)} + \frac{c_{r+1}}{s - s_{r+1}} + \cdots + \frac{c_i}{s - s_i} + \cdots + \frac{c_n}{s - s_n}\right] \\ &= \left[\frac{c_r}{(r-1)!} t^{r-1} + \frac{c_{r-1}}{(r-2)!} t^{r-2} + \cdots + c_2 t + c_1\right] e^{s_1 t} + \sum_{i=r+1}^{n} c_i e^{s_i t} \end{aligned} \quad \text{(A-6)}$$

附录 B z 变换定义及对照表

1. z 变换的定义

对连续函数 $f(t)$ 进行拉普拉斯变换,即

$$F(s) = \int_0^\infty f(t) e^{-st} dt$$

对离散函数 $f^*(t) = \sum_{k=0}^{+\infty} f(t)\delta(t-kT)$ 进行拉普拉斯变换,即

$$F^*(s) = \int_0^\infty \left[\sum_{k=0}^{+\infty} f(t)\delta(t-kT)\right] e^{-st} dt = \sum_{k=0}^{+\infty} f(kT) \int_0^\infty \delta(t-kT) e^{-st} dt$$

$$= \sum_{k=0}^{+\infty} f(kT) e^{-kTs}$$

引入 $z = e^{Ts}$,则有

$$F(z) = \sum_{k=0}^{+\infty} f(kT) z^{-k} \tag{B-1}$$

称 $F(z)$ 为 $f^*(t)$ 的 z 变换,并记作

$$F(z) = \mathcal{Z}[f^*(t)] \tag{B-2}$$

$F(z)$ 表示离散信号 $f^*(t)$ 的 z 变换,它表征了连续信号 $f(t)$ 在采样时刻的信息。由于习惯原因,人们也称 $F(z)$ 是 $f(t)$ 或 $F(s)$ 的 z 变换,但其含义是离散信号 $f^*(t)$ 的 z 变换。

2. z 变换的基本性质

z 变换的基本性质如表 B-1 所示。

表 B-1 z 变换的基本性质

序 号	定 理	性 质
1	线性定理	$\mathcal{Z}[\alpha_1 f_1(t) \pm \alpha_2 f_2(t)] = \alpha_1 F_1(z) \pm \alpha_2 F_2(z)$
2	滞后定理	$\mathcal{Z}[f(t-k_1 t)] = z^{-k_1} F(z)$
3	超前定理	$\mathcal{Z}[f(t+k_1 t)] = z^{k_1} F(z) - z^{k_1} \sum_{k=0}^{k_1-1} f(kT) z^{-k}$
4	位移定理	$\mathcal{Z}[f(t) e^{\mp at}] = F(z e^{\pm aT})$
5	初值定理	$\lim_{t \to 0} f(t) = \lim_{z \to \infty} F(z)$
6	终值定理	$\lim_{t \to \infty} f(t) = \lim_{z \to 1} (z-1) F(z)$

3. z变换和拉普拉斯变换

常用函数的拉普拉斯变换和 z 变换如表 B-2 所示。

表 B-2　常用函数的拉普拉斯变换和 z 变换

序号	拉普拉斯变换 $F(s)$	时间函数 $f(t)$	z 变换 $F(z)$
1	1	$\delta(t)$	1
2	$\dfrac{1}{1-e^{-Ts}}$	$\delta_T(t) = \sum_{n=0}^{\infty} \delta(t-nT)$	$\dfrac{z}{z-1}$
3	$\dfrac{1}{s}$	$u(t)$	$\dfrac{z}{z-1}$
4	$\dfrac{1}{s^2}$	t	$\dfrac{Tz}{(z-1)^2}$
5	$\dfrac{1}{s^3}$	$\dfrac{t^2}{2}$	$\dfrac{T^2 z(z+1)}{2(z-1)^3}$
6	$\dfrac{1}{s^{n+1}}$	$\dfrac{t^n}{n!}$	$\lim\limits_{a \to 0} \dfrac{(-1)^n}{n!} \dfrac{\partial^n}{\partial a^n}\left(\dfrac{z}{z-e^{-aT}}\right)$
7	$\dfrac{1}{s+a}$	e^{-at}	$\dfrac{z}{z-e^{-aT}}$
8	$\dfrac{1}{(s+a)^2}$	te^{-at}	$\dfrac{Tze^{-aT}}{(z-e^{-aT})^2}$
9	$\dfrac{a}{s(s+a)}$	$1-e^{-at}$	$\dfrac{(1-e^{-aT})z}{(z-1)(z-e^{-aT})}$
10	$\dfrac{b-a}{(s+a)(s+b)}$	$e^{-at}-e^{-bt}$	$\dfrac{z}{z-e^{-aT}} - \dfrac{z}{z-e^{-bT}}$
11	$\dfrac{\omega}{s^2+\omega^2}$	$\sin\omega t$	$\dfrac{z\sin\omega T}{z^2-2z\cos\omega T+1}$
12	$\dfrac{s}{s^2+\omega^2}$	$\cos\omega t$	$\dfrac{z(z-\cos\omega T)}{z^2-2z\cos\omega T+1}$
13	$\dfrac{\omega}{(s+a)^2+\omega^2}$	$e^{-at}\sin\omega t$	$\dfrac{ze^{-aT}\sin\omega T}{z^2-2ze^{-aT}\cos\omega T+e^{-2aT}}$
14	$\dfrac{s+a}{(s+a)^2+\omega^2}$	$e^{-at}\cos\omega t$	$\dfrac{z^2-ze^{-aT}\cos\omega T}{z^2-2ze^{-aT}\cos\omega T+e^{-2aT}}$
15	$\dfrac{1}{s-(1/T)\ln a}$	$a^{t/T}$	$\dfrac{z}{z-a}$

附录 C

常用校正网络

Operational-Amplifier circuits that may be used as compensators see Tab. C-1.

Tab. C-1 Operational-Amplifier Circuits That May Be Used as Compensators

	Control Action	$G(s)=\dfrac{u_o(s)}{u_i(s)}$	Operational-Amplifier Circuits
1	P	$\dfrac{R_4}{R_3}\dfrac{R_2}{R_1}$	
2	I	$\dfrac{R_4}{R_3}\dfrac{1}{R_1 C_2 s}$	
3	PD	$\dfrac{R_4}{R_3}\dfrac{R_2}{R_1}(R_1 C_1 s+1)$	

续表

	Control Action	$G(s) = \dfrac{u_o(s)}{u_i(s)}$	Operational-Amplifier Circuits
4	PI	$\dfrac{R_4}{R_3}\dfrac{R_2}{R_1}\dfrac{R_2C_2s+1}{R_2C_2s}$	
5	PID	$\dfrac{R_4}{R_3}\dfrac{R_2}{R_1}\dfrac{(R_1C_1s+1)(R_2C_2s+1)}{R_2C_2s}$	
6	Lead or lag	$\dfrac{R_4}{R_3}\dfrac{R_2}{R_1}\dfrac{R_1C_1s+1}{R_2C_2s+1}$	
7	Lag-lead	$\dfrac{R_6}{R_5}\dfrac{R_4}{R_3}\dfrac{[(R_1+R_3)C_1s+1](R_2C_2s+1)}{(R_1C_1s+1)[(R_2+R_4)C_2s+1]}$	

附录 D

Bode图的绘制规则

Tab. D-1 is the rules for making Bode plots.

Tab. D-1 Rules for Making Bode Plots

Term	Magnitude	Phase
Constant: K	$20\lg(\|k\|)$	$K>0: 0$ $K<0: \pm 180$
Real Pole: $\dfrac{1}{\frac{s}{\omega_0}+1}$	☐ Low freq. Asymptote at 0dB ☐ High freq. Asymptote at -20dB/dec ☐ Connect lines at break freq	☐ Low freq. Asymptote at $0°$ ☐ High freq. Asymptote at $-90°$ ☐ Connect with straight line from $0.1\omega_0$ to $10\omega_0$
Real Zero: $\dfrac{s}{\omega_0}+1$	☐ Low freq. Asymptote at 0dB ☐ High freq. Asymptote at $+20$dB/dec ☐ Connect lines at break freq	☐ Low freq. Asymptote at $0°$ ☐ High freq. Asymptote at $+90°$ ☐ Connect with line from $0.1\omega_0$ to $10\omega_0$
Plot at origin: $\dfrac{1}{s}$	-20dB/dec, through 0dB at $\omega=1$	$-90°$
Zero at origin: s	-20dB/dec, though 0dB at $\omega=1$	$+90°$
Underdamped Ploes: $\dfrac{1}{(\frac{s}{\omega_0})^2+2\zeta(\frac{s}{\omega_0})+1}$	☐ Low freq. Asymptote at 0dB ☐ High freq. Asymptote at -40dB/dec ☐ Draw peak at freq. $\omega_r=\omega_0\sqrt{1-2\zeta^2}$ with amplitude $H(j\omega_r)=20\lg(2\zeta\sqrt{1-\zeta^2})$ ☐ Connect lines	☐ Low freq. Asymptote at $0°$ ☐ High freq. Asymptote at $-180°$ ☐ Connect with straight line from $\omega=\omega_0\dfrac{\lg\left(\frac{2}{\zeta}\right)}{2}$ to $\omega=\omega_0\dfrac{2}{\lg\left(\frac{2}{\zeta}\right)}$

续表

Term	Magnitude	Phase
Underdamped Zero*: $\left(\dfrac{s}{\omega_0}\right)^2 + 2\zeta\left(\dfrac{s}{\omega_0}\right) + 1$	☐ Draw low freq. Asymptote at 0dB ☐ Draw high freq. Asymptote at +40dB/dec ☐ Draw dip at freq. $\omega_r = \dfrac{\omega_0}{\sqrt{1-2\zeta^2}}$ with amplitude $H(j\omega_r) = +20\lg(2\zeta\sqrt{1-\zeta^2})$ ☐ Connect lines	☐ Low freq. Asymptote at 0° ☐ Draw high freq. Asymptote at $-180°$ ☐ Connect with straight line from $\omega = \omega_0 \dfrac{\lg\left(\dfrac{2}{\zeta}\right)}{2}$ to $\omega = \omega_0 \dfrac{2}{\lg\left(\dfrac{2}{\zeta}\right)}$

Notes

☐ Rules for drawing zeros create the mirror image (around 0dB) of those for a pole with the same break freq.

☐ For underdamped poles and zeros peak exist for $0 < \zeta < 0.707 = \dfrac{1}{\sqrt{2}}$ and peak freq. is typically very near the break freq.

☐ For underdamped poles and zeros if $\zeta < 0.02$ draw phase vertically from 0 to -180 degrees at break freq. For nth order pole or zero make asymptotes and peaks n time higher than shown(i.e., second order asymptote is -40dB/dec, and phase goes from 0° to $-180°$). Don't change frequencies, only plot values and slopes.

附录 E 常用MATLAB命令

常用 MATLAB 命令如表 E-1 所示。

表 E-1 常用 MATLAB 命令

命 令	意 义	命 令	意 义
conv	多项式相乘	sym,syms	定义符号变量
roots	求多项式的根	eig	求特征值
poly	由根构造多项式	expm	求矩阵指数
residue	有理分式变为部分分式	lyap	连续 Lyapunov 方程求解
	部分分式变为有理分式	det	求行列式
zpk	传递函数零极点形式	ctrb	求可控性矩阵
tf	传递函数的有理分式	obsv	求可观测性矩阵
tf2zp	有理分式化为零极点形式	rank	求矩阵的秩
rlocus	绘根轨迹	place	求极点配置的反馈阵,多变量
rlocfind	确定根轨迹一点的增益	acker	求极点配置的反馈阵,多重极点
rltool	根轨迹分析	inv	矩阵求逆
nyquist	绘 Nyquist 图	lqr	求最优控制的反馈阵
bode	绘 Bode 图	int	积分
margin	求稳定裕度与穿越频率	size	求矩阵的大小

附录 F 自动控制理论中的概念解析

系统：系统泛指由一群有关联的个体组成，根据预先编排好的规则工作，能完成个别元件不能单独完成的工作的群体。

自动控制：是指在没有人直接参与的情况下，利用外加的设备或装置，使机器、设备或生产过程的某个工作状态或参数自动地按照预定的规律运行。

开环控制：开环控制是最简单的一种控制方式。它的特点是，按照控制信息传递的路径，控制量与被控制量之间只有前向通路而没有反馈通路，即控制作用的传递路径不是闭合的，故称为开环。

闭环控制：凡是将系统的输出量反送至输入端，对系统的控制作用产生直接的影响，都称为闭环控制系统或反馈控制系统。这种自成循环的控制作用，使信息的传递路径形成了一个闭合的环路，故称为闭环。

复合控制：是开、闭环控制相结合的一种控制方式。

自动控制系统：不需要有人干预就可按照期望规律或预定程序运行的控制系统。

被控对象：指需要给以控制的机器、设备或生产过程。被控对象是控制系统的主体，例如火箭、锅炉、机器人、电冰箱等。控制装置则指对被控对象起控制作用的设备总体，有测量变换部件、放大部件和执行装置。

开环控制系统：不将控制的结果反馈回来影响当前控制的系统，是没有输出反馈的一类控制系统。一般其结构简单，易维修；但精度低、易受干扰。

闭环控制系统：可以将控制的结果反馈回来与期望值比较，并根据它们的误差调整控制作用的系统，又称为反馈控制系统。一般其结构复杂，不易维修；但精度高，抗干扰能力强，动态特性好。

手动控制系统：必须在人的直接干预下才能完成控制任务的系统。例如：骑自行车——人工闭环系统，导弹——自动闭环系统，人打开灯——人工开环系统，自动门、自动路灯——自动开环系统。

被控量：指被控对象中要求保持给定值、要按给定规律变化的物理量。被控量又称输出量、输出信号。

给定值：是作用于自动控制系统的输入端并作为控制依据的物理量。给定值又称输入信号、输入指令、参考输入。

干扰：除给定值之外，凡能引起被控量变化的因素，都是干扰。干扰又称扰动。

精度：精度是测量值与真值的接近程度，包含精密度和准确度两方面。

反馈：指将系统的输出返回到输入端并以某种方式改变输入，进而影响系统功能的过程,即将输出量通过恰当的检测装置返回到输入端并与输入量进行比较的过程。

自动控制系统的性能要求：稳定性、快速性、准确性和鲁棒性。

过程控制：对生产过程的某一或某些物理参数进行的自动控制,在人直接参与的情况下,利用控制装置使被控制对象和过程按预定规律变化。

线性系统：用线性微分方程或线性差分方程描述的系统,也可指状态变量和输出变量对于所有可能的输入变量和初始状态都满足叠加原理的系统。一个由线性元部件所组成的系统必是线性系统。

非线性系统：用非线性微分方程或差分方程描述的系统。只要系统中包含一个或一个以上具有非线性特性的元件,即称为非线性系统。典型的非线性特性包括饱和特性、回环特性、死区特性、继电器特性。

连续系统：当系统中各元件的输入量和输出量均是连续量或模拟量时,就称此类系统是连续系统。

离散系统：当系统中某处或多处信号是脉冲序列或数字形式时,就称这类系统是离散系统。

恒值控制系统：控制系统在运行中被控量的给定值保持不变。

随动控制系统：控制系统被控量的值不是预先设定的,而是受外来的某些随机因素影响而变化,其变化规律是未知的时间函数。

程序控制系统：控制系统被控量的给定值是预定的时间函数,并要求被控量随之变化。

自动控制系统的组成如下。

- **测量元件**：其功能是测量被控制的物理量,如果这个物理量是非电量,一般要转换为电量。
- **给定元件**：其功能是给出与期望的被控量相对应的系统输入量(即参变量)。
- **比较元件**：把测量元件检测的被控量实际值与给定元件给出的参考量进行比较,求出它们之间的偏差。
- **放大元件**：将比较元件给出的偏差进行放大,用来推动执行元件去控制被控对象。
- **执行元件**：直接推动被控对象,使其被控量发生变化。
- **校正元件**：也称补偿元件,它是结构或参数便于调整的元件,用串联或反馈的方式连接在系统中,以改善系统性能。

数学模型：是描述系统内部物理量(或变量)之间动态关系的数学表达式。描述自动控制系统输入、输出变量以及内部各变量之间关系的数学表达式称为数学模型。

传递函数：线性定常系统在零初始条件下,输出量的拉普拉斯变换与输入量的拉普拉斯变换之比,称为传递函数。

零点/极点：分子多项式的零点(分子多项式的根)称为传递函数的零点；分母多项式的零点(分母多项式的根)称为传递函数的极点。

动态结构图：把系统中所有环节或元件的传递函数填在系统原理方块图的方块中,并把相应的输入、输出信号分别以拉普拉斯变换来表示,从而得到的传递函数方块图就称为动态结构图。

信号流图：是表示复杂控制系统中变量间相互关系的一种图解法,由节点和支路等组成。

梅逊增益公式：利用梅逊增益公式，可以直接得到系统输出量与输入变量之间的传递函数。

主导极点：如果闭环极点离虚轴很远，则它对应的暂态分量衰减得很快，只在响应的起始部分起微弱作用，而离虚轴最近的闭环极点（复极点或实极点）对系统瞬态过程性能的影响最大，在整个响应过程中起着主要的决定性作用，称它为主导极点。具体定义见教材。

偶极子：当极点与某零点靠得很近时，它们之间的模值很小，那么该极点的对应系数也就很小，对应暂态分量的幅值也很小，故该分量对响应的影响可忽略不计。我们将一对在复平面上位置很近的闭环零、极点称为偶极子。

输入节点（又称源点）：只有输出支路的节点叫输入节点或源点。

输出节点（又称陷点）：只有输入支路的节点叫输出节点，它对应于因变量或输出信号。

混合节点：既有输入支路又有输出支路的节点叫混合节点。

如果通路与任意一个节点相交不多于一次的称为开通路。如果通路的终点就是通路的起点，并且与任何其他节点相交不多于一次的，则称为闭通路。

闭环零点：闭环传递函数中分子多项式的根称为系统的闭环零点。

稳定性：所谓稳定性，就是指系统在扰动消失后，由初始偏差状态恢复到原来平衡状态的能力。若系统能恢复到平衡状态，则称系统是稳定的。

控制量：控制器的输出信号。

前馈控制系统：前馈控制系统直接根据扰动信号进行调节，扰动量是控制的依据，由于它没有被控量的反馈信号，故不形成闭合回路，所以它是一种开环控制系统。

程序控制系统：这种系统的给定量是按照一定的时间函数变化的，如程序控制机床的程序控制系统的输出量应与给定量的变化规律相同。

最小相位系统：如果系统的开环传递函数在右半 S 平面上没有极点和零点，则称为最小相位传递函数。

基本单元：信号线、分支点、加减点、方块。

开环传递函数：反馈信号与偏差信号之比。系统的主反馈回路接通以后，输出量与输入量之间的传递函数。

闭环传递函数：系统的主反馈回路接通以后，输出量与输入量之间的传递函数。

稳态误差：对单位负反馈系统，当时间趋于无穷大时，系统对输入信号响应的实际值与期望值（即输入量）之差的极限值，称为稳态误差，它反映系统复现输入信号的（稳态）精度。

时域：一种数学域，与频域相区别，用时间和时间响应来描述系统。

一阶系统：控制系统的运动方程为一阶微分方程，称为一阶系统。

二阶系统：控制系统的运动方程为二阶微分方程，称为二阶系统。

单位阶跃响应：系统在零状态条件下，在单位阶跃信号作用下的响应称为单位阶跃响应。

阻尼比 ζ：阻尼就是使自由振动衰减的各种摩擦和其他阻碍作用。在自动控制、土木、机械、航天等领域是结构动力学的一个重要概念。一般是指阻尼系数与临界阻尼系数之比，表达结构体标准化的阻尼大小。在经典控制理论中，阻尼比与二阶系统的特征根在 S 平面上的位置密切相关，不同阻尼比对应系统不同的运动规律。

性能指标：系统性能的定量度量。

上升时间：响应从终值 10% 上升到终值 90% 所需时间；对有振荡系统也可定义为响应从零第一次上升到终值所需时间。上升时间是响应速度的度量。

峰值时间：响应超过其终值到达第一个峰值所需时间。

调节时间：响应到达并保持在稳态误差要求范围内所需时间。

超调量σ%：响应的最大偏离量与终值之差的百分比。

劳斯判据：判断系统的闭环稳定性的一种代数判据。

稳态误差：是指系统达到稳态时，输出量的期望值与稳态值之间的差值。（由参考输入引起的稳态误差称为给定值稳态误差，由扰动输入引起的稳态误差，称为扰动稳态误差。）

根轨迹：是指开环系统某个参数由 0 变化到 ∞，闭环特征根在 S 平面上移动的轨迹。

根轨迹方程：根轨迹所应满足的方程，称为根轨迹方程。由相角方程和幅值方程组成。

参数根轨迹：如果系统的可变参数不是增益（根轨迹 $K*$ 或开环增益 K）而是系统的其他参数时，此时的根轨迹叫参数根轨迹。

零度根轨迹：在某些情况下，相角方程右边相角的主值将不再是 180°，而是 0°，将这种根轨迹叫零度根轨迹。

根轨迹的渐近线：如果开环零点数 m 小于开环极点数 n，则系统的开环增益为 K 时，趋向无穷远处的根轨迹共有 $(n-m)$ 条，决定这 $(n-m)$ 条根轨迹趋向无穷远处方位的渐近线称为根轨迹的渐近线。

根轨迹实轴上的会合点（或分离点）：几条根轨迹在 S 平面上相遇后又分开（或分开后又相遇）的点，称为根轨迹的分离点（或会合点）。

非最小相位系统：开环传递函数在右半 S 平面有一个或多个零极点的系统称为非最小相位系统。

最小相位（相角）系统：零点、极点均在 S 平面的左半平面的系统。

频域：一种数学域，与时域相区别，用频率和频率响应来描述系统。

频率特性：对于线性系统来说，当输入信号为正弦信号时，稳态时的输出信号是一个与输入信号同频率的正弦信号，不同的只是其幅值与相位，且幅值与相位随输入信号的频率不同而不同。输出与输入的幅值比随频率变化的函数称为幅频特性，输出与输入的相位差随频率变化的函数称为相频特性。两者合称频率特性。

幅相曲线：对于一个确定的频率，必有一个幅频特性的幅值和一个幅频特性的相角与之对应，幅值与相角在复平面上代表一个向量。当频率 ω 从零变化到无穷时，相应向量的矢端就描绘出一条曲线。这条曲线就是幅相频率特性曲线，简称幅相曲线。

对数频率特性曲线：又称为伯德图（曲线），其横坐标采用对数分度，对数幅频曲线的纵坐标的单位是分贝，记作 dB，对数相频曲线的纵坐标单位是度。

奈奎斯特稳定判据：简称奈氏判据，是根据开环频率特性曲线判断闭环系统稳定性的一种简便方法。

Nyquist 判据：当 ω 由 $-\infty$ 变化到 $+\infty$ 时，Nyquist 曲线（极坐标图）逆时针包围 $(-1, j0)$ 点的圈数 N，等于系统 $G(s)H(s)$ 位于 S 右半平面的极点数 P，即 $N=P$，则闭环系统稳定；否则 $(N \neq P)$ 闭环系统不稳定，且闭环系统位于 S 右半平面的极点数 Z 为：$Z=|P-N|$。

稳定裕度：表征系统稳定程度的指标，包括幅值裕度和相角裕度。

幅值裕度：对于闭环稳定系统，如果系统开环幅频特性再增大 h 倍，则系统将处于临界稳定状态。

相角裕度：对于闭环稳定系统，如果系统开环相频特性再滞后 Y 度，则系统将处于临界稳定状态。

低频段：在理想幅频特性曲线中，低频段表征系统的稳态性能。

中频段：在理想幅频特性曲线中，中频段表征系统的暂态性能高。

高频段：在理想幅频特性曲线中，高频段表征系统抗干扰性能。

尼柯尔斯图（Nichols 图）：将对数幅频特性和对数相频特性画在一个图上，即以（度）为线性分度的横轴，以 $l(\omega)=20\lg A(\omega)$（dB）为线性分度的纵轴，以 ω 为参变量绘制的 $\varphi(\omega)$ 曲线，称为对数幅相频率特性，或称作尼柯尔斯图（Nichols 图）。

系统校正：为了使系统达到要求，给系统加入特定的环节，这个过程叫系统校正。

超前网络：具有相角超前特性的网络称超前网络。

滞后网络：具有相角滞后特性的网络称滞后网络。

串联校正：将校正装置接在测量点之后和放大器之前，串接于系统前向通道中，称为串联校正。

校正类型：按照校正装置在系统中的连接方式，控制系统校正方式可分为串联校正、反馈校正、前馈校正和复合校正。

串联校正以及串联校正的种类：校正装置与系统不可变部分呈串联连接的方式称为串联校正。分为串联超前校正、串联滞后校正、串联滞后-超前校正。

反馈校正：校正装置与系统不可变部分或不可变部分中的一部分按反馈方式连接称为反馈校正。

连续控制系统：当控制系统的传递信号都是时间的连续函数，这种系统称为连续控制系统。

离散控制系统：控制系统在某处或几处传递的信号是脉冲系列或数字形式的时间上是离散的系统，称为离散控制系统。

描述函数：非线性元件稳态输出的基波分量与输入正弦信号的复数比定义为非线性环节的描述函数。

自激振荡：非线性系统在没有外界周期变化信号的作用下，系统中能产生具有固定振幅和频率的稳定的周期运动，称为自激振荡。

相平面：以状态变量为横坐标，以其一阶导数为纵坐标组成的直角坐标平面称为相平面。

零阶保持器：零阶保持器是将离散信号恢复到相应的连续信号的环节，它把采样时刻的采样值恒定不变地保持（或外推）到下一采样时刻。

采样与复现：把连续信号变换为脉冲序列的过程称为采样过程；将离散信号转换复原成连续信号的过程称为信号复现过程。

零阶保持器：零阶保持器把采样时刻 kT 的采样值恒定不变地保持到下一个采样周期 $(k+1)T$。

z 变换：从 s 域到 z 域的变换。

脉冲传递函数：线性定常离散系统在零初始条件下，离散输出信号的 z 变换与离散输入信号的 z 变换之比，称为离散系统的脉冲传递函数。

z 平面：水平轴为 z 的实部、垂直轴为 z 的虚部的复平面。

采样控制系统：通常把系统中的离散信号是脉冲序列形成的离散系统称为采样控制系统或脉冲控制系统。

采样开关：在采样控制系统中，把连续信号变换为脉冲序列的装置。

保持器：用数字计算机作为信息处理机构时，处理结果的输出如同原始信息的获取一样，一般有两种方式，其中一种需要把数字信号转换为连续信号。用于这种转换过程的装置，称为保持器。

采样定理：要由离散信号完全复现出采样前的连续信号，必须满足：采样角频率大于或等于两倍的采样器输入连续信号频谱中的最高频率。

脉冲传递函数：在线性离散系统中，把初始值为零时，系统离散输出信号的 z 变换与离散输入信号的 z 变换之比，定义为脉冲传递函数。

状态变量：足以表征系统的运动状态的最小个数的一组变量称为状态变量。

状态方程：由系统状态变量构成的一阶微分方程组。

输出方程：在指定系统输出的情况下，该输出与状态变量间的函数关系式。

状态空间表达式：状态方程和输出方程总和，构成对一个系统完整的动态描述。

系统的自由解：指系统零输入时，由初始状态引起的自由运动。

友矩阵：主对角线上方元素均为 1；最后一行元素可取任意值；其余元素均为 0。

状态转移矩阵：e^{At}，记作 $\boldsymbol{\Phi}(t)$，反映了从初始时刻的状态矢量到任意时刻的状态矢量的一种矢量变换关系。

能控：使系统由某一初始状态 $x(t_0)$ 转移到指定的任一终端状态 $x(t_f)$，称此状态是能控的。若系统的所有状态都是能控的，称系统是状态完全能控的。

线性系统的可控性：控制作用能够对线性系统所有状态产生影响，从而对系统的状态实现控制，称为线性系统的可控性。

最小实现问题：由描述系统输入输出动态关系的运动方程或传递函数，建立系统的状态空间表达式的过程称为实现过程。若原传递函数没有零极点对消的情况，则对该传递函数的实现称为最小实现。从工程的观点看，在无穷多个内部不同结构的系统中，其中维数最小的一类系统就是所谓的最小实现问题。

BIBO 稳定性：当系统受到外部的有界输入作用时，输出也是有界的，称为有界输入有界输出（BIBO）的稳定性。

渐近稳定性：系统没有输入作用，仅在初始条件作用下，输出能随时间的推延而趋于零（指系统的平衡状态），称为渐近稳定性。

可观测性：给定控制后，能在有限的时间间隔内根据系统输出唯一地确定系统的所有起始状态，则系统是完全可观测的。如果只能确定部分起始状态，则系统不完全可观测。

可控制性：当系统用状态方程描述时，给定系统的任意初始状态，可以找到允许的输入量，在有限的时间之内把系统的所有状态引向状态空间的原点（即零状态），则系统是完全可控制的。如果只有对部分状态变量可以做到这一点，则系统不完全可控。

注：附录"自动控制理论中的概念解析"索引并综述了相关概念，完整定义请参考教材中的具体内容。

附录 G 控制理论术语中英文对照

A

absolute error	绝对误差
absolute value	绝对值
absolute stability	绝对稳定性
acceleration error constant	加速度误差系数
across-variable	交叉变量
accumulated error	累积误差
accuracy	精确度
activate	启动,触发
active electric network	有源网络
actuating signal	作用信号,启动信号
actuator	执行机构,调节器,激励器
adjust	调整
adaptive control	自适应控制
additive perturbation	附加扰动
algebraic operations	代数运算
all-pass network	全通网络
amplifier	放大器
amplifying element	放大环节
amplitude	振幅,幅值
amplitude quantization error	幅值量化误差
analog computer	模拟计算机
analog-digital conversion	模拟转换
analog signal	模数信号
angle condition	相角条件
angle of arrival	入射角
angle of departure	出射角
angle of the asymptotes	渐近线夹角
angular acceleration	角加速度

analogous variables	模拟变量
argument	幅角
armature	电枢
assumptions	假设,假定
asymptotic stable	渐近稳定的
asymptote	渐近线
asymptote centroid	渐近线交叉点(与实轴)
automation	自动化
auxiliary polynomial	辅助多项式
automatic control	自动控制
autonomous system	自治系统
attenuation	衰减
auxiliary equation	辅助方程

B

backlash	间隙,回差
backward difference rule	向后差分规则
bandwidth	带宽
bang-bang control	开关式控制,继电控制
be proportional to	与……成比例
bilinear system	双线性系统
biocybernetics	生物控制论
block diagram	框图,方块图,结构图
bode plot	伯德图
branch	分支,支路
breakaway points	分离点
bump	撞击,扰动
by-pass	旁路

C

CACSD(Computer-Aided Control System Design)	控制系统计算机辅助设计
CACSE(Computer-Aided Control System Engineering)	控制系统计算机辅助工程
CAD(Computer Aided Design)	计算机辅助设计
cascade compensation	串联补偿(校正)
cascade control	串级控制
canonical form	标准型
cascade compensation network	串联补偿网络
Cauchy's theorem	柯西定理
channel	通道

characteristic equation	特征方程
characteristic gain locus	特征增益轨迹
circuit	电路
classical control theory	经典控制理论
closed-loop control	闭环控制
closed loop control system	闭环控制系统
closed loop frequency response	闭环频率响应
closed-loop feedback control system	闭环反馈控制系统
closed loop pole	闭环极点
closed-loop system	闭环系统
closed-loop transfer function	闭环传递函数
closed loop zero	闭环零点
combinational control system	复合控制系统
command following	跟随命令
comparator	比较器
complementary sensitivity function	互补灵敏度函数
comparing element	比较元件,比较环节
compatibility	相容性,兼容性
compound control	复合控制
components	部件
compensation	补偿,校正
complexity	复数
complex plane	复平面
complexity of design	设计复杂度
conditional stability	条件稳定
configuration	结构,配置,方案,组态
conformal mapping	保角映射
M loci	M 圆
constraint condition	约束条件
constant M loci	等 M 圆
continuous system	连续系统
contour map	围线图
control accuracy	控制精度
control law	控制律
controllable system	可控系统
controlled plant	被控对象
controlled variable	被控变量
controlling machine	控制机
control system	控制系统

control valve	调节阀
controllability	可控性,能控性
controllability matrix	可控矩阵
conveyor	传送器,传送带,传送装置
corner frequency	转折频率,交接频率
correcting unit	校正装置
correction	校正
coupling	耦合
coulomb damper	库仑阻尼
criterion	判据,准则
critical damping	临界阻尼
critical stability	临界稳定性
cross-over frequency	穿越频率
cut off rate	剪切率
cut off frequency	剪切频率
cybernetics	控制论

D

damped natural frequency	有阻尼自然频率
damped oscillation	阻尼振荡
damper	阻尼器
damping factor	阻尼系数
damping ratio	阻尼比
data acquisition	数据采集
DC motor	直流电机
dead band	死区
dead time	死区时间
deadbeat response	直进式的反应
decay	衰减,衰变
decade	十年,十倍
decibel (dB)	分贝
decomposition	分解
design	设计
design gap	设计差距
design of a control system	控制系统设计
design specifications	设计指标
delay	滞后
delay element	滞后环节
denominator	分母

describing function	描述函数
derivation action	微分作用
derivative control	微分控制
desired value	预期值,期望值
detectable	可测的
determinant	行列式
deviation	偏差
diagonal canonical form	对角标准型
difference equation	差分方程
differencing junction	比较点
differential action	微分作用
differential control	微分控制
differential equations	微分方程
differential feedback	微分反馈
digital filter	数字滤波器
digital computer	数字计算机
digital computer compensator	数字计算补偿器
digital control system	离散控制系统
digital signal processing	数字信号处理
discrete system	离散系统
discretization	离散化
distributed parameter control system	分布参数控制系统
disturbance	扰动,干扰
disturbance signal	扰动信号
disturbance compensation	扰动补偿
disturbance rejection property	抗干扰特性
dominant pole	主导极点
dominant roots	主导根
dominate	主导
drive control	传动控制
dual principle	对偶原理
duality	对偶性
duty ratio	负载比,占空率
dynamic characteristics	动态特性
dynamic equation	动态方程
dynamic error	动态误差
dynamic error coefficient	动态误差系数
dynamic process	动态过程

E

effectiveness	有效性
equilibrium state	平衡状态
eigenvalue	特征值
eigenvector	特征向量
element	元件,环节
embedded control	嵌入控制
engineering design	工程设计
error	误差
error coefficient	误差系数
error signal	误差信号
estimation error	估计误差
even symmetry	偶对称
expected value	期望值
exponential	指数,指数的,幂的
external description	外部描述
external disturbance	外扰
extremum	极值

F

fault diagnosis	故障诊断
feasibility	可行性,可能性,现实性
feedback	反馈
feedback compensation	反馈补偿
feedback control	反馈控制
feedback signal	反馈信号
feedback element	反馈环节
feedback path	反馈通道
feed forward	前馈
feed forward compensation	前馈补偿
final controlling element	执行器
final value	终值
final value theorem	终值定理
first-order holder	一阶保持器
first-order system	一阶系统
flyball governor	飞球调速器
focus	焦点
following device	随动装置

fourier transform	傅里叶变换
fourier transform pair	傅里叶变换对
forward path	前向通道
forward rectangular integration	前向矩形积分
fraction	分数
frequency	频率
frequency domain	频域
frequency response	频率响应
frequency response characteristic	频率响应特性
function	摩擦
full order observer	全阶观测器
full-state feedback control law	全状态反馈控制率
function	函数
fuzzy control	模糊控制

G

gain	增益
gain margin	增益裕度,幅值裕度
gap characteristics	间隙特性
gear backlash	齿轮间隙
general solution	通解
global asymptotic stability	全局渐进稳定性
global optimum	全局最优
graphical method	图解法
guidance system	制导系统
gravitation area	引力域
gyro	陀螺

H

harmonic	谐波,谐波量,谐振荡
harmonic response	谐波响应
hierarchical control	递阶控制
holder	保持器
homogeneity	齐次性
homogeneous equation	齐次方程
humidity control	湿度控制
hurwitz determinant	赫尔维茨行列式
hybrid fuel automobile	混合燃料汽车
hydraulic system	液压系统

hysteresis error	回差
hysteresis loop	磁滞回环

I

idealized system	理想化系统
identification	辨识
impulse response	脉冲响应
incompatibility principle	不相容原理
index of merit	品质因数
industrial automation	工业自动化
industrial robot	工业机器人
inertial	惯性的,惯量的,惰性的
inherent nonlinearity	固有非线性
inherent characteristic	固有特性
initial condition	初始条件
initial state	初始状态
initial value theorem	初值定理
inner loop	内环
input	输入
input node	输入节点
input signal	输入信号
input feedforward canonical Form	前馈输入标准型
instability	不稳定性
integral action	积分作用
integral control	积分控制
integral performance criterion	积分性能准则
integration network	积分网络
IAE(Integrated Absolute Error)	绝对误差积分
ISE(Integrated Square Error)	平方误差积分
integrity	整体性
internal model design	内部模型设计
internal model principle	内部模型原理
internal description	内部描述
internal disturbance	内扰
intelligent instrument	智能仪表
invariant	不变的,恒定的
inverse matrix	逆矩阵
inverse transformation	反变换
inverse Laplace transforms	拉普拉斯反变换

isometric method	等倾线法
iterative algorithm	迭代算法

J

Jordan block	约旦块
Jordan canonical form	约旦标准型
Jury stability criterion	朱利稳定判据

K

Kalman criterion	卡尔曼准则
Kalman filter	卡尔曼滤波器
Kalman state-space decomposition	卡尔曼状态空间分解

L

lag-lead compensation	滞后超前补偿
lag network	滞后网络
lag compensation	滞后补偿
Laplace transforms	拉普拉斯变换
Laplace transform pair	拉普拉斯变换对
large scale system	大系统
lead network	超前网络
lead-lag network	超前-滞后网络
least-mean-square	最小均方
limit cycle	极限环
linearization	线性化
linearity	线性度
linearized	线性化
linear approximation	线性估计
linear equation	线性方程
linear system	线性系统
linear quadratic regulator	线性二次型调节器
linear programming	线性规划
load	负载
load-response curve	负荷响应曲线
local asymptotic stability	局部渐近稳定性
local optimum	局部最优
locus	轨迹
logarithmic (decibel) measure	对数(分贝)测量
logarithmic magnitude	对数幅值

logarithmic sensitivity	对数敏感性
logic diagram	逻辑图
log magnitude	对数幅值
loop gain	回路增益
loss of gain	增益损失
low pass characteristic	低通特性
Lyapunov theorem of asymptotic stability	李雅普诺夫渐近稳定性定理

M

magnitude condition	幅值条件
magnitude-frequency characteristic	幅频特性
magnitude margin	幅值裕量
magnitude-versus-phase plot	幅相特性曲线
manipulated variable	操纵变量
manual PID tuning	手动 PID 调节
marginally stable	临界稳定
Mason rule	梅森公式
Mason loop rule	梅森公式
mathematical model	数学模型
matrix	矩阵
matrix exponential function	矩阵指数函数
maximum overshoot	最大超调量
maximum principle	极大值原理
maximum value of the frequency response	频率响应最大值
mean-square error criterion	均方误差准则
measurable	可测量的
measured variable	被测变量
measurement noise	测量噪声
measurement noise attenuation	测量噪声衰减
mechatronics	机电一体化
microcomputer	微型机
minimal realization	最小实现
minimum phase system	最小相位系统
minimum phase transfer function	最小相位传递函数
minimum step system	最小拍系统
minimum variance estimation	最小方差估计
model decomposition	模型分解
module library	模块库
modulus	模数

moment of inertia	转动惯量
MIMO(Multi-Input Multi-Output) system	多输入多输出系统
multiloop control	多回路控制
multiloop feedback control system	多回路反馈控制系统
multiplicative perturbation	乘性扰动
multivariable control system	多变量控制系统
multinomial	多项式(的)
multivariable system	多变量系统

N

natural frequency	自然频率
necessary condition	必要条件
negative feedback	负反馈
negative gain root locus	负增益根轨迹
Nichols chart	尼柯尔斯图线
N loci	N 圆
node	节点
noise	噪声
nonlinear control system	非线性控制系统
non-minimum phase system	非最小相位系统
non-minimum phase transfer functions	非最小相位传递函数
nonsingular	非奇异的
nontouching	非接触
norm	范数
number of separate loci	根轨迹分支数
numerator	分子
numerical control	数字控制,数控
Nyquist criterion	奈奎斯特判据
Nyquist contour	奈奎斯特轨线
Nyquist stability criterion	奈奎斯特稳定性判据

O

objective function	目标函数
observability	可观性,能观性
observability matrix	观测矩阵
observer	观测器
observable system	观测系统
octave	八度
odd symmetry	奇对称

off line	离线
offset	偏移,位移
on line	在线
on-off control	通断控制
open loop	开环
open-loop control	开环控制
open-loop pole	开环极点
open-loop transfer function	开环传递函数
optimal control	最优控制
optimization	最优化
origin	原点
oscillating loop	振荡回路
oscillation	振荡
oscillating period	振荡周期
oscillatory response	振荡响应
outer loop	外环
output	输出
output equation	输出等式
output signal	输出信号
over damped	过阻尼
over damping	过阻尼
overshoot	超调量

P

parameter	参数
parallel	并联
parameter design	参数设计
path	回路
peak overshoot	超调峰值
peak time	峰值时间
percent overshoot	超调量百分比
performance index	性能指标
perturbance	扰动,摄动
phase lag	相位滞后
phase locus	相轨迹
phase lead	相位超前
phase margin	相角裕量
phase modifier	相位调节器
phase variable canonical form	相变量标准形

phase variables	相变量
phase plane	相平面
phase-lag compensation	相位滞后补偿器
phase-lag network	相位滞后网络
phase lead compensation	相位超前补偿器
phase-lead network	相位超前网络
physical variables	物理变量
pickoff point	引出点
PID(Proportional plus Integral plus Derivative) controller	PID(比例、积分、微分)控制器
PID controller	PID 控制器
PID tuning	PID 调节
piece-wise linearization	分段线性化
pneumatic controller	气动调节器,气动控制器
polar plot	极坐标图
pole	极点
pole assignment	极点配置
pole placement	极点配置
pole-zero cancellation	零极点对消
polynomial	多项式
position error constant K_p	位置误差系数 K_p
position error	位置误差
positive definiteness	正定性
positive feedback	正反馈
positive feedback loop	正反馈回路
precision	精确度
prefilter	前置滤波器
pre-compensator	预补偿器
principle of superposition	叠加原理
process control	过程控制
productivity	生产率
proportional action	比例作用
proportional band	比例带
proportional control	比例控制
proportional plus derivative (PD) controller	比例加导数(PD)控制器
proportional plus integral (PI) controller	比例加积分(PI)控制器
prototype	原型,模型,样机
pulse	脉冲
pulse width	脉宽
pure delay	纯滞后

Q

quadratic	二次的
quadratic form	二次型
quarter amplitude decay	四等分振幅衰减
quality control	质量控制
quantizer	数字转换器

R

ramp function	斜坡函数
ramp input	斜坡输入
ramp response	斜坡响应
rate feedback	速度反馈
rate time	微分时间,预调时间
rational	有理(数)的,合理的
reaction curve	反作用曲线
realization	实现
reference order observer	降阶观测器
reference input	参考输入
reference variable	参考变量
regulator	调节器
relay	继电器
relative stability	相对稳定性
reliability	可靠性
remote control	遥控
reproducibility	再现性
resilience	弹性,弹性形变
resonance	谐振
resonant frequency	谐振频率
response	响应
response curve	响应曲线
reset time	再调时间,积分时间
residue	留数
rise time	上升时间
risk	风险
RMS(Root Mean Square)	均方根
robust control	鲁棒控制
robust control system	鲁棒控制系统
robustness	鲁棒性

roots loci	根轨迹
root contours	根轨迹围线
root locus	根轨迹
root locus method	根轨迹法
root locus segments on the real axis	实轴上的根轨迹
root sensitivity	根敏感性
Routh array	劳斯阵列
Routh-Hurwitz criterion	劳斯-赫尔维茨判据
Routh stability criterion	劳斯稳定判据

S

sampling control	采样控制
sampling frequency	采样频率
sampling period	采样周期
sampled data	采样数据
sampled-data system	采样系统
saturation	饱和
saturation characteristics	饱和特性
scalar function	标量函数
scaling factor	比例因子
S-domain	S 域
second order system	二阶系统
sensitivity	灵敏度
sensitivity function	灵敏度函数
sensor	传感器
series compensation	串联补偿
servo	伺服机构,伺服电机
servodrive	伺服传动,伺服传动装置
set point	设定点
set value	设定值
settling time	调节时间；稳定时间
separation principle	分离点原理
signal flow graph	信号流图
simulation block diagram	仿真框图
simulation experiment	仿真实验
simulation velocity	仿真速度
singularity	奇点
sinusoidal	正弦的
sinusoidal function	正弦函数

SISO(Single Input Single Output) system	单输入单输出系统
slope	斜率
specifications	规格,(产品等的)说明书
S-plane	S 平面
stability	稳定(性)
stability of a sampled-data system	采样系统稳定性
stability criterion	稳定判据
stability margin	稳定裕量
stabilizable	稳定的
stabilizing controller	稳定控制器
stable system	稳定系统
state of a system	系统状态
state transition matrix	状态转移矩阵
state differential equation	状态微分方程
state equations	状态方程
state space	状态空间
state variables	状态变量
state variable feedback	状态变量反馈
state vector	状态向量
stationary	稳态的
steady-state	稳态
steady-state deviation	稳态偏差
steady-state error	稳态误差
steady-state response	稳态响应
state-space representation	状态空间表达式
step-by-step control	步进控制
step function	阶跃函数
step signal	阶跃信号
step response	阶跃响应
stochastic process	随机过程
summing junction	相加点
superposition	叠加
supervise	监控,检测,操纵
synthesis	综合
system	系统
system sensitivity	系统灵敏性
systematic deviation	系统偏差
system engineering	系统工程
system identification	系统辨识

T

temperature control	温度控制
tangent	切线
Taylor series	泰勒序列
terminology	术语
test input signal	测试输入信号
threshold value	阈值
through-variable	同等变量
time delay	时间延迟
time constant	时间常数
time domain	时域
time response	时间响应
time-invariant system	常定(时不变)系统
time-varying system	时变系统
tolerated error	容许误差
trade-off	权衡
tracking error	跟踪误差
trajectory	轨迹
transducer	传感器,变换器
transfer function	传递函数
transfer function in the frequency domain	频域传递函数
transfer matrix	转移矩阵
transient response	暂态响应
transition matrix(t)	转移矩阵
transmitter	变送器
transportation	传输滞后
transpose	转置
type number	型号(0型、1型等系统)
typical element	典型环节

U

undamped natural frequency	无阻尼自然频率
underdamping	欠阻尼
uniform stability	一致稳定
uniformly asymptotic stability	一致渐近稳定性
unit circle	单位圆
unit impulse	单位脉冲
unit step function	单位阶跃函数

unit feedback	单位反馈
unity feedback	单位反馈
unit matrix	单位矩阵
unstable	不稳定的
unsymmetrical	不对称的
ultimate gain	最大增益
ultimate period	基本周期

V

value of quantity	量值
variable	变量
vector	向量
velocity feedback	速度反馈
velocity error constant	速度误差系数
viscous friction	黏摩擦
viscous damper	黏阻尼

W

wave	波
waveform	波形
weighting function	加权函数
white noise	白噪声

Z

zero	零点
zero input response	零输入响应
zero-order holder	零阶保持器
zero-state response	零状态响应
z-transfer function	z 传递函数
z-transformation	z 变换
z-transform	z 变换
Ziegler-Nichols PID tuning method	Ziegler-Nichols PID 调节方法
z-plane	z-平面

注：为提高本书作为双语教材的适用性，控制理论术语的编著，不局限于本书所讲的控制理论范围。

参 考 文 献

[1] 于长官.自动控制原理.哈尔滨:哈尔滨工业大学出版社,1996.
[2] 于长官.现代控制理论.哈尔滨:哈尔滨工业大学出版社,1997.
[3] 自动化名词审定委员会.自动化名词.北京:科学出版社,1991.
[4] 杨位钦,谢锡祺.自动控制理论基础.北京:北京理工大学出版社,1990.
[5] 李友善.自动控制原理.北京:国防工业出版社,1989.
[6] 姜春瑞等.自动控制原理与系统.北京.北京大学出版社,2005.
[7] 李友善,梅晓榕,王彤.自动控制原理360题.哈尔滨:哈尔滨工业大学出版社,2002.
[8] 吴祺.自动控制原理.北京:清华大学出版社,1990.
[9] 吴韫章.自动控制理论基础.西安:西安交通大学出版社,2002.
[10] 张铨.微计算机在自动控制中的应用.北京:国防工业出版社,1986.
[11] 陈小琳.自动控制原理例题习题集.北京:国防工业出版社,1982.
[12] 胡寿松.自动控制原理.4版.北京:科学出版社,2001.
[13] 施阳等.MATLAB语言精要及动态仿真工具SIMULINK.西安:西北工业大学出版社,1997.
[14] 夏德铃.近代控制理论引论.哈尔滨:哈尔滨工业大学出版社,1983.
[15] 梅晓榕.自动控制原理.2版.北京:科学出版社,2007.
[16] 傅佩琛.自动控制原理.哈尔滨:哈尔滨工业大学出版社,1988.
[17] 鄢景华.自动控制原理.哈尔滨:哈尔滨工业大学出版社,1996.
[18] 楼顺天,于卫.基于MATLAB的系统分析与设计——控制系统.西安:西安电子科技大学出版社,1996.
[19] 薛定宇.反馈控制系统设计与分析.北京:清华大学出版社,2000.
[20] 戴忠达.自动控制理论基础.北京:清华大学出版社,1991.
[21] Katsuhiko Ogata.现代控制工程.卢伯英,等译.北京:电子工业出版社,2000.
[22] Richard C Dorf,Robert H Bishop.现代控制系统.11版.北京:电子工业大学出版社,2009.
[23] 姜春瑞,槐春晶,刘丽.自动控制原理与系统.北京:北京大学出版社,2005.
[24] Control Tutorial for Matlab Simulink. http://www.library.cmu.edu/ctms/.

图书资源支持

感谢您一直以来对清华版图书的支持和爱护。为了配合本书的使用,本书提供配套的资源,有需求的读者请扫描下方的"书圈"微信公众号二维码,在图书专区下载,也可以拨打电话或发送电子邮件咨询。

如果您在使用本书的过程中遇到了什么问题,或者有相关图书出版计划,也请您发邮件告诉我们,以便我们更好地为您服务。

我们的联系方式:

清华大学出版社计算机与信息分社网站:https://www.shuimushuhui.com/

地　　址:北京市海淀区双清路学研大厦 A 座 714

邮　　编:100084

电　　话:010-83470236　010-83470237

客服邮箱:2301891038@qq.com

QQ:2301891038(请写明您的单位和姓名)

资源下载: 关注公众号"书圈"下载配套资源。

资源下载、样书申请
书圈

图书案例
清华计算机学堂

观看课程直播